CALCULUS

Volume II

MULTIVARIABLE

Giovanni Viglino

RAMAPO COLLEGE OF NEW JERSEY

September, 2021

CONTENTS
MULTIVARIABLE CALCULUS

SINGLE VARIABLE CALCULUS

APPENDIX A **CHECK YOUR UNDERSTANDING SOLUTIONS**
CHAPTERS 1 THROUGH 10

APPENDIX B **ADDITIONAL THEORETICAL DEVELOPMENT**
CHAPTERS 1 THROUGH 10

APPENDIX C **ANSWERS TO ODD-NUMBERED EXERCISES**
CHAPTERS 1 THROUGH 10

PREFACE

Acknowledgements typically appear at the end of a preface. In this case, however, my indebtedness to Professor Marion Berger for her invaluable input throughout the development of this text is such that I am compelled to express my gratitude for her contributions at the beginning: Thank you, dear colleague and friend.

That said:

Our text consists of two volumes. Volume I addresses those topics typically covered in standard Calculus I and Calculus II courses; which is to say, the Single Variable Calculus. Multivariable Calculus is covered in Volume II.

Our primary goal all along has been to write a readable text, without compromising mathematical integrity. Along the way you will encounter numerous Check Your Understanding boxes designed to challenge your understanding of each newly-introduced concept. Complete solutions to the problems in those boxes appear in Appendix A, but please don't be in too much of a hurry to look at those solutions. You should TRY to solve the problems on your own, for it is only through ATTEMPTING to solve a problem that one grows mathematically. In the words of Descartes:

> *WE NEVER UNDERSTAND A THING SO WELL, AND MAKE IT*
> *OUR OWN, WHEN WE LEARN IT FROM ANOTHER, AS WHEN*
> *WE HAVE DISCOVERED IT FOR OURSELVES.*

You will encounter a few graphing calculator glimpses in the text. In the final analysis, however, one can not escape the fact that:

> *MATHEMATICS DOES NOT RUN ON BATTERIES*

CHAPTER 11
FUNCTIONS OF SEVERAL VARIABLES

§1 Limits and Continuity

Just as the (understood) domain of a function $y = f(x)$ of one variable consists of those numbers for which $f(x)$ is defined, so then does the (understood) domain of the function $z = f(x, y)$ consist of all pairs of numbers (x, y) for which $f(x, y)$ is defined. For example:

$$f(x) = 2x + \frac{1}{\sqrt{x-5}} \text{ has domain } D_f = \{x \mid x > 5\}$$

$$\text{and } f(x, y) = 2xy + \frac{1}{\sqrt{x-y}} \text{ has domain } D_f = \{(x, y) \mid x > y\}$$

CHECK YOUR UNDERSTANDING 11.1

Answers:
(a) $D_f = \{x \mid x \neq \pm 3\}$
(b) $D_f = \{(x, y) \mid x \neq \pm y\}$
(c) $D_f = \{(x, y, z) \mid (z \neq -1)\}$

Determine the domain of the given function:

(a) $f(x) = \dfrac{1}{x^2 - 9}$ (b) $f(x, y) = \dfrac{1}{x^2 - y^2}$ (c) $f(x, y, z) = \dfrac{xy}{z + 1}$

LIMITS AND CONTINUITY

We begin by modifying the limit concept of a single-valued function

$$\lim_{x \to c} f(x) = L \text{ if:}$$

For any given $\varepsilon > 0$ there exists $\delta > 0$ such that:

$$0 < |x - c| < \delta \Rightarrow |f(x) - L| < \varepsilon$$

to accommodate a real-valued function of two variables:

DEFINITION 11.1
LIMIT

$$\lim_{(x, y) \to (x_0, y_0)} f(x, y) = L \text{ if:}$$

For any given $\varepsilon > 0$ there exists $\delta > 0$ such that for every (x, y) in the domain of f:

$$0 < \|(x, y) - (x_0, y_0)\| < \delta \Rightarrow |f(x, y) - L| < \varepsilon$$

Note that $|f(x, y) - L|$ denotes the distance between the two real numbers $f(x, y)$ and L, while the notation $\|(x, y) - (x_0, y_0)\|$ is used to represent the distance between the points (x, y) and (x_0, y_0) in the plane; specifically: $\|(x, y) - (x_0, y_0)\| = \sqrt{(x - x_0)^2 + (y - y_0)^2}$ (see margin).

For example:
$$|9 - 4| = 5$$
and:
$$\|(1, 3) - (4, -1)\|$$
$$= \sqrt{(1 - 4)^2 + [3 - (-1)]^2}$$
$$= \sqrt{9 + 16} = 5$$

Here is a geometrical interpretation of Definition 12.2:

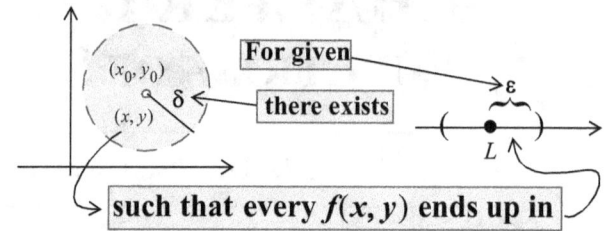

As was the case with functions of one variable:
The limit does not care about what happens at the point of interest.

EXAMPLE 11.1   (a) Let $f(x, y) = x + y$. Prove that:
$$\lim_{(x, y) \to (1, 1)} f(x, y) = 2$$

(b) Let $g(x, y) = \dfrac{x^2 - y^2}{x - y}$. Prove that:
$$\lim_{(x, y) \to (1, 1)} g(x, y) = 2$$

SOLUTION: (a) For given $\varepsilon > 0$ we are to find $\delta > 0$ such that:
$$0 < \|(x, y) - (1, 1)\| < \delta \Rightarrow |x + y - 2| < \varepsilon \quad (*)$$
Noting that $|x + y - 2| = |(x - 1) + (y - 1)| \le |x - 1| + |y - 1|$ we can conclude that (*) will be satisfied for any $\delta > 0$ for which:
$$0 < \|(x, y) - (1, 1)\| < \delta \Rightarrow |x - 1| < \frac{\varepsilon}{2} \text{ and } |y - 1| < \frac{\varepsilon}{2}$$

And $\delta = \dfrac{\varepsilon}{2}$ fits that bill:

For if (x, y) is within $\varepsilon/2$ units of $(1, 1)$, then certainly both x and y must be within $\varepsilon/2$ units of 1.

(b) Note that the function $g(x, y) = \dfrac{x^2 - y^2}{x - y}$ is not defined at $(1, 1)$. Nonetheless:

Reminiscent of Example 2.1, page 44.

$$\lim_{(x, y) \to (1, 1)} g(x, y) = \lim_{(x, y) \to (1, 1)} \frac{x^2 - y^2}{x - y} = \lim_{(x, y) \to (1, 1)} \frac{(x + y)(x - y)}{x - y}$$

$$= \lim_{(x, y) \to (1, 1)} (x + y) \underset{\substack{\uparrow \\ \text{by (a)}}}{=} 2$$

CHECK YOUR UNDERSTANDING 11.2

For $f(x, y) = x^2 + y^2$, indicate the value of $\displaystyle\lim_{(x, y) \to (0, 0)} f(x, y)$, and then use Definition 11.1 to justify your answer.

Answer: See page A-1.

Definition 1.3 of page 6 extends to real-valued functions of two (or more) variable, as does Theorem 2.3 of page 55:

THEOREM 11.1

LIMIT THEOREMS

If $\lim\limits_{(x,y)\to(x_0,y_0)} f(x,y) = L$ and $\lim\limits_{(x,y)\to(x_0,y_0)} g(x,y) = M$, then:

(a) $\lim\limits_{(x,y)\to(x_0,y_0)} [f(x,y)+g(x,y)] = L+M$.

(b) $\lim\limits_{(x,y)\to(x_0,y_0)} [f(x,y)-g(x,y)] = L-M$

(c) $\lim\limits_{(x,y)\to(x_0,y_0)} [f(x,y)\cdot g(x,y)] = LM$.

(d) $\lim\limits_{(x,y)\to(x_0,y_0)} \dfrac{f(x,y)}{g(x,y)} = \dfrac{L}{M}$ if $M\neq 0$.

(e) $\lim\limits_{(x,y)\to(x_0,y_0)} [cf(x,y)] = cL$ for any number c.

PROOF: We content ourselves by establishing (a). The proof is "identical" to that of Theorem 2.3(a), page 55 (see margin).

The only difference is that the distance notation $|x-c|$ on page 55 is adjusted to represent the distance between points in the plane; namely: $\|(x,y)-(x_0,y_0)\|$.

For a given $\varepsilon > 0$ we are to find $\delta > 0$ such that:

$$0 < \|(x,y)-(x_0,y_0)\| < \delta \Rightarrow |(f+g)(x,y)-(L+M)| < \varepsilon$$

$$0 < \|(x,y)-(x_0,y_0)\| < \delta \Rightarrow |f(x,y)+g(x,y)-L-M| < \varepsilon$$

$$0 < \|(x,y)-(x_0,y_0)\| < \delta \Rightarrow |[f(x,y)-L]+[g(x,y)-M]| < \varepsilon \ (*)$$

By virtue of the triangle inequality we have:

$$|[f(x,y)-L]+[g(x,y)-M]| \leq |f(x,y)-L|+|g(x,y)-M|$$

It follows that $(*)$ will hold for any $\delta > 0$ for which $0 < \|(x,y)-(x_0,y_0)\| < \delta$ implies that **BOTH** $|f(x,y)-L| < \dfrac{\varepsilon}{2}$ and $|g(x,y)-M| < \dfrac{\varepsilon}{2}$. Let's find such a δ:

Since $\lim\limits_{(x,y)\to(x_0,y_0)} f(x,y) = L$, there is a $\delta_1 > 0$ such that

$$0 < \|(x,y)-(x_0,y_0)\| < \delta_1 \Rightarrow |f(x,y)-L| < \dfrac{\varepsilon}{2}.$$

Since $\lim\limits_{(x,y)\to(x_0,y_0)} g(x,y) = M$ there is a $\delta_2 > 0$ such that

$$0 < \|(x,y)-(x_0,y_0)\| < \delta_2 \Rightarrow |g(x,y)-M| < \dfrac{\varepsilon}{2}.$$

It follows that for δ the smaller of δ_1 and δ_2:

$$0 < \|(x,y)-(x_0,y_0)\| < \delta \Rightarrow |f(x,y)-L| < \dfrac{\varepsilon}{2} \text{ and } |g(x,y)-M| < \dfrac{\varepsilon}{2}.$$

In the exercises you are invited to verify that for

$$f(x, y) = x: \quad \lim_{(x,y) \to (x_0, y_0)} f(x, y) = x_0$$

$$f(x, y) = y: \quad \lim_{(x,y) \to (x_0, y_0)} f(x, y) = y_0$$

$$\lim_{(x,y) \to (x_0, y_0)} c = c \text{ for any constant } c.$$

It then follows, from Theorem 11.1, that for any polynomial $p(x, y)$ of two variables (see margin):

$$\lim_{(x,y) \to (x_0, y_0)} p(x, y) = p(x_0, y_0)$$

Thus: $\lim_{(x,y) \to (2, 3)} 2x^2 y + x - 3y = 2(2^2 \cdot 3) + 2 - 3 \cdot 3 = 17$

Moreover, for any polynomials $p(x, y)$, $q(x, y)$:

$$\lim_{(x,y) \to (x_0, y_0)} \frac{p(x, y)}{q(x, y)} = \frac{p(x_0, y_0)}{q(x_0, y_0)} \text{ (providing } q(x_0, y_0) \neq 0 \text{)}.$$

Thus: $\lim_{(x,y) \to (3, -4)} \frac{x^2 + y}{xy - 1} = \frac{9 - 4}{-12 - 1} = -\frac{5}{13}$

> A **polynomial function of two variables** (or simply a polynomial function) is a sum of terms of the form $cx^n y^m$, where c is a constant and n and m are nonnegative integers. Moreover, a **rational function** is a ratio of two polynomials.

CHECK YOUR UNDERSTANDING 11.3

(a) Prove Theorem 11.1(e).

(b) Use Theorem 11.1, along with Example 11.1, to evaluate

$$\lim_{(x,y) \to (1, 1)} 5(x + y)\left(\frac{x^2 - y^2}{x - y}\right)$$

> Answers: (a) See page A-1
> (b) 20

CONTINUITY

As might be anticipated (see page 19):

DEFINITION 11.2
CONTINUITY

The real-valued function $f(x, y)$ is **continuous** at (x_0, y_0) if:

$$\lim_{(x,y) \to (x_0, y_0)} f(x, y) = f(x_0, y_0)$$

A **function that** is continuous at every point in its domain is said to be a **continuous function**.

As is the case with functions of one variable:

THEOREM 11.2

If f and g are continuous at (x_0, y_0) then so are the functions:

(a) $f + g$ (b) $f - g$ (c) fg

(d) $\dfrac{f}{g}$ [providing $g(x_0, y_0) \neq 0$]

(e) cf (for any real number c)

PROOF: We establish (c), and relegate the others to the exercises. By definition, the product function fg is defined by:

$$(fg)(x, y) = f(x, y)g(x, y)$$

We proceed to show that $\displaystyle\lim_{(x,y) \to (x_0, y_0)} (fg)(x, y) = (fg)(x_0, y_0)$:

The "same" as the proof of Theorem 2.4(c), page 20.

$$\lim_{(x,y) \to (x_0, y_0)} (fg)(x, y) = \lim_{(x,y) \to (x_0, y_0)} [f(x, y) \cdot g(x, y)]$$

Theorem 11.1(c):
$$= \lim_{(x,y) \to (x_0, y_0)} f(x, y) \cdot \lim_{(x,y) \to (x_0, y_0)} g(x, y)$$

Since f and g are continuous at (x_0, y_0):
$$= f(x_0, y_0)g(x_0, y_0) = (fg)(x_0, y_0)$$

Here are a couple of "new" continuity theorems:

THEOREM 11.3 (a) If f and g are real-valued continuous functions of one variable, then:

$$H(x, y) = f(x) + g(y) \text{ and } K(x, y) = f(x)g(y)$$

are continuous functions of two variables.

(b) If g is a real-valued continuous function of two variables, and if f is a real-valued continuous function of one variable, then:

$$H(x, y) = f[g(x, y)]$$

is a continuous function of two variables.

Proof: See Appendix B, page B-1.

CHECK YOUR UNDERSTANDING 11.4

(a) Prove Theorem 11.2 (c).
(b) Use Theorem 11.3 to establish the continuity of

$$H(x, y) = \sqrt{y^2(\sin x + e^x)}, \text{ for } x > 0.$$

Answer: See page A-2.

The previous discussion involving functions of two variables, readily extends to functions of three (or more) variables. In particular:

DEFINITION 11.3

AS IN
DEFINITION 12.2

$$\lim_{(x,y,z) \to (x_0, y_0, z_0)} f(x, y, z) = L \text{ if:}$$

For any given $\varepsilon > 0$ there exists $\delta > 0$ such that:

$$0 < \|(x, y, z) - (x_0, y_0, z_0)\| < \delta$$
$$\Rightarrow |f(x, y, z) - L| < \varepsilon$$

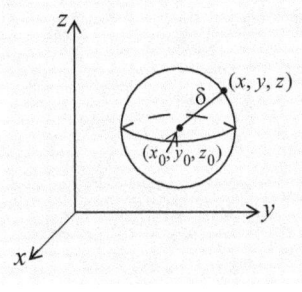

In the above:

$$\|(x, y, z) - (x_0, y_0, z_0)\| = \sqrt{(x - x_0)^2 + (y - y_0)^2 + (z - z_0)^2}$$

represents the distance between the points (x, y, z) and (x_0, y_0, z_0) in three-space (see margin).

DEFINITION 11.4
AS IN
DEFINITION 11.3

The function $f(x, y, z)$ is **continuous** at (x_0, y_0, z_0) if:

$$\lim_{(x, y, z) \to (x_0, y_0, z_0)} f(x, y, z) = f(x_0, y_0, z_0)$$

We note that Theorems 11.1, 11.2, and 11.3 generalize to functions of three (or more) variables. In particular:

THEOREM 11.4
AS IN
THEOREM 11.3

If f, g, and h are real-valued continuous functions of one variable, then the following are continuous real-valued functions of three variables:

$$H(x, y, z) = f(x) + g(y) + h(z) \text{ and }$$
$$K(x, y, z) = f(x)g(y)h(z)$$

If f is a real-valued continuous function of one variable, and if g is a real-valued continuous function of three variables, then $H(x, y, z) = f[g(x, y, z)]$ is also a real-valued continuous function of three variables.

Definitions 11.2, 11.3, and 11.4, along with Theorems 12.2, 11.3, and 11.4, assure us that polynomial functions of two or three variables are continuous everywhere, and that rational functions of two or three variables are continuous throughout their domains (which exclude only those points where the polynomial in the denominator is zero).

	EXERCISES	

Exercises 1-4. Determine the domain of the given function.

1. $f(x, y) = \dfrac{xy - 100}{x^2 + 9}$

2. $f(x, y) = \dfrac{\sqrt{x - 100}}{x^2 - y^2}$

3. $f(x, y) = \tan xy$

4. $f(x, y) = \dfrac{\sqrt{xy}}{\sin x - \cos y}$

5. $f(x, y, z) = \dfrac{1}{x + y + z}$

6. $f(x, y, z) = \dfrac{1}{\sin xyz}$

Exercises 7-9. Use Definition 11.1 to verify that:

7. $\displaystyle\lim_{(x, y) \to (1, 1)} 2x + y = 3$

8. $\displaystyle\lim_{(x, y) \to (2, 3)} 4y - 3x = 6$

9. $\displaystyle\lim_{(x, y) \to (2, 1)} \dfrac{x^2 - 4xy + 4y^2}{x - 2y} = 0$

Exercises 10-20. Evaluate:

10. $\displaystyle\lim_{(x, y) \to (0, 1)} (2x + y)^3$

11. $\displaystyle\lim_{(x, y) \to (0, \pi)} \dfrac{y + x \sin x}{y^2}$

12. $\displaystyle\lim_{(x, y) \to (0, 0)} \dfrac{e^{x + y}}{1 - xy}$

13. $\displaystyle\lim_{(x, y) \to (2, 2)} \dfrac{x^4 - y^4}{x^2 - y^2}$

14. $\displaystyle\lim_{(x, y) \to (2, 2)} \dfrac{x^4 - y^4}{x^2 + y^2}$

15. $\displaystyle\lim_{(x, y) \to (1, e)} \ln(xy)$

16. $\displaystyle\lim_{(x, y) \to (0, 4)} \dfrac{4x - xy}{4xy - xy^2}$

17. $\displaystyle\lim_{(x, y) \to (1, 2)} \dfrac{x^2 + 2xy + y^2 - 9}{x + y - 3}$

18. $\displaystyle\lim_{(x, y, z) \to (0, 1, 0)} \dfrac{x + y}{2y - z}$

19. $\displaystyle\lim_{(x, y, z) \to (1, 1, 1)} \dfrac{zx - zy}{x - y}$

20. $\displaystyle\lim_{(x, y, z) \to (1, 1, 1)} (xy - z, -x)$

Exercises 21-28. Find the set of discontinuities of the given function.

21. $f(x, y) = \dfrac{x^4 - y^4}{x^2 - y^2}$

22. $f(x, y) = \dfrac{4x - xy}{4xy - xy^2}$

23. $f(x, y) = \dfrac{x^2 e^{xy}}{\ln e^x}$

24. $f(x, y) = \dfrac{y^2}{x^2 + y^2}$

25. $f(x, y, z) = \dfrac{zx - zy}{x - y}$

26. $f(x, y, z) = \dfrac{x + y}{2y - z}$

27. $f(x, y) = \begin{cases} \dfrac{x^2 - y^2}{x - y} & \text{if } x \neq y \\ x + y & \text{if } x = y \end{cases}$

28. $f(x, y) = \begin{cases} \dfrac{x^2 - y^2}{x - y} & \text{if } x \neq y \\ 0 & \text{if } x = y \end{cases}$

29. Construct a function $f: \Re^2 \to \Re$ with domain $\Re^2$ which is discontinuous only at $(0, 1)$.

30. Construct a function $f: \Re^2 \to \Re$ with domain $\Re^2$ which is discontinuous only at $(0, 1)$ and $(1, 0)$.

31. Construct a function $f: \Re^3 \to \Re$ with domain $\Re^3$ which is discontinuous only at $(0, 1, 0)$.

32. Show that the function $f(x, y) = \dfrac{x^2 y}{x^4 + y^2}$ approaches 0 as $(x, y) \to (0, 0)$ along any line $y = mx$. Does $\lim\limits_{(x, y) \to (0, 0)} f(x, y)$ exist?

Exercises 33-41. Prove:

33. Theorem 11.1(b)   34. Theorem 11.1(c)

35. Theorem 11.2(a)   36. Theorem 11.2(b)

37. Theorem 11.2(d)   38. Theorem 11.2(e)

39. If $f(x, y) = x$, then $\lim\limits_{(x, y) \to (x_0, y_0)} f(x, y) = x_0$.

40. If $f(x, y) = y$, then $\lim\limits_{(x, y) \to (x_0, y_0)} f(x, y) = y_0$.

41. For any constant c, $\lim\limits_{(x, y) \to (x_0, y_0)} c = c$

§2. GRAPHING FUNCTIONS OF TWO VARIABLES

A general discussion of planes takes place in Section 12.3.

Just as the graph of the function $y = f(x) = ax + b$ is a line, so then is the graph of $z = f(x, y) = ax + by + c$ a plane in three-space. And just as a line is determined by two points, so then is a plane determined by three points (that do not lie on a common line). Consider the following example.

EXAMPLE 11.2 (a) Sketch the graph of the plane
$$z = -2x - 4y + 4$$

(b) Find the equation of the plane containing the points $(-1, 1, 1)$, $(1, 0, 6)$, $(0, 1, 3)$.

SOLUTION: (a) We choose three "nice" points on the plane; specifically, points of the form $(x, 0, 0)$, $(0, y, 0)$, and $(0, 0, z)$:

Setting y and z to 0 in $z = -2x - 4y + 4$ we have $0 = -2x + 4$ or $x = 2$. Thus $(2, 0, 0)$ lies on the plane.

Setting x and z to 0 in $z = -2x - 4y + 4$ we have $0 = -4y + 4$ or $y = 1$. Thus $(0, 1, 0)$ lies on the plane.

Setting x and y to 0 in $z = -2x - 4y + 4$ we have $z = 4$. Thus $(0, 0, 4)$ lies on the plane.

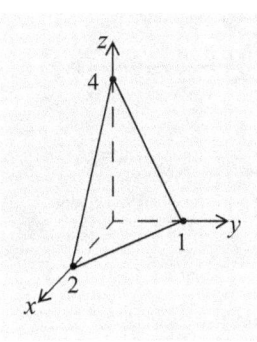

One can envision the plane containing those points (a sheet of paper sitting upon them). The first octant portion of the plane is depicted in the margin.

(b) In order for the point $(x, y, z) = (-1, 1, 1)$ to lie on the plane

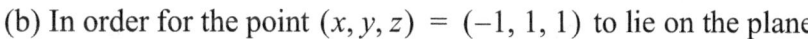

$z = ax + by + c$ we must have: $\underset{z}{1} = a\underset{x}{(-1)} + b \cdot \underset{y}{1} + c$

Similarly, $(1, 0, 6)$ and $(0, 1, 3)$ lead us to:
$$6 = a \cdot 1 + b \cdot 0 + c \quad \text{and} \quad 3 = a \cdot 0 + b \cdot 1 + c$$
And so we have three equations in three unknowns:

(1): $-a + b + c = 1$ \qquad (2): $a = 6 - c$ \qquad (3): $b = 3 - c$

Substituting for a and b in (1): $-(6 - c) + (3 - c) + c = 1 \Rightarrow \boxed{c = 4}$.

Substituting 4 for c in (2) and (3): $\boxed{a = 2}$ and $\boxed{b = -1}$.

Conclusion: The plane $z = 2x - y + 4$ contains $(-1, 1, 1)$, $(1, 0, 6)$, and $(0, 1, 3)$.

While the graph of a function $y = f(x)$ of one variable resides in two-dimensional space (the plane), that of a function $z = f(x, y)$ of two variables lives in three-dimensional space; which can be difficult to draw on a two-dimensional surface. One may, however, get a sense of its graph by considering the two-dimensional **traces** resulting from the intersection of the surface with planes parallel to the coordinate planes. Consider the following example.

EXAMPLE 11.3 Sketch the graph of the function:
$$f(x, y) = x^2 + y^2$$

SOLUTION: Setting $z = 4, z = 9, z = 16$ in $z = f(x, y) = x^2 + y^2$ yields the concentric circles of Figure 11.1(a). Hoisting those circles 4, 9, and 16 units up the z axis brings us to Figure 11.1(b). Imagining that those hoisted circles are connected by some sort of elastic membrane takes us to the graph depicted in Figure 11.1(c), called a paraboloid as its traces on the xz- and yz-axis are parabolas.

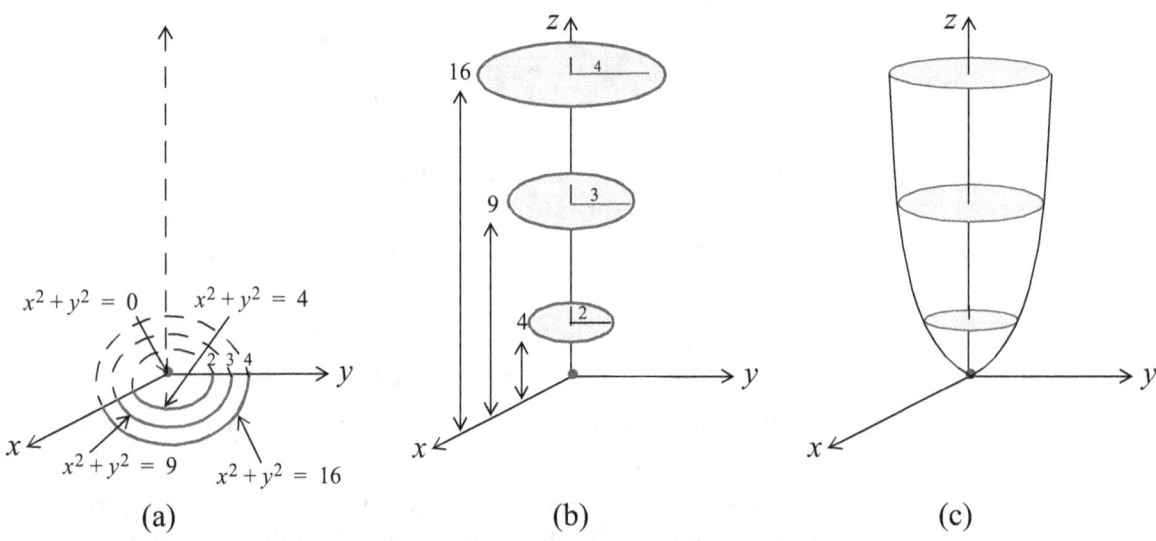

Figure 11.1

EXAMPLE 11.4 Sketch the graph of the function:
$$f(x, y) = x^2 - y$$

SOLUTION: Setting $z = 0, z = 2, z = 4$ in $z = f(x, y) = x^2 - y$ yields the parabolas of Figure 11.2(a). Hoisting those parabolas 0, 2, and 4 units up the z axis brings us to Figure 11.2(b). Figure 11.2(c) "solidifies" the construction process. Note that its trace on the xz-plane is the parabola $z = x^2$, while its trace on the yz-plane is the line $z = -y$.

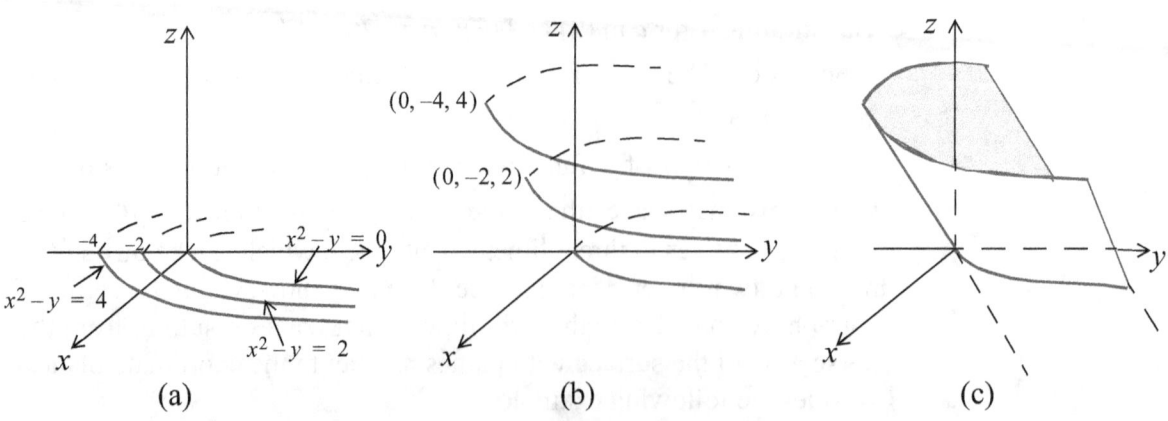

Figure 11.2

Answer: See page A-2.

CHECK YOUR UNDERSTANDING 11.5
Sketch the graph of: $$f(x, y) = |x| - y$$

DEFINITION 11.5
CYLINDER

A **cylinder** in $\Re^3$ is composed of lines (called **rulings**) that are parallel to a given line and pass through a given plane curve (called a **directrix**).

We will restrict our attention to cylinders with directrix residing in a coordinate plane and with rulings perpendicular to the directrix. The equation of such a cylinder contains only two variables, and the surface extends forever in the direction of the missing variable. Consider the following example.

EXAMPLE 11.5 Sketch the graph of the parabolic cylinder:
$$z = x^2$$

SOLUTION: Note that the equation $z = x^2$ does not involve the variable y. This is an indication that we are dealing with a cylinder with directrix $z = x^2$ residing in the xz-plane. Since the y-variable is unrestricted, the rulings run parallel to the y-axis (see adjacent figure).

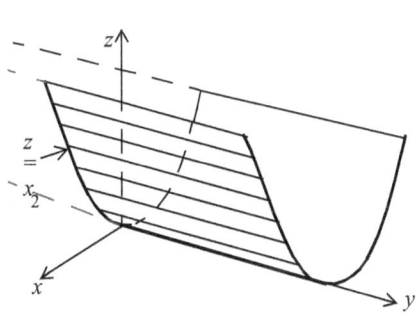

Answer: See page A-2.

CHECK YOUR UNDERSTANDING 11.6
Sketch the graph of the elliptical cylinder $x^2 + \dfrac{y^2}{4} = 1$.

Conic sections (circles, ellipses, parabolas, and hyperbolas) consist of those points in $\Re^2$ satisfying second degree equations in the variables x and y. We now turn our attention to **quadric surfaces** in $\Re^3$ which satisfy second-degree equations in the variables x, y, and z of the form:

$$Ax^2 + By^2 + Cz^2 + Dx + Ey + Fz + G = 0$$

(ellipsoids, paraboloids, cones, and hyperboloids)

General quadratic equations also contain xy, yz, and xz terms.

ELLIPTIC PARABOLOIDS

The quadric surface $x^2 + y^2 - z = 0$ of Example 11.3 is said to be a circular paraboloid [its trace on the planes $z = k$ (for $k \geq 0$) are circles, while those with the planes $x = k$ and $y = k$ are parabolas]. It is a special case of an elliptic paraboloid — one of which is featured below.

In general:

$$\frac{z}{c} = \frac{x^2}{a^2} + \frac{y^2}{b^2}$$

$$\frac{y}{c} = \frac{x^2}{a^2} + \frac{z^2}{b^2}$$

$$\frac{x}{c} = \frac{y^2}{a^2} + \frac{z^2}{b^2}$$

represent **elliptic paraboloids**.

EXAMPLE 11.6   Sketch the graph of the elliptic paraboloid:

$$z = x^2 + \frac{y^2}{9}$$

SOLUTION: Projecting the traces of the surface with the planes $z = 1, z = 4$ and $z = 9$ onto the xy-plane brings us to the ellipses

$x^2 + \dfrac{y^2}{9} = 1$, $x^2 + \dfrac{y^2}{9} = 4$ and $x^2 + \dfrac{y^2}{9} = 9$ appearing in Figure 11.3(a). Lifting those ellipses back up the z-axis by 1, 4, and 9 units brings us to Figure 11.3(b). Note that the traces in the planes $x = k$ and $y = k$ are parabolas. The associated elliptic paraboloid appears In Figure 11.3(c).

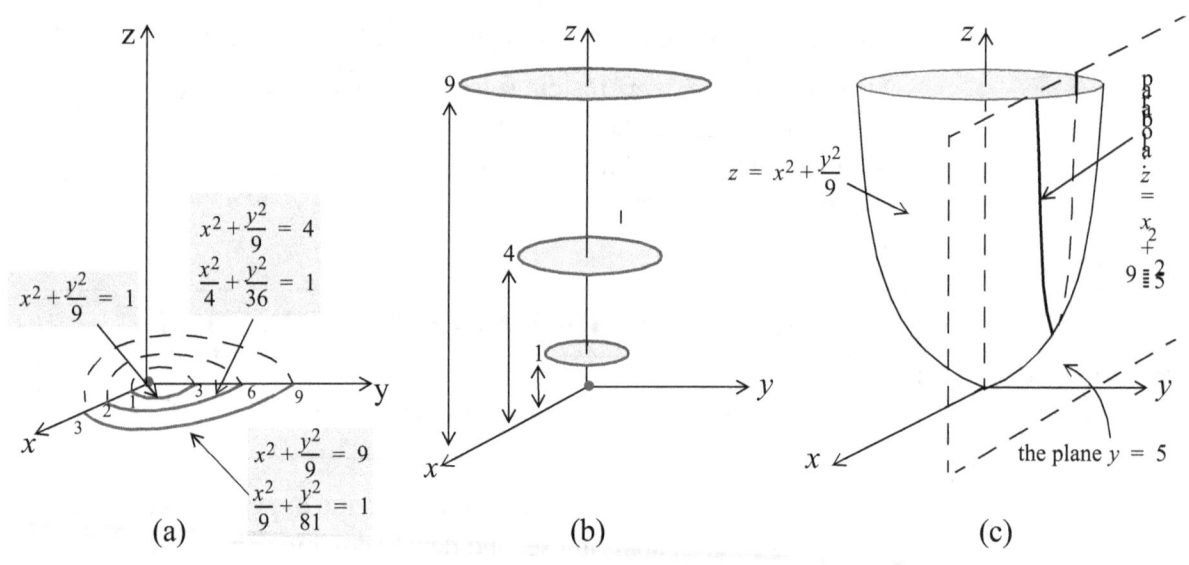

(a)   (b)   (c)

Figure 11.3

CHECK YOUR UNDERSTANDING 11.7

Sketch the graph of the elliptic paraboloid:

$$y = x^2 + \frac{z^2}{9}$$

Answer: See page A-2.

ELLIPSOIDS

Spheres are "special ellipsoids." In particular:
$$\frac{x^2}{r^2} + \frac{y^2}{r^2} + \frac{z^2}{r^2} = 1$$
denotes the sphere centered at the origin of radius r.

Just as equations of the form $\dfrac{x^2}{a^2} + \dfrac{y^2}{b^2} = 1$ represent ellipses centered at the origin, so then do equations of the form $\dfrac{x^2}{a^2} + \dfrac{y^2}{b^2} + \dfrac{z^2}{c^2} = 1$ represent ellipsoids centered at the origin. Consider the following example:

EXAMPLE 11.7 Sketch the ellipsoid:
$$\frac{x^2}{4} + \frac{y^2}{1} + \frac{z^2}{9} = 1$$

SOLUTION: Portions of the elliptical traces of the ellipsoid on the planes $x = 0$, $y = 0$, and $z = 0$ appear in Figure 11.4(a), as do those on the coordinate planes $z = 1$ and $z = 2$. Traces on the planes $x = k$ or $y = k$ **are also ellipses**. The associated ellipsoid appears in Figure 11.4(b).

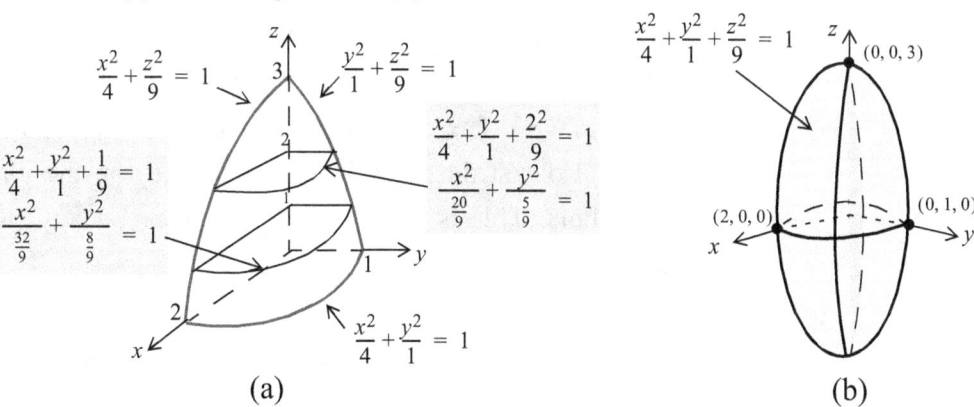

(a)   (b)

Figure 11.4

CHECK YOUR UNDERSTANDING 11.8

Sketch the graph of the ellipsoid:
$$\frac{x^2}{9} + \frac{y^2}{25} + \frac{z^2}{4} = 1$$

Answer: See page A-3.

In general:

$$\frac{z^2}{c^2} = \frac{x^2}{a^2} + \frac{y^2}{b^2}$$

$$\frac{y^2}{c^2} = \frac{x^2}{a^2} + \frac{z^2}{b^2}$$

$$\frac{x^2}{c^2} = \frac{y^2}{a^2} + \frac{z^2}{b^2}$$

represent **elliptic cones**.

ELLIPTIC CONES

As noted in Example 11.6, the equation $z = x^2 + \frac{y^2}{9}$ is that of an elliptic paraboloid. Replacing z with z^2 we arrive at the equation of an elliptic cone:

EXAMPLE 11.8 Sketch the graph of the elliptic cone:

$$z^2 = x^2 + \frac{y^2}{9}$$

SOLUTION: Projecting the trace on the planes $z = 1, z = 2$, and $z = 3$ onto the xy-plane yields the ellipses $x^2 + \frac{y^2}{9} = 1$, $x^2 + \frac{y^2}{9} = 4$ and $x^2 + \frac{y^2}{9} = 9$ that appear in Figure 11.5(a). Since $z = \pm\sqrt{x^2 + \frac{y^2}{9}}$, those three ellipses were both raised and lowered 1, 2, and 3 units in Figure 11.5(b). Note that the traces on the coordinate planes are lines. For example the traces on the plane $y = 0$ are the lines $z = \pm x$. The associated elliptic cone appears in Figure 11.5(c). Note also that the trace in Figure 11.5(c) on the plane $y = 3$ is a hyperbola. This is the case for the trace on any plane $y = k$ (or $x = k$).

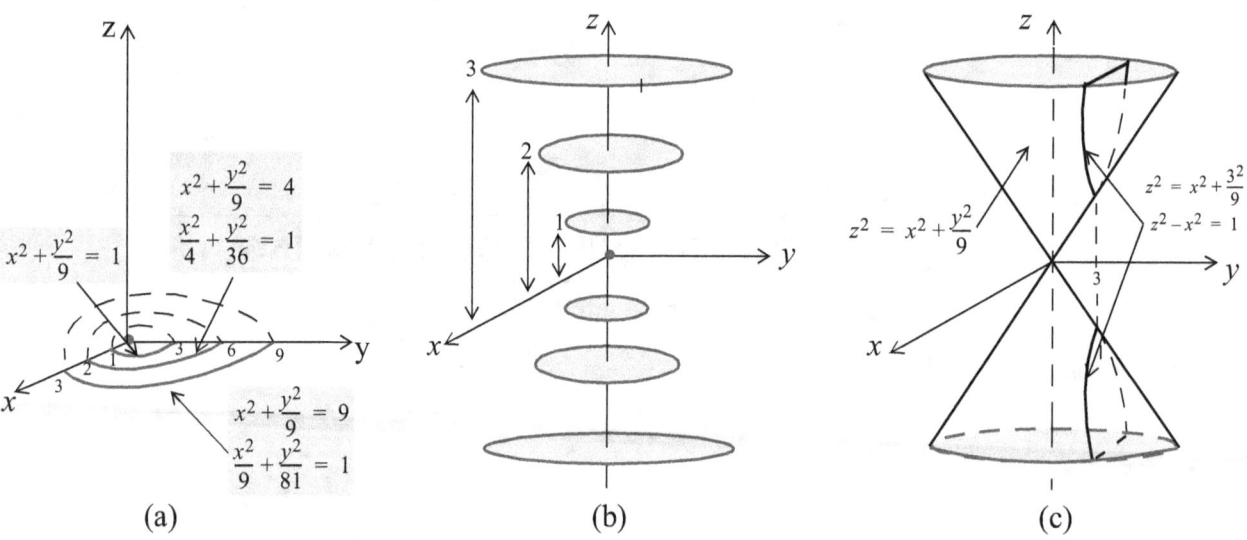

(a)   (b)   (c)

Figure 11.5

CHECK YOUR UNDERSTANDING 11.9

Sketch the graph of the elliptic cone:

$$y^2 = x^2 + \frac{z^2}{9}$$

Answer: See page A-3.

HYPERBOLOIDS OF ONE SHEET AND OF TWO SHEETS

In general:

$$\frac{x^2}{a^2}+\frac{y^2}{b^2}-\frac{z^2}{c^2}=1$$

$$\frac{x^2}{a^2}-\frac{y^2}{b^2}+\frac{z^2}{c^2}=1$$

$$-\frac{x^2}{a^2}+\frac{y^2}{b^2}+\frac{z^2}{c^2}=1$$

represent **hyperbolas of one sheet**.

The equation $\dfrac{x^2}{a^2}+\dfrac{y^2}{b^2}-\dfrac{z^2}{c^2}=1$ represents a hyperboloid of one sheet. Its trace on the xy-plane ($z=0$) is the ellipse $\dfrac{x^2}{a^2}+\dfrac{y^2}{b^2}=1$; that on the xz-plane ($y=0$) is the hyperbola $\dfrac{x^2}{a^2}-\dfrac{z^2}{c^2}=1$; and that on the yz-plane ($x=0$) is the hyperbola $\dfrac{y^2}{b^2}-\dfrac{z^2}{c^2}=1$.

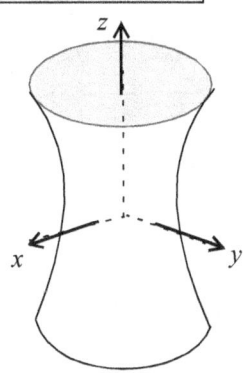

Traces parallel to the xy-plane are again ellipses which expand as one moves further up or down along the z-axis. Traces parallel to the other coordinate planes are hyperbolas with vertices's moving further from the origin as you move away from the x or y-axis.

The equation $-\dfrac{x^2}{a^2}-\dfrac{y^2}{b^2}+\dfrac{z^2}{c^2}=1$ or: $\dfrac{x^2}{a^2}+\dfrac{y^2}{b^2}=\dfrac{z^2}{c^2}-1$

In general:

$$\frac{x^2}{a^2}-\frac{y^2}{b^2}-\frac{z^2}{c^2}=1$$

$$-\frac{x^2}{a^2}-\frac{y^2}{b^2}+\frac{z^2}{c^2}=1$$

$$-\frac{x^2}{a^2}+\frac{y^2}{b^2}-\frac{z^2}{c^2}=1$$

represent **hyperbolas of two sheets**,

represents a hyperboloid of two sheets. The traces on the planes $z=h$ for $|h|\geq|c|$ are ellipses which expand as $|h|$ increases. Traces parallel to the other coordinate planes are hyperbolas with vertices's moving further from the origin as you move away from the x or y-axis.

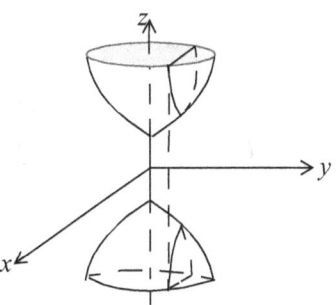

In general:

$$\frac{z}{c}=\frac{x^2}{a^2}-\frac{y^2}{b^2}$$

$$\frac{y}{c}=\frac{x^2}{a^2}-\frac{z^2}{b^2}$$

$$\frac{x}{c}=\frac{y^2}{a^2}-\frac{z^2}{b^2}$$

represent **hyperbolic paraboloids**.

HYPERBOLIC PARABOLOIDS

Equations of the form $z=\dfrac{x^2}{a^2}-\dfrac{y^2}{b^2}$ represent hyperbolic paraboloids. Taking the easy way out:

saddle point

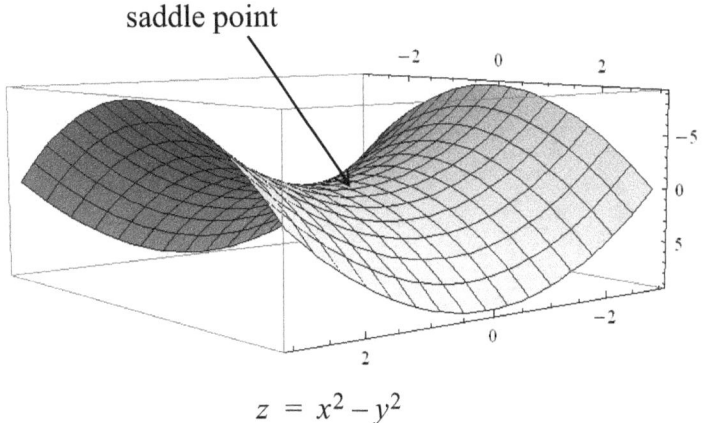

$$z=x^2-y^2$$

COURTESY OF *MATHEMATICA*			
Ellipsoid $$\frac{x^2}{a^2}+\frac{y^2}{b^2}+\frac{z^2}{c^2}=1$$		**Elliptic Paraboloid** $$\frac{z}{c}=\frac{x^2}{a^2}+\frac{y^2}{b^2}$$ $$\frac{y}{c}=\frac{x^2}{a^2}+\frac{z^2}{b^2}$$ $$\frac{x}{c}=\frac{y^2}{a^2}+\frac{z^2}{b^2}$$	
Elliptic Cone $$\frac{z^2}{c^2}=\frac{x^2}{a^2}+\frac{y^2}{b^2}$$ $$\frac{y^2}{c^2}=\frac{x^2}{a^2}+\frac{z^2}{b^2}$$ $$\frac{x^2}{c^2}=\frac{y^2}{a^2}+\frac{z^2}{b^2}$$		**Hyperboloid of One Sheets** $$\frac{x^2}{a^2}+\frac{y^2}{b^2}-\frac{z^2}{c^2}=1$$ $$\frac{x^2}{a^2}-\frac{y^2}{b^2}+\frac{z^2}{c^2}=1$$ $$-\frac{x^2}{a^2}+\frac{y^2}{b^2}+\frac{z^2}{c^2}=1$$	
Hyperboloid of Two Sheets $$\frac{x^2}{a^2}-\frac{y^2}{b^2}-\frac{z^2}{c^2}=1$$ $$-\frac{x^2}{a^2}-\frac{y^2}{b^2}+\frac{z^2}{c^2}=1$$ $$-\frac{x^2}{a^2}+\frac{y^2}{b^2}-\frac{z^2}{c^2}=1$$		**Hyperbolic Paraboloid** $$\frac{z}{c}=\frac{x^2}{a^2}-\frac{y^2}{b^2}$$ $$\frac{y}{c}=\frac{x^2}{a^2}-\frac{z^2}{b^2}$$ $$\frac{x}{c}=\frac{y^2}{a^2}-\frac{z^2}{b^2}$$	

	EXERCISES	

Exercises 1-2. Sketch the graph of the given plane in the first octant.

1. $z = -4x - 2y + 2$ 2. $z = -x - y + 1$

Exercises 1-3. Find the equation of the plane containing the given points.

3. $(0, 0, 0)$, $(1, 0, 2)$, $(0, 2, 5)$ 4. $(-1, 1, 1)$, $(1, 0, 0)$, $(0, 1, 2)$

Exercises 5-10. Sketch the given cylinder in $\Re^3$.

5. $x + 2y = 1$ 6. $4x^2 + z = 4$ 7. $y^2 + z = 0$

8. $25x^2 - 9y^2 - 1 = 0$ 9. $x^2 - 2y^2 = 1$ 10. $x - 2xy = 1$

Exercises 11-26. Identify and sketch the given quadratic surface.

11. $z = x^2 + y^2$ 12. $x = y^2 + 4z^2$

13. $36x^2 + 9y^2 + 4z^2 = 36$ 14. $x - y^2 - z^2 = 0$

15. $x^2 - y^2 + z^2 = 0$ 16. $x^2 + 2y + z^2 = 0$

17. $25x^2 - 4y^2 + 25z^2 + 100 = 0$ 18. $x^2 + 4y + z^2 = 0$

19. $x^2 + 4y - z^2 = 0$ 20. $16x^2 - 9y^2 - 9z^2 = 0$

21. $x^2 + y^2 - 4z^2 = 4$ 22. $9x^2 - 36y^2 + 16z^2 + 144 = 0$

23. $25x^2 - 4y^2 + 25z^2 = 100$ 24. $z = (x + 2)^2 + (y - 3)^2 - 9$

25. $9x^2 - 4y^2 = 36z$ 26. $x^2 - y^2 - 9z = 0$

Why discuss integrals prior to derivatives? Because, as you will see, several variations of the derivative concept come into play within the realm of multivariable functions, while integration is more straightforward.

§3. DOUBLE INTEGRALS

We take the definition of an integrable function $y = f(x)$ of one variable appearing on page 178:

DEFINITION 5.3

DEFINITE INTEGRAL

A function $y = f(x)$ is said to be **integrable** over the interval $[a, b]$ if $\lim\limits_{\Delta x \to 0} \sum\limits_a^b f(x)\Delta x$ exists. In this case, we write:

$$\int_a^b f(x)dx = \lim_{\Delta x \to 0} \sum_a^b f(x)\Delta x$$

and modify it to accommodate a function of two variables:

DEFINITION 11.6

A function $z = f(x, y)$ is said to be **integrable** over the region R in the plane (see margin) if $\lim\limits_{\substack{\Delta x \to 0 \\ \Delta y \to 0}} \sum\limits_R f(x, y)\Delta A$ exists. In this case, we write:

$$\iint\limits_R f(x, y)dA = \lim_{\substack{\Delta x \to 0 \\ \Delta y \to 0}} \sum_R f(x, y)\Delta A$$

$$= \lim_{\substack{\Delta x \to 0 \\ \Delta y \to 0}} \sum_R f(x, y)\Delta x \Delta y$$

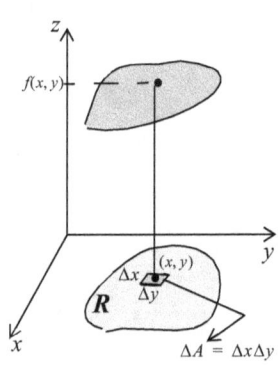

Quite a modification! But just as the "geometrical-area" approach lead us to Definition 5.3:

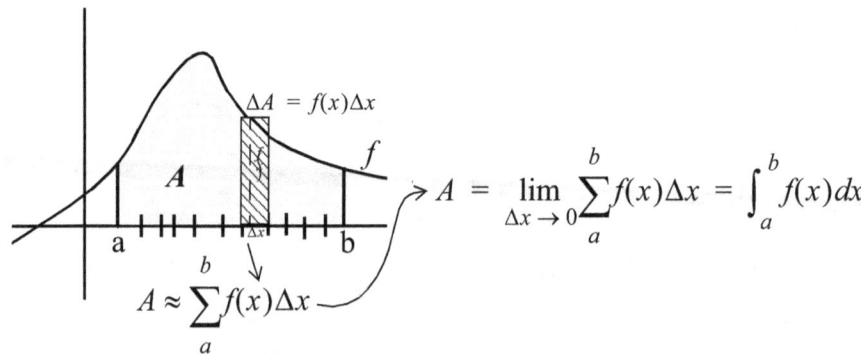

$$A = \lim_{\Delta x \to 0} \sum_a^b f(x)\Delta x = \int_a^b f(x)dx$$

$$A \approx \sum_a^b f(x)\Delta x$$

so then will the geometric approach highlighted in Figure 11.6 lead us to a better understanding of Definition 116:

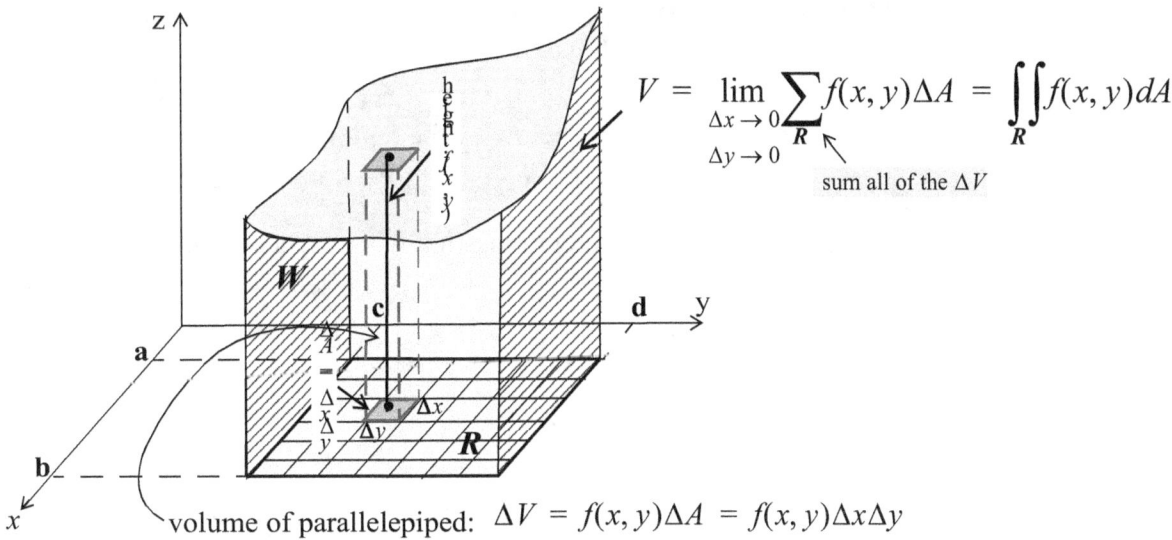

$$V = \lim_{\substack{\Delta x \to 0 \\ \Delta y \to 0}} \sum_R f(x,y)\Delta A = \iint_R f(x,y)dA$$

sum all of the ΔV

volume of parallelepiped: $\Delta V = f(x,y)\Delta A = f(x,y)\Delta x \Delta y$

Figure 11.6

Fine, but how does one go about evaluating $\iint_R f(x,y)dA$? Like this:

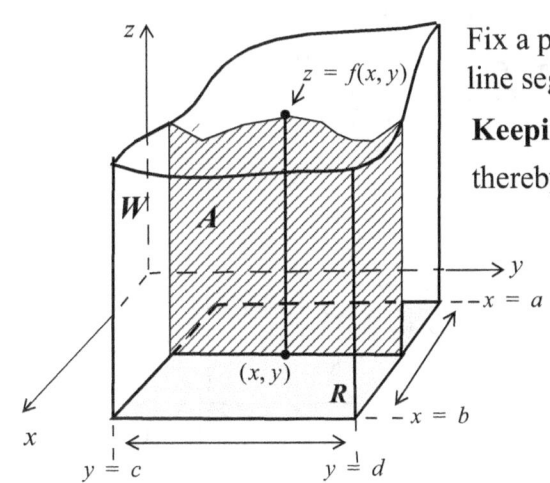

Fix a point (x, y) in **R**, and consider the depicted vertical line segment from (x, y) to $z = f(x, y)$.

Keeping x fixed, slide that line from $y = c$ to $y = d$,

thereby sweeping out the hashed two-dimensional region A.

Next, slide A from $x = a$ to $x = b$, thereby filling in the entire solid **W**.

This brings us to the following so-called iterated double integral representation:

integrating first with respect to y holding x fixed.

$$V = \iint_R f(x,y)dA = \int_a^b \left[\int_c^d f(x,y)dy \right] dx$$

Figure 11.7

One can also arrive at the volume of the above solid **W** by first sliding the vertical line from $x = a$ to $x = b$, and then sliding the resulting planar region, now perpendicular to the y-axis, from $y = c$ to $y = d$. In other words:

$$V = \iint_R f(x,y)dA = \int_c^d \int_a^b f(x,y)dxdy$$

in this integral, y is held fixed

EXAMPLE 11.9 Calculate the volume V of the solid bounded above by the surface

$$z = f(x, y) = x^2 + 6xy$$

and below by the region:

$$R = \{(x, y) | 0 \le x \le 2, 1 \le y \le 3\}$$

SOLUTION: One way (Integrating first with respect to y):

$$\overset{\text{x is held fixed and y runs from 1 to 3}}{\uparrow}$$

$$V = \iint_R (x^2 + 6xy) dA = \int_0^2 \int_1^3 (x^2 + 6xy) dy dx$$

$$= \int_0^2 (x^2 y + 3xy^2) \Big|_{y=1}^{y=3} dx$$

$$= \int_0^2 [(3x^2 + 27x) - (x^2 + 3x)] dx$$

$$= \int_0^2 [2x^2 + 24x] dx = \left(2 \cdot \frac{x^3}{3} + 12x^2\right) \Big|_0^2 = \frac{160}{3}$$

Another way (Integrating first with respect to x):

$$\overset{\text{y is held fixed and x runs from 0 to 2}}{\uparrow}$$

$$V = \iint_R (x^2 + 6xy) dA = \int_1^3 \int_0^2 (x^2 + 6xy) dx dy$$

$$= \int_1^3 \left(\frac{x^3}{3} + 3x^2 y\right) \Big|_{x=0}^{x=2} dy$$

$$= \int_1^3 \left[\left(\frac{8}{3} + 12y\right)\right] dy = \left(\frac{8}{3}y + 6y^2\right) \Big|_1^3 = \frac{160}{3}$$

CHECK YOUR UNDERSTANDING 11.10

Determine the volume of the solid W bounded above by the surface $z = f(x, y) = 2x + y + 3xy$ and below by the region $R = \{(x, y) | 1 \le x \le 2, 2 \le y \le 4\}$. Do so by integrating first with respect to y, and then again by integrating first with respect to x.

Answer: 39

Our next goal is to set up an integrated integral representing the volume V of the solid in Figure 11.8 that lies above the region R, and below the surface $z = h(x, y)$. Note that, unlike the situation in Figure 11.7, the region R is no longer a rectangle.

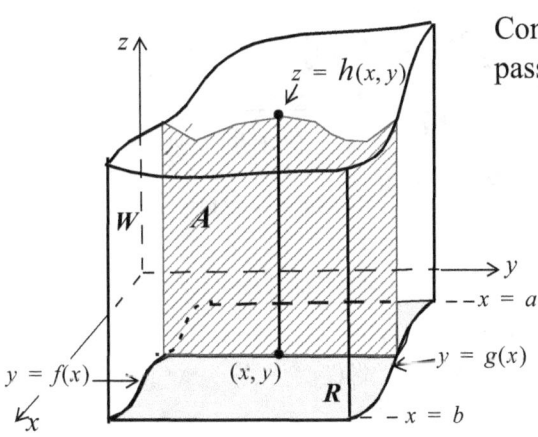

Consider, again, the vertical line of length $z = h(x, y)$ passing through the point (x, y) in R.

Keeping x fixed, slide the line from the curve $y = f(x)$ to the curve $y = g(x)$, sweeping out the hashed region A.

Finally, slide that two dimenstional region from $x = a$ to $x = b$ to fill in the entire solid W.

Bringing us to:

$$V = \iint_R h(x, y)dA = \int_a^b \int_{f(x)}^{g(x)} h(x, y)dydx$$

Figure 11.8

Leading us to (proof omitted):

THEOREM 11.5   Let $h(x, y)$ be continuous on a region R.

(a) For R defined by $a \le x \le b$ and $f(x) \le y \le g(x)$:

$$\iint_R h(x, y)dA = \int_a^b \int_{f(x)}^{g(x)} h(x, y)dydx$$

[see Figure 11.9(a)]

(b) For R defined by $c \le y \le d$ and $f(y) \le x \le g(y)$:

$$\iint_R h(x, y)dA = \int_c^d \int_{f(y)}^{g(y)} h(x, y)dxdy$$

[see Figure 11.9(b)]

Fact: If h is continuous on R then h is integrable on R,

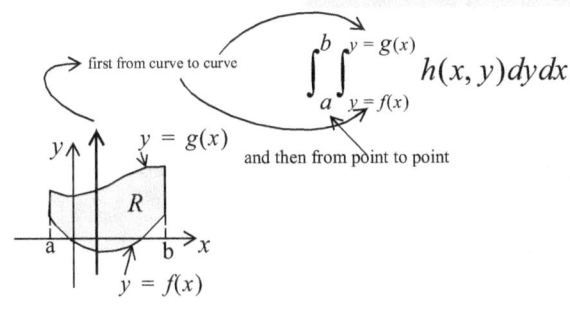

Integrating first with respect to y
(holding x fixed)

(a)

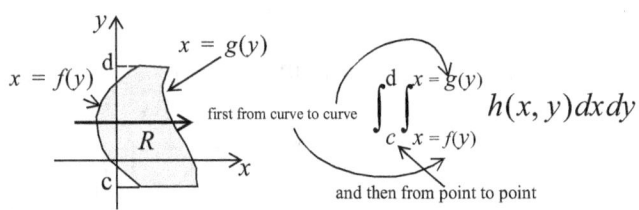

Integrating first with respect to x
(holding y fixed)

(b)

Figure 11.9

EXAMPLE 11.10   Let R be the finite region bounded by the lines $x = 0$, $y = x$, and $y = -x + 2$. Evaluate $\iint_R 2xy dA$ by integrating first with:

(a) respect to y.   (b) respect to x.

SOLUTION:

(a)

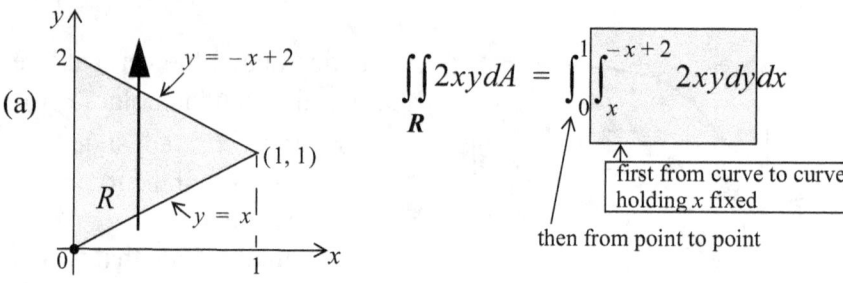

Bringing us to:

$$\iint_R 2xy\,dA = \int_0^1 \int_x^{-x+2} 2xy\,dy\,dx = \int_0^1 2x\frac{y^2}{2}\Big|_{y=x}^{y=-x+2} dx = \int_0^1 x[(-x+2)^2 - x^2]\,dx$$

(x is held fixed)

$$= \int_0^1 (-4x^2 + 4x)\,dx$$

$$= \left(-4\frac{x^3}{3} + 4\frac{x^2}{2}\right)\Big|_0^1$$

$$= -\frac{4}{3} + 2 = \frac{2}{3}$$

(b)

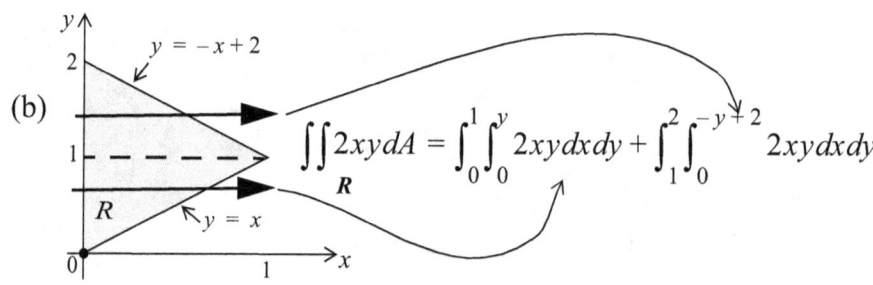

Bringing us to:

$$\iint_R 2xy\,dA = \int_0^1 \int_0^y 2xy\,dx\,dy + \int_1^2 \int_0^{-y+2} (2xy)\,dx\,dy$$

y is held fixed: $= \int_0^1 \left(2\frac{x^2}{2}y\right)\Big|_{x=0}^{x=y} dy + \int_1^2 \left(2\frac{x^2}{2}y\right)\Big|_{x=0}^{x=-y+2} dy$

$$= \int_0^1 y^3\,dy + \int_1^2 (-y+2)^2 y\,dy = \int_0^1 y^3\,dy + \int_1^2 (y^3 - 4y^2 + 4y)\,dy$$

$$= \frac{y^4}{4}\Big|_0^1 + \left(\frac{y^4}{4} - 4\frac{y^3}{3} + 2y^2\right)\Big|_1^2$$

$$= \frac{1}{4} + \left[\left(4 - \frac{32}{3} + 8\right) - \left(\frac{1}{4} - \frac{4}{3} + 2\right)\right] = \frac{2}{3}$$

CHECK YOUR UNDERSTANDING 11.11

Let R denote the finite region R in the xy-plane bounded by $y = -x + 1$, $y = \sqrt{x - 1}$ and $y = 2$.

(a) Express $\iint\limits_R 2xy\,dA$ in both iterated integral forms.

(b) Evaluate $\iint\limits_R 2xy\,dA$.

Answer for (b): 20

Employing the "dominant-minus-subordinate" approach used is Example5.17, page197, we now calculate the volume V of the solid W depicted below.

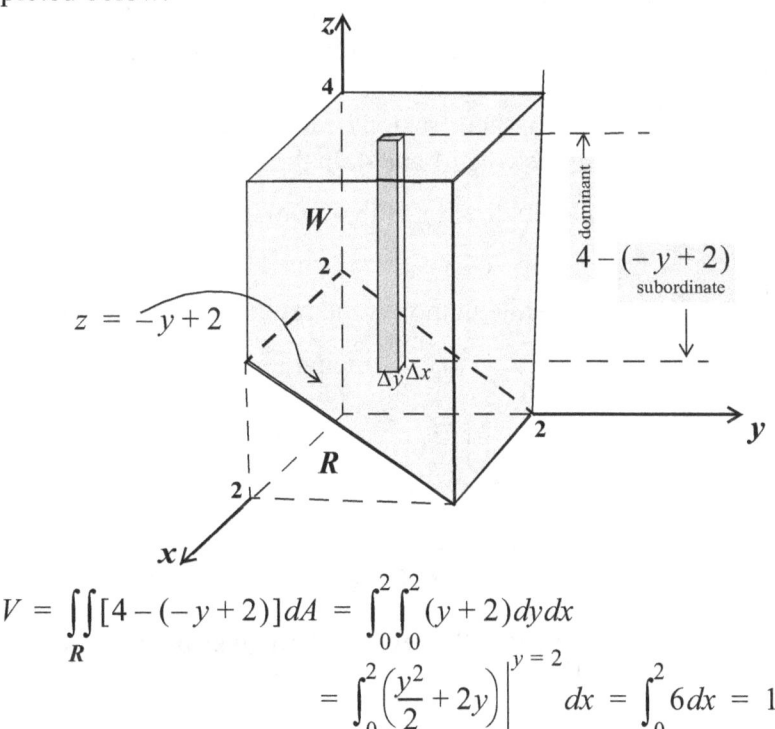

$$V = \iint\limits_R [4 - (-y + 2)]\,dA = \int_0^2 \int_0^2 (y + 2)\,dy\,dx$$

$$= \int_0^2 \left(\frac{y^2}{2} + 2y\right)\Bigg|_{y=0}^{y=2} dx = \int_0^2 6\,dx = 12$$

CHECK YOUR UNDERSTANDING 11.12

Determine the volume V of the solid W in the adjacent figure.

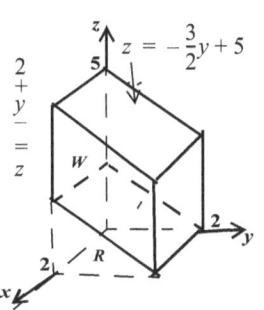

Answer: 10

Let's turn things around. Instead of going from a region R to an iterated integral, we now reconstruct R from a given integral:

EXAMPLE 11.11   Sketch the region R of integration for the integral:

$$\iint_R h(x,y)\,dA = \int_{-1}^{2} \int_{x^2}^{x+2} h(x,y)\,dy\,dx$$

and then write an equivalent integral with the order of integration reversed.

SOLUTION: Fixing x between -1 and 2, we go from the curve $y = x^2$ to the curve $y = x + 2$ (see R on the right).

As is suggested in the margin figure, two integrals are needed in order to reverse the order of integration: one if we fix y between 0 and 1 (hashed region), and another if we fix y between 1 and 4. In the hashed region, the line labeled (1) will go from the curve $x = -\sqrt{y}$ to the curve $x = \sqrt{y}$ $(y = x^2 \Rightarrow x = \pm\sqrt{y})$; while in the second region, the line labeled (2) will go from $x = y - 2$ to $x = \sqrt{y}$. Bringing us to:

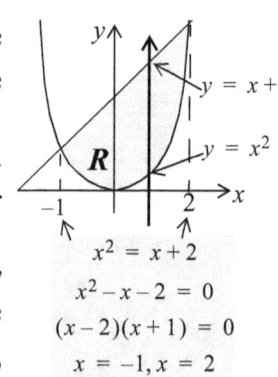

$$x^2 = x + 2$$
$$x^2 - x - 2 = 0$$
$$(x-2)(x+1) = 0$$
$$x = -1, x = 2$$

$$\iint_R h(x,y)\,dA = \int_0^1 \int_{-\sqrt{y}}^{\sqrt{y}} h(x,y)\,dx\,dy + \int_1^4 \int_{y-2}^{\sqrt{y}} h(x,y)\,dx\,dy$$

CHECK YOUR UNDERSTANDING 11.13

Sketch the region **of integration R** for the integral

$$\iint_R h(x,y)\,dA = \int_0^1 \left[\int_0^{x^2} h(x,y)\,dy \right] dx + \int_1^2 \left[\int_0^{-x+2} h(x,y)\,dy \right] dx$$

and then write an equivalent integral with the order of integration reversed.

Answer:
$$\int_0^1 \left[\int_{\sqrt{y}}^{-y+2} h(x,y)\,dx \right] dy$$

MASS

An idealized thin flat object is called a **lamina**. A **homogeneous lamina** is a lamina with constant density throughout, where:

The **density**, δ, of a homogeneous lamina of mass M and area A is defined to be its mass per unit Area:

$$\delta = \frac{M}{A}. \quad \text{Leading us to: } M = \delta A$$

A lamina that is not homogeneous is said to be **inhomogeneous**.

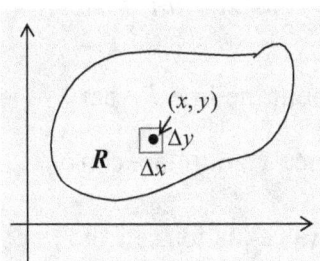

The **density function**, $\delta(x, y)$, at a point (x, y) in an inhomogeneous lamina R in the xy-plane is defined to be the limit of the masses of rectangular regions containing (x, y) divided by the area of those regions as the dimensions of the rectangular regions tend to zero:

$$\delta(x, y) = \lim_{\substack{\Delta x \to 0 \\ \Delta y \to 0}} \frac{\Delta M}{\Delta A}. \quad \text{Leading us to: } \Delta M \approx \delta(x, y)\Delta A$$

Bringing us to:

> The total **mass** M of a lamina R with density function $\delta(x, y)$ is given by:
>
> $$M = \iint_R \delta(x, y)dA$$

EXAMPLE 11.12 A triangular lamina R in the xy-plane bounded by the lines $y = 0$, $y = x$ and $y = -x + 2$ has density function $\delta(x, y) = 2xy$. Find its total mass.

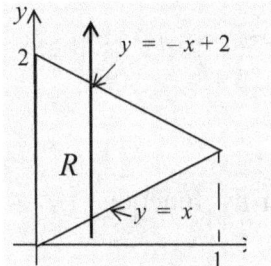

SOLUTION:

$$M = \iint_R \delta(x, y)dA = \iint_R 2xydA$$

$$= \int_0^1 \int_x^{-x+2} (2xy) \; dydx = \frac{2}{3}$$

See Example 11.10

CHECK YOUR UNDERSTANDING 11.14

Answer: $\frac{1}{24}$

Find the mass of a triangular lamina with vertices at $(0, 0)$, $(0, 1)$, and $(1, 0)$ with density function $\delta(x, y) = xy$.

CENTER OF MASS

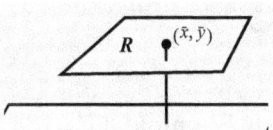

Roughly speaking, the center of mass, or center of gravity, of a lamina R is the point $(\bar{x}, \bar{y})$ in R about which the lamina is "horizontally balanced" (see margin).

To be horizontally balanced at $(\bar{x}, \bar{y})$, R must surely be horizontally balanced when positioned on the line $L_{\bar{x}}$ parallel to the y-axis at $\bar{x}$. Partitioning the region into small rectangular regions of area ΔA_i, we conclude (see margin) that the lamina R with density function $\delta(x, y)$ will nearly balance about $L_{\bar{x}}$ if:

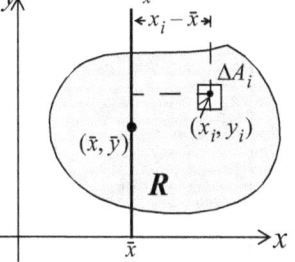

If masses $m_1, m_2, ..., m_n$ are positioned at $x_1, x_2, ..., x_n$:

then the seesaw will balance if and only if:

$$\sum_{i=1}^{n} (x_i - p)m_i = 0$$

$$\sum_R (x_i - \bar{x})\delta(x_i, y_i)\Delta A_i = 0$$

It follows that R will balanced about $L_{\bar{x}}$ if: $\iint\limits_R (x - \bar{x})\delta(x, y)dA = 0$.

A similar argument reveals that R will balance about the line $L_{\bar{y}}$ parallel to the x-axis at $\bar{y}$: $\iint\limits_R (y - \bar{y})\delta(x, y)dA = 0$. Since $\bar{x}$ and $\bar{y}$ are constant, we can express the above two double integral equations in the form:

$$\iint\limits_R x\delta(x, y)dA = \bar{x}\iint\limits_R \delta(x, y)dA \text{ and } \iint\limits_R y\delta(x, y)dA = \bar{y}\iint\limits_R \delta(x, y)dA$$

$$\bar{x} = \frac{\iint\limits_R x\delta(x, y)dA}{\iint\limits_R \delta(x, y)dA} \quad \text{and} \quad \bar{y} = \frac{\iint\limits_R y\delta(x, y)dA}{\iint\limits_R \delta(x, y)dA}$$

The expression $\iint\limits_R x\delta(x, y)dA$, denoted by M_y, is said to be the **moment of R about the y-axis**, while $M_x = \iint\limits_R y\delta(x, y)dA$ is the **moment of R about the x-axis**.

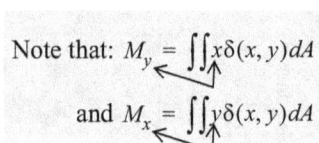
Note that: $M_y = \iint\limits_R x\delta(x, y)dA$
and $M_x = \iint\limits_R y\delta(x, y)dA$

To summarize:

Center of mass $(\bar{x}, \bar{y})$ of a lamina R with density function $\delta(x, y)$ is given by:

$$\bar{x} = \frac{M_y}{M} = \frac{\iint\limits_R x\delta(x, y)dA}{\iint\limits_R \delta(x, y)dA} \qquad \bar{y} = \frac{M_x}{M} = \frac{\iint\limits_R y\delta(x, y)dA}{\iint\limits_R \delta(x, y)dA}$$

EXAMPLE 11.13 Find the center of mass of the triangular lamina in the xy-plane bounded by the lines $y = 0$, $y = \frac{x}{2}$ and $y = -\frac{x}{2} + 4$ with density function $\delta(x, y) = 11 - x - y$.

SOLUTION: We already encountered this lamina in Example 11.14 where we found that $M = \iint\limits_R \delta(x, y)dA = \frac{184}{3}$.

We also have:

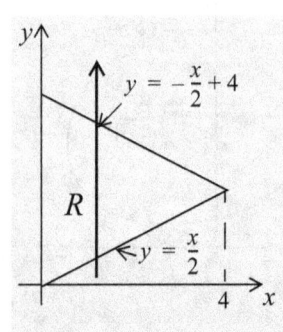

$$M_y = \iint\limits_R x\delta(x,y)dA = \iint\limits_R (11x - x^2 - xy)dA$$

$$= \int_0^4 \left[\int_{x/2}^{-x/2+4} (11x - x^2 - xy)dy \right] dx$$

$$= \int_0^4 \left[\left(11xy - x^2y - \frac{1}{2}xy^2 \right) \Bigg|_{x/2}^{-x/2+4} \right] dx$$

$$= \int_0^4 (x^3 - 13x^2 + 36x)dx = \left(\frac{1}{4}x^4 - \frac{13}{3}x^3 + 18x^2 \right) \Bigg|_0^4 = \frac{224}{3}$$

$$M_x = \iint\limits_R y\delta(x,y)dA = \iint\limits_R (11y - xy - y^2)dA$$

$$= \int_0^4 \int_{x/2}^{-x/2+4} (11y - xy - y^2)dydx$$

$$= \int_0^4 \left(\frac{11}{2}y^2 - \frac{1}{2}xy^2 - \frac{1}{3}y^3 \right) \Bigg|_{x/2}^{-x/2+4} dx$$

$$= \int_0^4 \left(\frac{1}{12}x^3 + x^2 - 22x + \frac{200}{3} \right) dx$$

$$= \left(\frac{1}{48}x^4 + \frac{1}{3}x^3 - 11x^2 + \frac{200}{3}x \right) \Bigg|_0^4 = \frac{352}{3}$$

$$\text{Conclusion:} = \left(\frac{224/3}{184/3}, \frac{352/3}{184/3} \right) = \left(\frac{28}{23}, \frac{44}{23} \right) \approx (1.22, 1.91)$$

CHECK YOUR UNDERSTANDING 11.15

Find the center of mass of a triangular lamina with vertices at $(0,0)$, $(0,1)$, and $(1,0)$ with density function $\delta(x,y) = xy$.

Answer: $\left(\frac{2}{5}, \frac{2}{5} \right)$

	EXERCISES	

Exercises 1-8. Determine the volume V of the solid bounded above by the function $h(x, y)$ and below by the region R by integrating first with respect to y, and also by integrating first with respect to x.

1. $h(x, y) = xy^2$; $R: 0 \leq x \leq 1, 1 \leq y \leq 2$.

2. $h(x, y) = x^2 y^3$; $R: 1 \leq x \leq 2, 2 \leq y \leq 3$.

3. $h(x, y) = y - 2x$; $R: 1 \leq x \leq 2, 3 \leq y \leq 5$.

4. $h(x, y) = 2xy - y^2$; $R: 0 \leq x \leq 4, 0 \leq y \leq 2$.

5. $h(x, y) = \sin x + \cos y$; $R: 0 \leq x \leq \dfrac{\pi}{2}, 0 \leq y \leq \dfrac{\pi}{2}$.

6. $h(x, y) = xy \ln xy$; $R: 1 \leq x \leq 2, 2 \leq y \leq 4$.

7. $h(x, y) = 3xy^2$; R is enclosed by $y = x^2$ and $y = 2x$.

8. $h(x, y) = x^2 + y^2$; R is enclosed by $y = x^2$ and $y = 2x$.

9. Find the volume of the solid enclosed in the first octant by the plane $x + y + z = 1$.

10. Find the volume of the solid bounded by the coordinate planes and the plane $3x + 2y + z = 6$.

11. Find the volume of the solid below the graph of $z = xy$ and above the triangle with vertices $(1, 1)$, $(4, 1)$, and $(1, 2)$.

12. Find the volume of the solid below the graph of $z = x^2 + y^2$ and inside $x^2 + y^2 = 1$.

13. Find the volume of the solid bounded above by the plane $2x + y - z = -2$, and below by the region in the xy-plane bounded by $x = 0$ and $y^2 + x = 1$.

14. Find the volume of the solid bounded by the cylinder $x^2 + y^2 = 4$ and the planes $y + z = 4$ and $z = 0$.

Exercises 15-26. Evaluate.

15. $\displaystyle\int_{-1}^{1}\int_{-x^2}^{x^2} (x^2 - y)\,dy\,dx$

16. $\displaystyle\int_{0}^{1}\int_{x^2}^{x^{1/4}} (x^{1/2} - y^2)\,dy\,dx$

17. $\displaystyle\int_{0}^{1}\int_{-1}^{y} (xy - y^3)\,dx\,dy$

18. $\displaystyle\int_{0}^{1}\int_{y^4}^{\sqrt{y}} (x^{1/2} - y^2)\,dx\,dy$

19. $\displaystyle\int_{0}^{2}\int_{0}^{\sqrt{4-x^2}} (x + 2y)\,dy\,dx$

20. $\displaystyle\int_{0}^{4}\int_{y/2}^{2} e^{8x}\,dx\,dy$

21. $\int_0^2 \int_0^{x/2} e^{x^2}\,dy\,dx$

22. $\int_0^\pi \int_0^1 e^x \sin y\,dy\,dx$

23. $\int_1^2 \int_y^{\sqrt{2y}} xy^2\,dx\,dy$

24. $\int_0^\pi \int_0^{\sin x} x\,dy\,dx$

25. $\int_0^{\sqrt{\pi}} \int_0^{x^2} x\sin y\,dy\,dx$

26. $\int_0^1 \int_x^1 \frac{1}{x}\frac{1}{x^2+1}\,dy\,dx$

Exercises 27-38. Express the iterated integral as an equivalent iterated integral or integrals with the order of integration reversed.

27. $\int_{-3}^2 \int_0^1 h(x,y)\,dy\,dx$

28. $\int_a^b \int_c^d h(x,y)\,dx\,dy$

29. $\int_0^1 \int_{x^4}^{x^2} h(x,y)\,dy\,dx$

30. $\int_0^1 \int_y^1 h(x,y)\,dx\,dy$

31. $\int_{-1}^1 \int_{1/e}^{e^x} h(x,y)\,dy\,dx$

32. $\int_0^{\pi/4} \int_{\sin x}^{\cos x} h(x,y)\,dy\,dx$

33. $\int_0^{2/3} \int_{3y}^{\sqrt{6y}} h(x,y)\,dx\,dy$

34. $\int_1^e \int_0^{\ln x} h(x,y)\,dy\,dx$

35. $\int_0^3 \int_{-\sqrt{9-y^2}}^{\sqrt{9-y^2}} h(x,y)\,dx\,dy$

36. $\int_1^{e^2} \int_{\ln x}^2 h(x,y)\,dy\,dx$

37. $\int_0^1 \int_{1-y}^{1+y} h(x,y)\,dx\,dy$

38. $\int_1^4 \int_x^{2x} h(x,y)\,dy\,dx$

Exercises 39-41. Are you able to evaluate the given iterated integral as stated? If not, then give it a shot after reversing the order of integration.

39. $\int_0^4 \int_{y/2}^2 e^{x^2}\,dx\,dy$

40. $\int_0^1 \int_{\sin^{-1}y}^{\pi/2} \sec^2(\cos x)\,dx\,dy$

41. $\int_0^1 \int_x^1 \sin y^2\,dy\,dx$

Exercises 42-45. Express the double integral as one iterated integral.

42. $\iint_R h(x,y)\,dA = \int_{-1}^0 \int_{-x}^1 h(x,y)\,dy\,dx + \int_0^1 \int_{x^2}^1 h(x,y)\,dy\,dx$

43. $\iint_R h(x,y)\,dA = \int_0^1 \int_0^x h(x,y)\,dy\,dx + \int_1^3 \int_0^1 h(x,y)\,dy\,dx$

44. $\iint_R h(x,y)\,dA = \int_0^1 \int_0^y h(x,y)\,dx\,dy + \int_1^2 \int_0^{2-y} h(x,y)\,dx\,dy$

45. $\iint_R h(x,y)\,dA = \int_1^2 \int_1^y h(x,y)\,dx\,dy + \int_2^4 \int_{y/2}^y h(x,y)\,dx\,dy + \int_4^8 \int_{y/2}^4 h(x,y)\,dx\,dy$

Exercises 46-54. Find the mass of the lamina R with density function $\delta(x, y)$.

46. R is the triangular region with vertices $(0, 0)$, $(0, 2)$, $(1, 0)$, and $\delta(x, y) = 1 + 3x + y$.

47. R is the region bounded by $y = x^3$, $y = 0$, $x = 1$, and $\delta(x, y) = xy$.

48. R is the triangular region with vertices $(2, 1)$, $(4, 3)$, $(6, 1)$, and $\delta(x, y) = 6x + 9y$.

49. R is the triangular region with vertices $(0, 0)$, $(1, 2)$, $(0, 1)$, and $\delta(x, y) = 6x + 6y + 6$.

50. R is the region bounded by $y = x^2$, $y = 0$, $x = 1$, and $\delta(x, y) = 1 + x + y$.

51. R is the region bounded by $y = 2x - 4$, $y = \sqrt{x^2 - 4}$, and $\delta(x, y) = 4x$.

52. R is the region inside $x^2 + 4y^2 = 4$, and $\delta(x, y) = x^2 + y^2$.

53. R is the region bounded by $y = \sin x$, $x = 0$, $x = \pi$, and $\delta(x, y) = |\cos x|$.

54. R is the region bounded by $y = \sin x$, $y = 0$, $y = \pi$, and $\delta(x, y) = x + y$.

Exercises 55-64. Find the center of mass of the lamina R with density function $\delta(x, y)$.

55. R is the rectangular region with vertices $(0, 0)$, $(0, 2)$, $(3, 0)$, $(3, 2)$ and $\delta(x, y) = 6x + 6y + 6$.

56. R is the triangular region with vertices $(0, 0)$, $(2, 1)$, $(0, 3)$, and $\delta(x, y) = xy^2$

57. R is the triangular region with vertices $(0, 0)$, $(2, 1)$, $(0, 3)$, and $\delta(x, y) = x + y$.

58. R is the triangular region enclosed by the lines $y = 6 - 2x$, $y = x$, $x = 0$, and $\delta(x, y) = x^2$.

59. R is the region bounded by $y = \sqrt{x}$, $y = 0$, $x = 1$, and $\delta(x, y) = x + y$.

60. R is the region bounded by $y = \sqrt{x}$, $y = 0$, $x = 1$, and $\delta(x, y) = x$.

61. R is the region bounded by $y = e^x$, $y = 0$, $0 \le x \le 1$, and $\delta(x, y) = y$.

62. R is the region above the x-axis and inside the circle $x^2 + y^2 = 1$, and $\delta(x, y) = x^2 + y^2$.

63. R is the region bounded by, $0 \le x \le \pi$, $0 \le y \le \sin x$, and $\delta(x, y) = 8y$.

64. R is the region bounded by $0 \le x \le e$, $0 \le y \le \ln x$, and $\delta(x, y) = x$.

§4. DOUBLE INTEGRALS IN POLAR COORDINATES

The area ΔA, between the circles of radius r and $r + \Delta r$, and the radial lines of angles θ and $\theta + \Delta\theta$, is the difference of two circular sectors:

$$\Delta A = \frac{1}{2}(r + \Delta r)^2 \Delta\theta - \frac{1}{2}r^2 \Delta\theta$$

$$= r\Delta r \Delta\theta + \frac{1}{2}(\Delta r)^2 \Delta\theta$$

Note: The Area of a sector in a circle of radius r that is subtended by an angle θ is to the area of the circle, as θ is to the circumference of the circle: Since, for small increments, $\frac{1}{2}(\Delta r)^2\Delta\theta$ is negligible compared to $r\Delta r\Delta\theta$, we have:

$$\Delta A \approx r\Delta r\Delta\theta.$$

At times, a region **R** in the xy-plane can more efficiently be represented using polar coordinates[1]. And just as one can partition the region in Figure 11.10(a) into rectangles of area $\Delta A = \Delta x \Delta y$, so then can the polar region in Figure 11.10(b) be partitioned into "polar rectangles" of area $\Delta A \approx r\Delta r\Delta\theta$ (see margin).

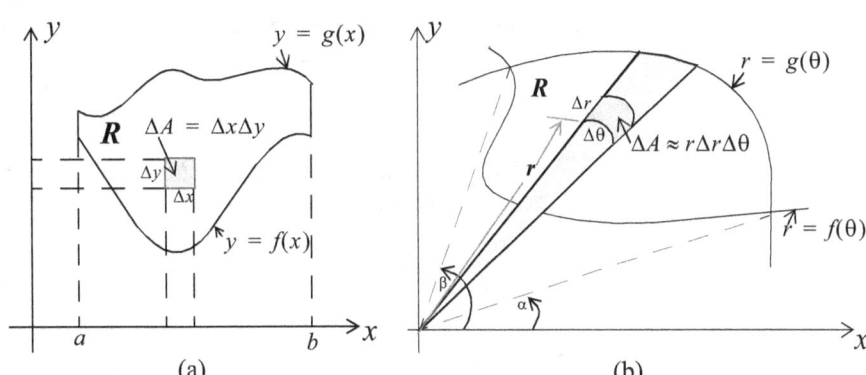

Figure 11.10

Bringing us to:

THEOREM 11.6 If R is a region of the type shown in Figure 11.10(b), and if $h(r, \theta)$ is integrable over **R**, then:

$$\iint_R h(r, \theta)dA = \int_\alpha^\beta \int_{f(\theta)}^{g(\theta)} h(r, \theta)r\,dr\,d\theta$$

EXAMPLE 11.14 For $R = \{(x, y)\,|\,x^2 + y^2 \le 4\}$, evaluate:

$$\iint_R (x^2 + y^2)dA$$

SOLUTION: If you choose to go with the Cartesian coordinate system, then you will be faced with a tedious double integral (see margin):

$$\iint_R (x^2 + y^2)dy\,dx = \int_{-2}^2 \int_{-\sqrt{4-x^2}}^{\sqrt{4-x^2}} (x^2 + y^2)dy\,dx$$

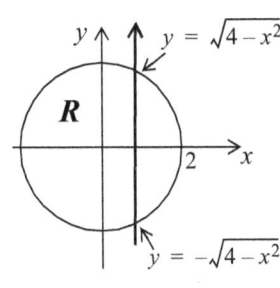

Recall (see page 405):
$$x = r\cos\theta \qquad y = r\sin\theta$$
$$r^2 = x^2 + y^2 \qquad \tan\theta = \frac{y}{x}$$

Turning to polar coordinates:

$$\iint_R (x^2 + y^2)dA = \int_0^{2\pi}\int_0^2 r^2 r\,dr\,d\theta = \int_0^{2\pi} \left.\left(\frac{r^4}{4}\right)\right|_0^2 d\theta$$

$$= 4 \cdot 2\pi = 8\pi$$

---

1. Polar coordinates were first introduced in Section 10.2, page 405.

If a positive function $z = h(x, y)$ is defined on the region R of Figure 11.10(a), then the volume of the surface bounded above by h and below by the region R is given by:

$$V = \iint_R h(x, y)dA = \int_a^b \int_{f(x)}^{g(x)} h(x, y)dydx$$

(see Figure 11.8, page 447)

Similarly, if a positive function $z = h(r, \theta)$ is defined on the region R in Figure 11.10(b), then the volume V of the resulting solid can be approximated by the Riemann sum:

$$V \approx \sum_R h(r, \theta)\Delta A \approx \sum_R h(r, \theta)r\Delta r\Delta \theta$$

Moreover, in a fashion analogous to that developed in the previous section, we have:

$$V = \lim_{\substack{\Delta r \to 0 \\ \Delta \theta \to 0}} \sum_R h(r, \theta)\Delta A = \iint_R h(r, \theta)dA = \int_\alpha^\beta \int_{f(\theta)}^{g(\theta)} h(r, \theta)rdrd\theta$$

Note: Observe that if $h(r, \theta) = 1$ for every (r, θ) in R, then V above is also the area A of R; which is to say:

$$A = \int_\alpha^\beta \int_{f(\theta)}^{g(\theta)} rdrd\theta = \int_\alpha^\beta \frac{r^2}{2}\Big|_{f(\theta)}^{g(\theta)} d\theta = \frac{1}{2}\int_\alpha^\beta ([f(\theta)]^2 - [g(\theta)]^2)d\theta \text{ (sam}$$

(same as formula in Figure 10.10(b), page418)

EXAMPLE 11.15 Use polar coordinates to find the volume of the solid bounded above by

$$z = f(x, y) = (x^2 + y^2)^{3/2}$$

and below by $R = \{(x, y)|x^2 + y^2 \le 1\}$.

SOLUTION: The approach of the previous section brings us to:

$$V = \iint_R (x^2 + y^2)^{3/2}dA = \int_{-1}^1 \left(\int_{-\sqrt{1-x^2}}^{\sqrt{1-x^2}} (x^2 + y^2)^{3/2}dy\right)dx$$

Not a nice integral! There is a better way:

$$V = \iint_R (x^2 + y^2)^{3/2}dA = \iint_R (r^2)^{3/2}dA = \int_0^{2\pi}\left(\int_0^1 r^3 rdr\right)d\theta$$

$$= \int_0^{2\pi}\left(\frac{r^5}{5}\right)\Big|_0^1 d\theta$$

$$= \int_0^{2\pi}\frac{1}{5}d\theta = \frac{2\pi}{5}$$

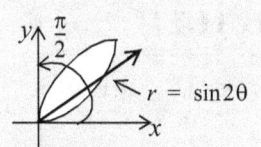

As the arrow indicates, r varies from 0 to $\sin 2\theta$, as

θ varies from 0 to $\dfrac{\pi}{2}$.

EXAMPLE 11.16 Find the area A of one petal of the four-leaf rose $r = \sin 2\theta$ of Example 10.8(b), page 408.

SOLUTION:

$$A = \int_0^{\frac{\pi}{2}}\left(\int_0^{\sin 2\theta} r\,dr\right)d\theta = \int_0^{\frac{\pi}{2}}\left(\frac{r^2}{2}\right)\Bigg|_0^{\sin 2\theta} d\theta = \frac{1}{2}\int_0^{\frac{\pi}{2}}\sin^2 2\theta\,d\theta$$

Theorem 1.5(viii), page 37: $= \dfrac{1}{2}\displaystyle\int_0^{\frac{\pi}{2}}\dfrac{(1-\cos 4\theta)}{2}\,d\theta$

$$= \frac{1}{4}\left(\theta - \frac{1}{4}\sin 4\theta\right)\Bigg|_0^{\frac{\pi}{2}} = \frac{\pi}{8}$$

CHECK YOUR UNDERSTANDING 11.16

(a) Find the volume of the solid bounded by the plane $z = 0$ and the paraboloid $z = 1 - x^2 - y^2$.

(b) Using a double integral, find the area of the region R enclosed by the cardioid $r = 1 - \cos\theta$ (compare with Example 10.10, page 417).

Answer: (a) $\dfrac{\pi}{2}$ (b) $\dfrac{3\pi}{2}$

Recall that:
$$M = \iint_R \delta(x,y)\,dA$$
(page 450)

EXAMPLE 11.17 Find the mass of the region enclosed by the lemniscate $r^2 = 4\cos 2\theta$ of Example 10.8(c), page 370, with $\delta(r,\theta) = r^2$.

SOLUTION: Noting that the region and density are both symmetrical about the x- and y-axis, we find the mass of the region by quadrupling its mass in the first quadrant [see Figure 10.8(a), page 408]:

$$M = \iint_R \delta(r,\theta)\,dA = 4\int_0^{\frac{\pi}{4}}\int_0^{2\sqrt{\cos 2\theta}} r^2 \cdot r\,dr\,d\theta$$

$$= 4\int_0^{\frac{\pi}{4}}\left[\left(\frac{r^4}{4}\right)\Bigg|_0^{2\sqrt{\cos 2\theta}}\right]d\theta = 4\int_0^{\frac{\pi}{4}}\left(\frac{2^4}{4}\cos^2 2\theta\right)d\theta$$

$$= 16\int_0^{\frac{\pi}{4}}\frac{1+\cos 4\theta}{2}\,d\theta$$

$$= 8\left(\theta + \frac{\sin 4\theta}{4}\right)\Bigg|_0^{\frac{\pi}{4}} = 8\left[\frac{\pi}{4} - 0\right] = 2\pi$$

CHECK YOUR UNDERSTANDING 11.17

Find the mass of the region enclosed by the cardioid $r = 1 + \sin\theta$ of Example 10.9, page 411, with $\delta(r, \theta) = r$.

Recall that:

$$\bar{x} = \frac{M_y}{M} = \frac{\displaystyle\iint_R x\delta(x, y)dA}{\displaystyle\iint_R \delta(x, y)dA}$$

$$\bar{y} = \frac{M_x}{M} = \frac{\displaystyle\iint_R y\delta(x, y)dA}{\displaystyle\iint_R \delta(x, y)dA}$$

(page452)

EXAMPLE 11.18   Find the center of mass $(\bar{x}, \bar{y})$ of the semicircular region R given by $r = 2$ for $0 \le \theta \le \pi$, with $\delta(r, \theta) = r$.

SOLUTION: Determining the mass of R:

$$M = \iint_R \delta(x, y)dA = \int_0^\pi \left(\int_0^2 (r \cdot rdr)\right)d\theta$$

$$= \int_0^\pi \left(\frac{r^3}{3}\right)\Big|_0^2 d\theta = \int_0^\pi \frac{8}{3}d\theta = \frac{8\pi}{3}$$

Finding M_y and M_x:

$$M_y = \iint x\delta(x, y)dA = \int_0^\pi \left(\int_0^2 r\cos\theta \cdot r \cdot rdr\right)d\theta$$

$$= \int_0^\pi \left(\frac{r^4}{4}\cos\theta\right)\Big|_0^2 d\theta = 4\int_0^\pi \cos\theta d\theta$$

$$= 4\sin\theta\Big|_0^\pi = 0$$

$$M_x = \iint y\delta(x, y)dA = \int_0^\pi \left(\int_0^2 r\sin\theta \cdot r \cdot rdr\right)d\theta$$

$$= \int_0^\pi \left(\frac{r^4}{4}\sin\theta\right)\Big|_0^2 d\theta = 4\int_0^\pi \sin\theta d\theta$$

$$= -4\cos\theta\Big|_0^\pi = 8$$

Conclusion: $(\bar{x}, \bar{y}) = \left(\dfrac{M_y}{M}, \dfrac{M_x}{M}\right) = \left(\dfrac{0}{8\pi/3}, \dfrac{8}{8\pi/3}\right) = \left(0, \dfrac{3}{\pi}\right)$

CHECK YOUR UNDERSTANDING 11.18

Find the center of mass $(\bar{x}, \bar{y})$ of the semicircular region R given by $r = \cos\theta$ for $0 \le \theta \le \dfrac{\pi}{2}$, with $\delta(r, \theta) = \sin\theta$.

	EXERCISES	

Exercises 1-12. Evaluate.

1. $\int_0^{\frac{\pi}{4}} \int_0^{\sqrt{4\cos 2\theta}} r\,dr\,d\theta$

2. $\int_0^{2\pi} \int_0^2 (8r - 2r^3)\,dr\,d\theta$

3. $\int_{-\frac{\pi}{4}}^{\frac{\pi}{4}} \int_0^{\cos 2\theta} r\,dr\,d\theta$

4. $\int_{-\frac{\pi}{2}}^{\frac{\pi}{2}} \int_0^{1+\sin\theta} r^2\,dr\,d\theta$

5. $\int_0^{\pi} \int_0^1 e^{r^2} r\,dr\,d\theta$

6. $\int_0^{2\pi} \int_0^a e^{-r^2} r\,dr\,d\theta$

7. $\int_0^{\frac{\pi}{2}} \int_0^{\sin\theta} r\cos\theta\,dr\,d\theta$

8. $\int_0^{\pi} \int_0^{1-\sin\theta} r^2\cos\theta\,dr\,d\theta$ 0

9. $\int_0^{\frac{\pi}{6}} \int_{\sin\theta}^{\sec\theta} r\,dr\,d\theta$

10. $\int_0^{\frac{\pi}{4}} \int_0^{\tan\theta} \frac{r}{(1+r^2)^{3/2}}\,dr\,d\theta$

11. $\int_0^{2\pi} \int_0^1 r(1-r^2)^{3/2}\,dr\,d\theta$

12. $\int_{-\frac{\pi}{6}}^{\frac{\pi}{6}} \int_0^{\cos 3\theta} r^2\sin^2 3\theta\,dr\,d\theta$

Exercises 13-24. Sketch the region over which the integration occurs, and then evaluate the iterated integral using polar coordinates.

13. $\int_0^2 \int_0^{\sqrt{2x-x^2}} \sqrt{x^2+y^2}\,dy\,dx$

14. $\int_0^4 \int_0^{\sqrt{4y-y^2}} (x^2+y^2)\,dx\,dy$

15. $\int_0^6 \int_0^y x\,dx\,dy$

16. $\int_0^1 \int_0^{\sqrt{1-y^2}} e^{\sqrt{x^2+y^2}}\,dx\,dy$

17. $\int_{-1}^0 \int_{-\sqrt{1-x^2}}^0 \frac{2}{1+\sqrt{x^2+y^2}}\,dy\,dx$

18. $\int_{-1}^1 \int_{-\sqrt{1-x^2}}^{\sqrt{1-x^2}} \frac{2}{1+x^2+y^2}\,dy\,dx$

19. $\iint\limits_R e^{-(x^2+y^2)}\,dA$ where R is the region enclosed by the circle $x^2+y^2 = 1$.

20. $\iint\limits_R e^{-(x^2+y^2)}\,dA$ where R is in the first quadrant bounded by $y = 0$, $y = x$, and $x^2+y^2 = 4$.

21. $\iint\limits_R \sqrt{9-x^2-y^2}\,dA$ where R is the region in the first quadrant enclosed by the circle $x^2+y^2 = 1$.

22. $\iint\limits_R y^2\,dA$ where R is the region enclosed by the circle $x^2+y^2 = 2y$.

23. $\iint\limits_{R} y \, dA$ where R is the region enclosed by the circle $x^2 + y^2 = y$.

24. $\iint\limits_{R} \sqrt{x^2 + y^2} \, dA$ where R is the region enclosed by $r = 3 + \cos\theta$.

Exercises 25-34. Use polar coordinates and double integrals to determine the area of the given region.

25. The region lies inside the circle $r = \dfrac{3}{2}$ and to the right of the line $x = \dfrac{3}{4}$ (i.e. $r\cos\theta = \dfrac{3}{4}$).

26. The region lies inside the circle $r = 2$ and above the line $y = 1$ (i.e. $r\sin\theta = 1$).

27. The region common to the circles $r = 2\cos\theta$ and $r = 2\sin\theta$.

28. The region that lies outside the circle $r = 3\cos\theta$ and inside the circle $r = \dfrac{3}{2}$.

29. The region that lies inside the circle $r = 3\cos\theta$ and outside the circle $r = \dfrac{3}{2}$.

30. The region inside the cardioid $r = 2(1 + \cos\theta)$.

31. The region inside the circle $r = 1$ and outside the cardioid $r = 1 - \cos\theta$.

32. One leaf of the rose $r = 3\sin 4\theta$.

33. The region in the first quadrant that is inside the circle $r = \sqrt{3}\cos\theta$ and outside the circle $r = \sin\theta$.

34. The region common to the cardioids $r = 2(1 + \cos\theta)$ and $r = 2(1 - \cos\theta)$.

Exercises 35-44. Use polar coordinates to find the volume of the given solid.

35. A sphere of radius a.

36. The ellipsoid $\dfrac{x^2}{4} + \dfrac{y^2}{4} + \dfrac{z^2}{3} = 1$.

37. The solid that is under the cone $z = \sqrt{x^2 + y^2}$ and above the disk $x^2 + y^2 \le 4$.

38. The solid that lies below by $z = 1 - x^2 - y^2$, above the xy-plane, and inside the cylinder $x^2 + y^2 - x = 0$.

39. The solid that lies below the paraboloid $z = x^2 + y^2$, above the xy-plane, and inside the cylinder $x^2 + y^2 = 2x$.

40. The solid that is bounded by the paraboloids $z = 3x^2 + 3y^2$ and $z = 4 - x^2 - y^2$.

41. The solid that is bounded below by the xy-plane, above by the spherical surface
$x^2 + y^2 + z^2 = 4$, and on the sides by the cylinder $x^2 + y^2 = 1$.

42. The solid that is bounded above by the cone $z^2 = x^2 + y^2$, and below by the region which lies inside the circle $x^2 + y^2 = 2a$.

43. The solid that is inside the cylinder $x^2 + y^2 = 4$ and the ellipsoid $4x^2 + 4y^2 + z^2 = 64$.

44. The solid that is bounded above by the surface $z = \dfrac{1}{\sqrt{x^2 + y^2}}$, below by the xy-plane, and
enclosed between the cylinders $x^2 + y^2 = 1$ and $x^2 + y^2 = 9$.

Exercises 45-49. Find the mass of the region R.

45. R is the cardioid $r = 1 + \sin\theta$, and $\delta(x, y) = r$.

46. R is the region outside the circle $r = 3$, inside the circle $r = 6\sin\theta$, and $\delta(x, y) = r$.

47. R is the region outside the circle $r = 3$ and inside the circle $r = 6\sin\theta$, and $\delta(x, y) = \dfrac{1}{r}$.

48. R is the region inside the circle $r = 3\cos\theta$, outside $r = 2 - \cos\theta$, and $\delta(x, y) = \dfrac{1}{r}$.

49. R is the region inside the circle $r = a > 0$, outside the circle $r = 2a\sin\theta$, and
$\delta(x, y) = \dfrac{1}{r}$.

Exercises 50-55. Find the center of mass of the region R.

50. R is the washer between the circles $r = 2$, $r = 4$, if $\delta(r, \theta) = r^2$.

51. R is the cardioid $r = 1 + \sin\theta$, and $\delta(x, y) = r$.

52. R is the smaller region cut from the circle $r = 6$ by the line $r\cos\theta = 3$ if $\delta(r, \theta) = \cos^2\theta$.

53. R is the region outside the circle $r = 3$, inside the circle $r = 6\sin\theta$, and $\delta(r, \theta) = \dfrac{1}{r}$.

54. R is the region bounded by $r = \cos2\theta$, $0 \le \theta \le \dfrac{\pi}{4}$, and $\delta(r, \theta) = r\theta$.

55. R is the region bounded by $r = \cos\theta$, $-\dfrac{\pi}{4} \le \theta \le \dfrac{\pi}{4}$, and $\delta(r, \theta) = r$.

§5. TRIPLE INTEGRALS

In defining $\int_a^b f(x)dx$ for $y = f(x)$, we had the luxury of being able to represent the graph of a function $y = f(x)$ in the xy-plane (see page 177). Though considerably more challenging, when defining $\iint_R h(x, y)dA$, we were still able to depict a function $z = h(x, y)$ in three space (see page 445). But when it comes to the next task, that of defining the triple integral $\iiint_W k(x, y, z)dW$ over a three-dimensional region W, we must abandon all hope of geometrically representing the function $w = k(x, y, z)$ in four-dimensional space, as we are three-dimensional creatures.

That's fine, for the eyes of mathematics are in the mind. We'll just generalize the Riemann sum procedure that lead us to the integral of a function $y = f(x)$ over an interval $[a, b]$ and of the integral a function $z = h(x, y)$ over a region R in the plane to arrive at the definition of the integral of a function $w = k(x, y, z)$ over the region W in Figure 11.11. Specifically, we chop that region into "small boxes" of lengths $\Delta x, \Delta y, \Delta z$, pick a point (x, y, z) in each of those boxes, and then define $\iiint_W k(x, y)dV$ to be the limit of the Riemann sums

$\sum_W k(x, y, z)\Delta x\Delta y\Delta z$ as all three dimensions of the boxes tend to zero

(providing the limit exists):.

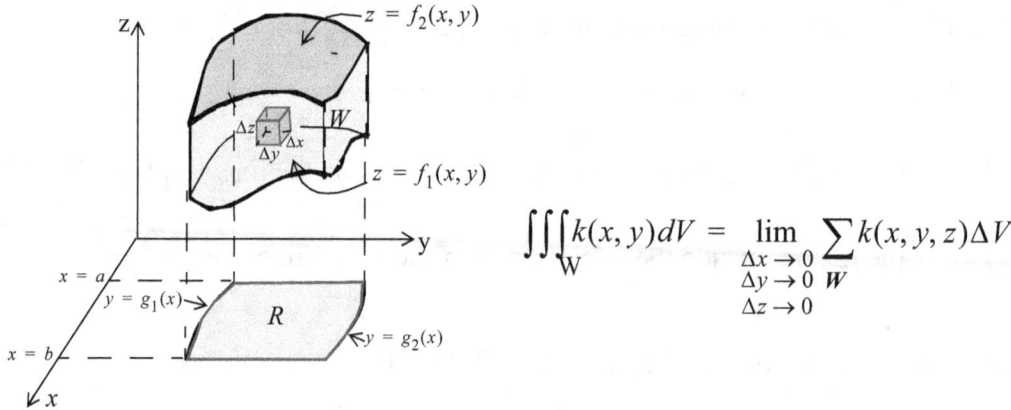

$$\iiint_W k(x, y)dV = \lim_{\substack{\Delta x \to 0 \\ \Delta y \to 0 \\ \Delta z \to 0}} \sum_W k(x, y, z)\Delta V$$

Figure 11.11

As is the case with double integrals, the triple integral can be evaluated using integrated integrals. Roughly speaking, for a given $w = k(x, y, z)$:

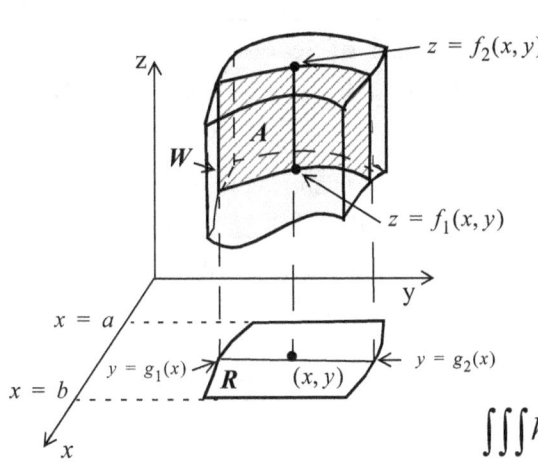

Fix a point (x, y) in $\boldsymbol{R}$, and start with the vertical line segment from $z = f_1(x, y)$ to $z = f_2(x, y)$. Keeping x fixed, slide that line from the curve $y = g_1(x)$ to the curve $y = g_2(x)$, sweeping out the hashed region $\boldsymbol{A}$.

Finally, slide that two-dimensional region $\boldsymbol{A}$ from $x = a$ to $x = b$ to fill in the entire solid $\boldsymbol{W}$.

Bringing us to:

$$\iiint h(x, y, z)\,dV = \int_{x=a}^{x=b}\int_{y=g_1(x)}^{y=g_2(x)}\int_{z=f_1(x,y)}^{z=f_2(x,y)} h(x, y, z)\,dz\,dy\,dx$$

first from surface to surface
then from curve to curve
finally from point to point

Needless to say, certain conditions have to be met in order for all of the above to work, and work it will if all of the functions g_1, g_2, f_1, f_2, and h are continuous over their associated domains.

EXAMPLE 11.19 Evaluate $\iiint_W x\,dz\,dy\,dx$, where $\boldsymbol{W}$ is the tetrahedron depicted in the margin.

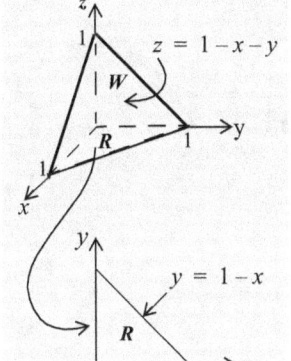

SOLUTION:

From surface to surface, holding x and y fixed.

$$\iiint_W x\,dz\,dy\,dx = \int_0^1\int_0^{1-x}\int_0^{1-x-y} x\,dz\,dy\,dx$$

From curve to curve, holding x fixed.

$$= \int_0^1\int_0^{1-x} (xz)\Big|_0^{1-x-y}\,dy\,dx = \int_0^1\int_0^{1-x} (x - x^2 - xy)\,dy\,dx$$

$$= \int_0^1 \left(xy - x^2 y - x\cdot\frac{y^2}{2}\right)\Bigg|_0^{1-x}\,dx$$

From point to point

$$= \int_0^1\left[x(1-x) - x^2(1-x) - x\frac{(1-x)^2}{2}\right]dx$$

$$= \int_0^1\left(\frac{x^3}{2} - x^2 + \frac{x}{2}\right)dx = \left(\frac{x^4}{8} - \frac{x^3}{3} + \frac{x^2}{4}\right)\Bigg|_0^1 = \frac{1}{24}$$

CHECK YOUR UNDERSTANDING 11.19

Use a triple Integral to find the volume of the tetrahedron of the previous example.

The concepts of mass and center of mass of Section 11.3 (pages 450 and 451, respectively) extend naturally to three-dimensional objects. Consider the following example.

EXAMPLE 11.20 Find the mass M and center of gravity $(\bar{x}, \bar{y}, \bar{z})$ of the cylinder depicted in the margin, with density at each point proportional to the distance from its base.

SOLUTION: Noting that $\delta(x, y, z) = kz$ for a constant k we have:

$$M = \iiint_V \delta(x, y, z)\,dV$$

$$= \int_{-r}^{r}\int_{-\sqrt{r^2-x^2}}^{\sqrt{r^2-x^2}}\int_{0}^{h} kz\,dz\,dy\,dx \overset{\text{by symmetry}}{=} 4\int_{0}^{r}\int_{0}^{\sqrt{r^2-x^2}}\int_{0}^{h} kz\,dz\,dy\,dx$$

$$= 4\int_{0}^{r}\int_{0}^{\sqrt{r^2-x^2}}\left(k\frac{z^2}{2}\right)\Big|_{0}^{h}$$

$$= 2kh^2\int_{0}^{r} y\Big|_{0}^{\sqrt{r^2-x^2}}dx$$

$$= 2kh^2\int_{0}^{r}\sqrt{r^2-x^2}\,dx$$

$$= 2kh^2\frac{\pi r^2}{4} = \frac{1}{2}kh^2 r^2\pi$$

$$\int_{0}^{r}\sqrt{r^2-x^2}\,dx = \int_{0}^{\frac{\pi}{2}}\sqrt{r^2\cos^2\theta}\,r\cos\theta\,d\theta$$

$$\begin{array}{|l|}\hline x = r\sin\theta \\ dx = r\cos\theta\,d\theta \\ x = 0 \Rightarrow \theta = 0 \\ x = r \Rightarrow \theta = \frac{\pi}{2} \\ \hline \end{array}$$

$$= r^2\int_{0}^{\frac{\pi}{2}}\cos^2\theta\,d\theta$$

$$= r^2\int_{0}^{\frac{\pi}{2}}\frac{1+\cos 2\theta}{2}\,d\theta$$

$$= \frac{r^2}{2}\left(\theta + \frac{\sin 2\theta}{2}\right)\Big|_{0}^{\frac{\pi}{2}} = \frac{\pi r^2}{4}$$

By symmetry we anticipate that the x and y coordinates of the center of mass of the cylindrical solid are both 0, a fact that you are invited to establish in the following CYU. That being the case, the center will lie on the z axis; but not at the midpoint of its altitude, as the material gets heavier the further up from its base.

Generalizing the situation depicted on page 452 to accommodate the three-dimensional object W we have:

$$\bar{z} = \frac{\iiint_W z\delta(x,y,z)dV}{\iiint_W \delta(x,y,z)dV} = \frac{\iiint_W z\delta(x,y,z)dV}{M} = \frac{\iiint_W z\delta(x,y,z)dV}{\frac{1}{2}kh^2r^2\pi}$$

where:

$$\iiint_W z\delta(x,y,z)dV = \int_{-r}^{r}\int_{-\sqrt{r^2-x^2}}^{\sqrt{r^2-x^2}}\int_0^h z(kz)\,dz\,dy\,dx$$

$$\text{by symmetry:} \quad = 4\int_0^r\int_0^{\sqrt{r^2-x^2}}\int_0^h z(kz)\,dz\,dy\,dx$$

$$= 4k\int_0^r\int_0^{\sqrt{r^2-x^2}}\frac{h^3}{3}\,dy\,dx$$

$$= \frac{4kh^3}{3}\int_0^r\sqrt{r^2-x^2}\,dx = \frac{4kh^3}{3}\cdot\frac{\pi r^2}{4} = \frac{1}{3}kh^3r^2\pi$$

And so: $(\bar{x},\bar{y},\bar{z}) = \left(0,0,\dfrac{\frac{1}{3}kh^3r^2\pi}{\frac{1}{2}kh^2r^2\pi}\right) = \left(0,0,\dfrac{2}{3}h\right)$ independent of k

---

CHECK YOUR UNDERSTANDING 11.20

Answer: See page A-8.

Verify that $\bar{x} = \bar{y} = 0$ in the previous example.

Up to now we have systematically evaluated triple integrals by integrating first with respect to z, then y, and then x:

> There are six possible orders of integration: $dzdydx$, $dzdxdy$, $dydzdx$ $dydxdz$, $dxdzdy$, $dxdydz$

$$\iiint k(x,yz)dV = \int_a^b\int_{g_1(x)}^{g_2(x)}\int_{f_1(x,y)}^{f_2(x,y)} k(x,y,z)\,dz\,dy\,dx$$

It may be advantageous to choose a different order. Consider the following example.

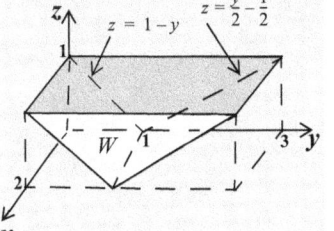

EXAMPLE 11.21 Find the mass of the solid wedge W lying between the planes $z=1, y=0, y=3$, $z=1-y$, $x=0$, $x=2$, and $z=\dfrac{y}{2}-\dfrac{1}{2}$ (see margin), if $\delta(x,y,z) = z$.

SOLUTION: Integrating first with respect to z would require two integrals (why?). On the other hand:

$$M = \iiint_W \delta(x, y, z)dV = \int_0^1 \int_0^2 \int_{1-z}^{2z+1} z\,dy\,\boxed{dx\,dz} \nearrow \text{hold } x \text{ and } z \text{ fixed}$$

$$= \int_0^1 \int_0^2 (zy)\Big|_{y=1-z}^{y=2z+1} dx\,dy = \int_0^1 \int_0^2 3z^2\,dx\,dz$$

$$= \int_0^1 (3z^2 x)\Big|_{x=0}^{x=2} dz$$

$$= 6\int_0^1 z^2\,dz = 2$$

CHECK YOUR UNDERSTANDING 11.21

The order of integration for M in the above example is $dy\,dx\,dz$. Express M in terms of the remaining five possible orders of integration.

Answer: See page A-8.

	EXERCISES	

Exercises 1-10. Evaluate.

1. $\int_0^1 \int_0^{x^2} \int_{x+y}^{x^2+y^2} dz\,dy\,dx$

2. $\int_0^1 \int_0^1 \int_0^1 (x^2+y^2+z^2)dz\,dy\,dx$

3. $\int_0^1 \int_0^{2-3x} \int_0^{x+y} x\,dz\,dy\,dx$

4. $\int_0^1 \int_0^{1-x^2} \int_0^{4-x^2-y} x\,dz\,dy\,dx$

5. $\int_0^3 \int_{-1}^1 \int_0^1 xyz^2 dz\,dy\,dx$

6. $\int_0^1 \int_0^{4-x} \int_0^{2x+y} z\,dz\,dy\,dx$

7. $\int_1^2 \int_x^{2x} \int_{y-x}^y (4x-12z)dz\,dy\,dx$

8. $\int_2^4 \int_{-1}^1 \int_1^3 (x^2y+y^2x)dz\,dy\,dx$

9. $\int_0^1 \int_0^{x^2} \int_0^{xy^3} xy^2z^3 dz\,dy\,dx$

10. $\int_0^{\frac{\pi}{2}} \int_x^{\frac{\pi}{2}} \int_0^{xy} \sin\frac{z}{y}dz\,dy\,dx$

Exercises 11-18. Evaluate. Note the specified order of integration.

11. $\int_0^1 \int_0^z \int_0^{z^2+y^2} dx\,dy\,dz$

12. $\int_0^1 \int_0^y \int_0^{x^2y} (x+y)dz\,dx\,dy$

13. $\int_0^1 \int_0^\pi \int_0^\pi y\sin z\,dx\,dy\,dz$

14. $\int_1^e \int_1^e \int_1^e \frac{1}{xyz}dx\,dy\,dz$

15. $\int_0^1 \int_0^{x^2} \int_0^{x^3y} x^3y^2z\,dz\,dy\,dx$

16. $\int_0^1 \int_0^x \int_0^{x^2y} x^3y^4z^2 dz\,dy\,dx$

17. $\int_0^1 \int_{-x}^x \int_{-x-z}^{x+z} e^{x+y+z}\,dy\,dz\,dx$

18. $\int_0^1 \int_0^{\sqrt{3}z} \int_0^{\sqrt{3(y^2+z^2)}} xyz\sqrt{x^2+y^2+z^2}dx\,dy\,dz$

Exercises 19-26. Find the volume of the solid W,

19. W is the solid in the first octant that lies between the planes $x+y+2z = 2$ and $2x+2y+z = 4$,

20. W is the solid bounded above by the paraboloid $z = 4-x^2-y^2$ and below by the plane $z = 4-2x$.

21. W is the solid enclosed between the cylinder $x^2+y^2 = 9$ and the planes $z = 1$ and $x+z = 5$.

22. W is the solid bounded by the cylinders $z = x^2$, $z = 4 - x^2$, and the planes $y = 0$ and $z + 2y = 4$.

23. W is the tetrahedron bounded by the planes $z = 0$, $x = 0$, $x = 2y$, and $x + 2y + z = 2$.

24. W is the solid enclosed by the cylinders $x^2 + y^2 = 1$ and $x^2 + z^2 = 1$.

25. W is the solid enclosed by the paraboloids $z = 5x^2 + 5y^2$ and $z = 6 - 7x^2 - y^2$.

26. W is the solid enclosed by the surface $z = \dfrac{y}{1 + x^2}$ and the planes $x = 0$, $y = 0$, $z = 0$, and $x + y = 1$.

Exercises 27-32. Find the mass of the object W with density function $\delta(x, y, z)$.

27. W is the cube given by $0 \le x \le a$, $0 \le y \le a$, $0 \le z \le a$; $\delta(x, y, z) = x^2 + y^2 + z^2$.

28. W is the solid bounded by the parabolic cylinder $z = 1 - y^2$ and the planes $x = 0$, $z = 0$, and $x + z = 1$; $\delta(x, y, z) = k$, for a constant k.

29. W is the solid bounded by the cylinder $x^2 + y^2 = 1$ and the planes $z = 0$, $z = 1$; $\delta(x, y, z) = |xyz|$.

30. W is the cube given by $0 \le x \le a$, $0 \le y \le a$, $0 \le z \le a$; $\delta(x, y, z) = xyz$

31. W is the solid bounded by the parabolic cylinder $x = y^2$ and the planes $z = 0$, $x = 1$, and $x = z$; $\delta(x, y, z) = k$, for a constant k.

32. W is the solid bounded by the coordinate planes and the plane $x + y + z = 1$; $\delta(x, y, z) = 1 - x - y$.

Exercises 33-40. Find the center of mass of the object W with density function $\delta(x, y, z)$.

33. W is the solid bounded above by the paraboloid $z = 4 - x^2 - y^2$ and below by the plane $z = 0$; $\delta(x, y, z) = a$.

34. W is the solid enclosed by the surface $z = 1 - y^2$, for $y \ge 0$, and the planes $x = -1$, $x = 1$, $z = 0$; $\delta(x, y, z) = yz$.

35. W is the solid bounded by the parabolic cylinder $x = y^2$ and the planes $z = 0$, $x = 1$, and $x = z$; $\delta(x, y, z) = k$, for a constant k.

36. W is the solid enclosed by the cylinder $x^2 + y^2 = a^2$ and the planes $z = 0$, $z = h$, for $h > 0$; $\delta(x, y, z) = h - z$,

37. W is the cube given by $0 \le x \le a$, $0 \le y \le a$, $0 \le z \le a$; $\delta(x, y, z) = x^2 + y^2 + z^2$.

38. $W = \{(x, y, z)|-1 \le x \le 0, x \le y \le 1, x + y \le z \le 2\}$; $\delta(x, y, z) = x + 3$.

39. $W = \{(x, y, z)|1 \le z \le 2, 0 \le x \le z, x \le y \le z\}$, $\delta(x, y, z) = y - x$.

40. $W = \{(x, y, z)|9 \le z \le 4, x^2 + y^2 \le 4\}$; $\delta(x, y, z) = x^2 + y^2 + z^2$.

Exercises 41-44. Express the integral $\iiint\limits_W k(x, y, z)dV$ in six different ways, where W is the solid bounded by the given surfaces.

41. $y = x^2, y = -2z + 4, z = 0$.

42. $z = 0, x = 2, y = 2, x + y - 2z = 2$

43. $y = 0, y = 4 - x^2 - 4z^2$

44. $x = -2, x = 2, y^2 + z^2 = 9$

Exercises 45-48. Rewrite the given integral in five additional forms by changing the order of integration.

45. $\displaystyle\int_0^1 \int_y^1 \int_0^y k(x, y, z)dz\,dx\,dy$

46. $\displaystyle\int_{-1}^1 \int_{x^2}^1 \int_0^{1-y} k(x, y, z)dz\,dy\,dx$

47. $\displaystyle\int_0^1 \int_{-1}^0 \int_0^{y^2} k(x, y, z)dz\,dy\,dx$

48. $\displaystyle\int_0^1 \int_0^{x^2} \int_0^y k(x, y, z)dz\,dy\,dx$

§6. CYLINDRICAL AND SPHERICAL COORDINATES

Just as certain planar regions are best described using polar coordinates, some three-dimensional surfaces are best identified using non-rectangular coordinate systems; one of which is the **cylindrical coordinate system**. In this system, a point P with projection Q on the xy-plane, is represented by (r, θ, z), where (r, θ) is a polar representation of Q, and z is the rectangular vertical coordinate of P (see Figure 11.12). Bridges between the cylindrical and rectangular coordinates also appear in the figure.

See the *Rectangular to Polar and Vice Versa* discussion on page 405.

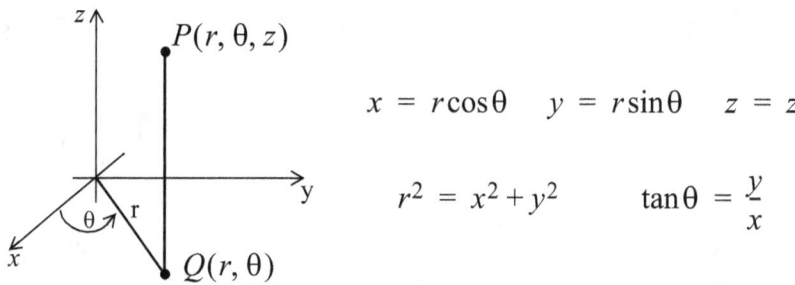

$$x = r\cos\theta \quad y = r\sin\theta \quad z = z$$

$$r^2 = x^2 + y^2 \qquad \tan\theta = \frac{y}{x}$$

Figure 11.12

EXAMPLE 11.22   (a) Express $P(r, \theta, z) = \left(2, \dfrac{3\pi}{4}, 5\right)$ in rectangular coordinates.

(b) Express $P(x, y, z) = (-1, \sqrt{3}, 2)$ in cylindrical coordinates.

SOLUTION: (a) For $P(r, \theta, z) = \left(2, \dfrac{3\pi}{4}, 5\right)$, $z = 5$.

Since $Q = (r, \theta) = \left(2, \dfrac{3\pi}{4}\right)$ (see Figure 11.12):

Taken directly from page 406.

$$x = 2\cos\frac{3\pi}{4} = 2\left(-\cos\frac{\pi}{4}\right) = -2\left(\frac{1}{\sqrt{2}}\right) = -\sqrt{2}$$

$$y = 2\sin\frac{3\pi}{4} = 2\left(\sin\frac{\pi}{4}\right) = 2\left(\frac{1}{\sqrt{2}}\right) = \sqrt{2}$$

Conclusion: $\left(2, \dfrac{3\pi}{4}, 5\right)$ has rectangular coordinates $(-\sqrt{2}, \sqrt{2}, 5)$.

(b) For $P(x, y, z) = (-1, \sqrt{3}, 2)$, $z = 2$.

Since $Q = (r, \theta) = (-1, \sqrt{3})$ (see Figure 11.12):

$$r^2 = (-1)^2 + (3^{1/2})^2 = 4 \quad \text{and} \quad \tan\theta = \frac{\sqrt{3}}{-1} = -\sqrt{3}$$

As shown in Example 10.6(b), page 405, there are infinitely many solutions of the above two equations; **one of which** is: $r = 2$ and $\theta = \dfrac{2\pi}{3}$.

Conclusion: $(-1, \sqrt{3}, 2)$ has cylindrical coordinates $\left(2, \frac{2\pi}{3}, 2\right)$.

CHECK YOUR UNDERSTANDING 11.22

(a) Express $\left(4, \frac{\pi}{6}, -1\right)$ in rectangular coordinates.

(b) Express $(2, 2, 4)$ in cylindrical coordinates with $r > 0$ and $0 < \theta < \frac{\pi}{2}$.

Answer: (a) $(2\sqrt{3}, 2, -1)$

(b) $\left(2\sqrt{2}, \frac{\pi}{4}, 4\right)$

Some volume building blocks are featured below:

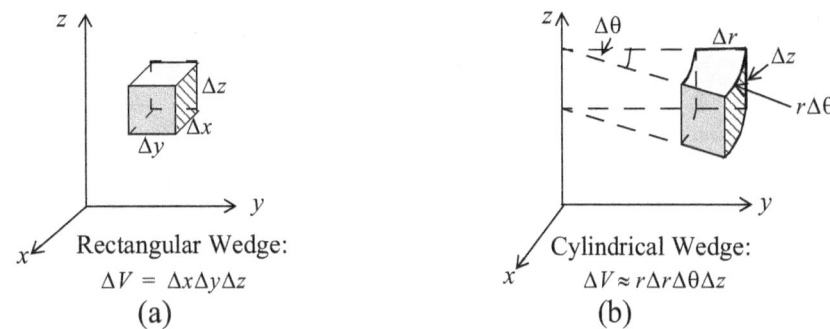

Rectangular Wedge:
$\Delta V = \Delta x \Delta y \Delta z$
(a)

Cylindrical Wedge:
$\Delta V \approx r \Delta r \Delta \theta \Delta z$
(b)

Figure 11.13

Figure 11.13(a) displays a rectangular building block. Such blocks were used to construct an iterated triple integral for $\iiint_W f(x, y, z)dV$. In a similar fashion, using the cylindrical building block in Figure 11.13(b)], we now formulate an iterated triple integral for $\iiint_W f(r, \theta, z)dV$:

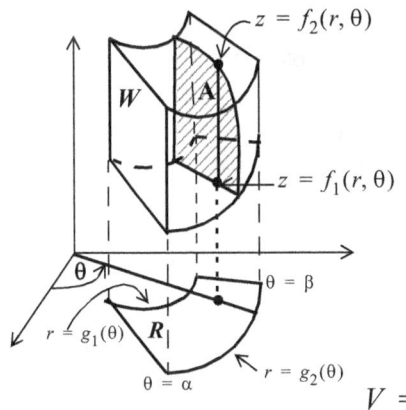

Fix a point (r, θ) in R, and consider the vertical line segment from $z = f_1(r, \theta)$ to $z = f_2(r, \theta)$.

Keeping θ fixed, vary r from $g_1(\theta)$ to $g_2(\theta)$ to sweep out the hashed region A.

Finally, rotate A from $\theta = \alpha$ to $\theta = \beta$ to fill in the entire solid W.

Bringing us to:

$$V = \iiint_W h(r, \theta, z)dV = \int_{\theta = \alpha}^{\theta = \beta} \int_{r = g_1(\theta)}^{r = g_2(\theta)} \int_{z = f_1(r, \theta)}^{z = f_2(r, \theta)} h(r, \theta, z) r \, dz \, dr \, d\theta$$

Once more: from surface to surface then from curve to curve (polar form) and now from angle to angle

Figure 11.14

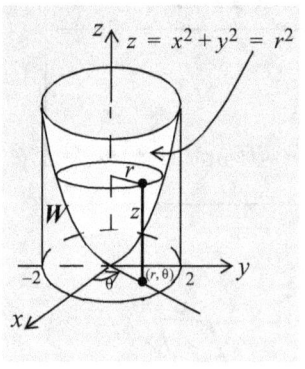

EXAMPLE 11.23 Find the volume, mass, and center of mass of the solid W contained within the cylinder $x^2 + y^2 = 4$, that is bounded above by the paraboloid $z = x^2 + y^2$, and below by the xy-plane. Assume that the density at each point is its distance from the axis of the cylinder.

SOLUTION: The solid W and density function δ can be expressed as:
$$W = \{(r, \theta, z) \mid 0 \le r \le 2, 0 \le \theta \le 2\pi, 0 \le z \le r^2\} \text{ and } \delta(r, \theta, z) = r$$
We then have:

$$V = \iiint_W 1\, dV = \int_0^{2\pi}\int_0^2\int_0^{r^2} r\, dz\, dr\, d\theta = \int_0^{2\pi}\int_0^2 (rz)\Big|_{z=0}^{z=r^2}\, dr\, d\theta$$

$$\underbrace{}_{\substack{\text{first}\\ \text{holding } r \text{ and } \theta \text{ constant}}}\quad = \int_0^{2\pi}\int_0^2 r^3\, dr\, d\theta = \int_0^{2\pi}\frac{r^4}{4}\Big|_{r=0}^{r=2}\, d\theta$$

$$= \int_0^{2\pi} 4\, d\theta = 8\pi$$

$$M = \iiint_W \delta(x, y, z)\, dV = \int_0^{2\pi}\int_0^2\int_0^{r^2} r \cdot r\, dz\, dr\, d\theta$$

$$= \int_0^{2\pi}\int_0^2 (r^2 z)\Big|_0^{r^2}\, dr\, d\theta$$

$$= \int_0^{2\pi}\int_0^2 r^4\, dr\, d\theta = \int_0^{2\pi}\frac{r^5}{5}\Big|_0^2\, d\theta = \frac{64\pi}{5}$$

Symmetry dictates that the x and y coordinates of the center of mass are 0. Let's find its z coordinate:

$$\bar{z} = \frac{1}{M}\iiint_W z\,\delta(x, y, z)\, dV = \frac{1}{M}\int_0^{2\pi}\int_0^2\int_0^{r^2} z \cdot r \cdot r\, dz\, dr\, d\theta$$

$$= \frac{1}{M}\int_0^{2\pi}\int_0^2 \left(r^2 \cdot \frac{z^2}{2}\right)\Big|_0^{r^2}\, dr\, d\theta$$

$$= \frac{1}{M}\int_0^{2\pi}\int_0^2 \frac{r^6}{2}\, dr\, d\theta = \frac{1}{M}\int_0^{2\pi}\frac{r^7}{14}\Big|_0^2\, d\theta$$

$$= \frac{2^7}{14\left(\dfrac{64\pi}{5}\right)} \cdot 2\pi = \frac{10}{7}$$

Conclusion: $(\bar{x}, \bar{y}, \bar{z}) = \left(0, 0, \dfrac{10}{7}\right)$

Answer:

$V = \frac{2\pi}{3}(40\sqrt{5} - 76)$

$M = \frac{2\pi k}{3}(40\sqrt{5} - 76)$

$\left(0, 0, \dfrac{19}{20\sqrt{5} - 38}\right)$

CHECK YOUR UNDERSTANDING 11.23

Find the volume, mass, and center of mass of the solid W contained within the paraboloid $z = x^2 + y^2$, that lies below the sphere $x^2 + y^2 + z^2 = 20$. Assume a constant density function.

SPHERICAL COORDINATES

Three-dimensional objects that are symmetrical with respect to the origin, such as spheres and cones, are often best described in terms of spherical coordinates; where:

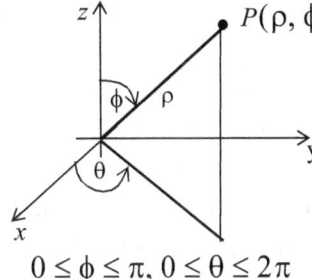

$P(\rho, \phi, \theta)$

$0 \leq \phi \leq \pi, 0 \leq \theta \leq 2\pi$

In the exercises you are invited to establish the following coordinate relations:

$$\rho^2 = x^2 + y^2 + z^2$$

$$x = \rho \sin\phi \cos\theta \quad y = \rho \sin\phi \sin\theta$$

$$z = \rho \cos\phi$$

Figure 11.15

EXAMPLE 11.24

(a) Represent $P(\rho, \phi, \theta) = \left(2, \dfrac{\pi}{4}, \dfrac{\pi}{3}\right)$ in rectangular coordinates.

(b) Represent $P(x, y, z) = (1, 1, \sqrt{6})$ in spherical coordinates.

SOLUTION: (a) From the equations in Figure 11.15 we have:

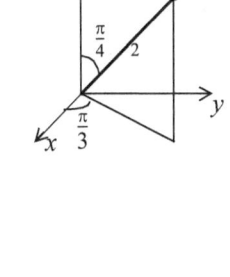

$x = 2\sin\dfrac{\pi}{4}\cos\dfrac{\pi}{3} = 2\left(\dfrac{1}{\sqrt{2}}\right)\left(\dfrac{1}{2}\right) = \dfrac{1}{\sqrt{2}}$

$y = 2\sin\dfrac{\pi}{4}\sin\dfrac{\pi}{3} = 2\left(\dfrac{1}{\sqrt{2}}\right)\left(\dfrac{\sqrt{3}}{2}\right) = \sqrt{\dfrac{3}{2}}$

$z = 2\cos\dfrac{\pi}{4} = 2\left(\dfrac{1}{\sqrt{2}}\right) = \sqrt{2}$

Conclusion: $P(x, y, z) = \left(\dfrac{1}{\sqrt{2}}, \sqrt{\dfrac{3}{2}}, \sqrt{2}\right) = \left(\dfrac{\sqrt{2}}{2}, \dfrac{\sqrt{6}}{2}, \sqrt{2}\right)$

(b) Using the equations in Figure 11.15, we have:

$$\rho = \sqrt{x^2 + y^2 + z^2} = \sqrt{1 + 1 + 6} = \sqrt{8} = 2\sqrt{2}$$

$$z = \rho\cos\phi \Rightarrow \cos\phi = \frac{z}{\rho} = \frac{\sqrt{6}}{2\sqrt{2}} = \frac{\sqrt{3}}{2} \Rightarrow \phi = \frac{\pi}{6}$$

While $\sin\dfrac{3\pi}{4}$ is also equal to $\dfrac{1}{\sqrt{2}}$, we exclude $\dfrac{3\pi}{4}$ as $(1, 1, \sqrt{6})$ lies in the first octant.

$$y = \rho\sin\phi\sin\theta \Rightarrow \sin\theta = \frac{y}{\rho\sin\phi} = \frac{1}{2\sqrt{2}\sin\dfrac{\pi}{6}} = \frac{1}{2\sqrt{2}\cdot\dfrac{1}{2}} = \frac{1}{\sqrt{2}}$$

$$\Rightarrow \theta = \frac{\pi}{4} \quad \text{(see margin)}$$

Conclusion: $P(\rho, \phi, \theta) = \left(2\sqrt{2}, \dfrac{\pi}{6}, \dfrac{\pi}{4}\right)$

Answers:

(a) $P(\rho, \phi, \theta) = \left(4, \dfrac{2\pi}{3}, \dfrac{\pi}{2}\right)$

(b) $P(x, y, z) = \left(-\dfrac{3}{2}, \dfrac{\sqrt{3}}{2}, -1\right)$

$P(r, \theta, z) = \left(\sqrt{3}, \dfrac{5\pi}{6}, -1\right)$

CHECK YOUR UNDERSTANDING 11.24

(a) Represent $P(x, y, z) = (0, 2\sqrt{3}, -2)$ in spherical coordinates.

(b) Represent $P(\rho, \phi, \theta) = \left(2, \dfrac{2\pi}{3}, \dfrac{5\pi}{6}\right)$ in both rectangular and cylindrical coordinates.

Our next concern is to unearth an integration method utilizing spherical coordinates. The first step is to find an approximation for the volume ΔV of the spherical wedge depicted in Figure 11.16. As is noted in the figure, that wedge can be approximated by a box with depth $\Delta\rho$. The remaining two dimensions are determined by arcs of circles of radius ρ; one of which is easily seen to be $\rho\Delta\phi$. The length of the remaining side is that of an arc on a circle of radius $\rho\sin\phi$ that is subtended by the arc $\Delta\theta$. As such, it has a length of $\rho\sin\phi\Delta\theta$. Bringing us to:

$$\Delta V \approx \Delta\rho \cdot \rho\sin\phi\Delta\theta \cdot \rho\Delta\phi = \rho^2\sin\phi\Delta\rho\Delta\phi\Delta\theta$$

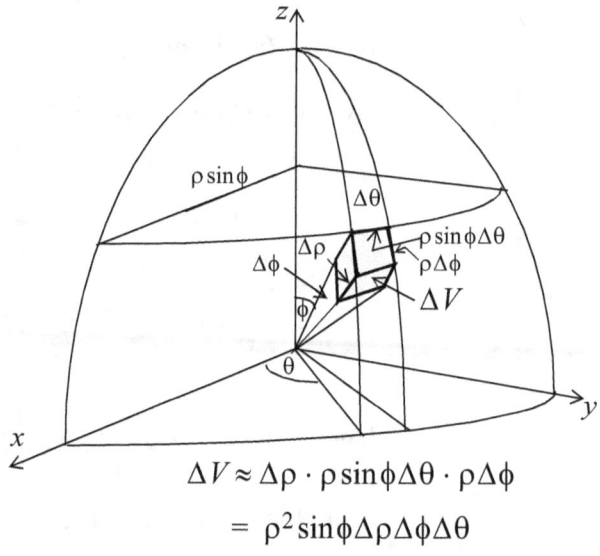

$$\Delta V \approx \Delta\rho \cdot \rho\sin\phi\Delta\theta \cdot \rho\Delta\phi$$

$$= \rho^2\sin\phi\Delta\rho\Delta\phi\Delta\theta$$

Figure 11.16

We now translate a familiar story, invoking the language of spherical coordinates:

For $h(\rho, \phi, \theta)$ continuous on the solid $W = \{(\rho, \phi, \theta) | \rho_1 \le \rho \le \rho_2, \phi_1 \le \phi \le \phi_2, \alpha \le \theta \le \beta\}$

Fix the angles ϕ and θ and construct the radial line l from the point $\rho_1(\phi, \theta)$ to the point $\rho_2(\phi, \theta)$.

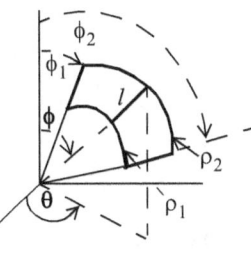

Holding θ fixed rotate the line from ϕ_1 to ϕ_2 sweeping out the shaded planar region A.

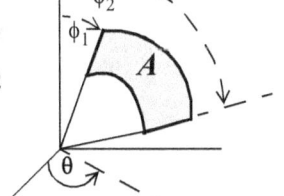

Finally, sweep that two-dimensional region A from α to β, thereby filling in the entire region W.

Bringing us to: $\iiint h(\rho, \phi, \theta)dV = \int_\alpha^\beta \int_{\phi_1(\theta)}^{\phi_2(\theta)} \int_{\rho_1(\phi, \theta)}^{\rho_2(\phi, \theta)} h(\rho, \phi, \theta)\rho^2 \sin\phi \, d\rho \, d\phi \, d\theta$

Figure 11.17

EXAMPLE 11.25   Find the volume, mass, and center of mass of the solid W bounded above by the sphere $\rho = 4$ and below by the cone $\phi = \dfrac{\pi}{3}$.

Assume that density is proportional to the point's distance from the origin.

Solution: $V = \int_0^{2\pi} \int_0^{\frac{\pi}{3}} \int_0^4 \rho^2 \sin\phi \, d\rho \, d\phi \, d\theta$

$= \int_0^{2\pi} \int_0^{\frac{\pi}{3}} \left(\dfrac{\rho^3}{3} \sin\phi \right) \Bigg|_{\rho=0}^{\rho=4} d\phi \, d\theta$

$= \dfrac{4^3}{3} \int_0^{2\pi} (-\cos\phi) \Big|_{\phi=0}^{\phi=\frac{\pi}{3}} d\theta$

$= \dfrac{4^3}{3} \int_0^{2\pi} \left(-\cos\dfrac{\pi}{3} + 1 \right) d\theta = \dfrac{64\pi}{3}$

We are given that $\delta(\rho, \phi, \theta) = c\rho$ for a constant c. Thus:

$$M = \int_0^{2\pi} \int_0^{\frac{\pi}{3}} \int_0^4 (c\rho)\rho^2 \sin\phi \, d\rho \, d\phi \, d\theta = c \int_0^{2\pi} \int_0^{\frac{\pi}{3}} \int_0^4 \rho^3 \sin\phi \, d\rho \, d\phi \, d\theta$$

$$= c \int_0^{2\pi} \int_0^{\frac{\pi}{3}} \left(\frac{\rho^4}{4}\right)\Big|_0^4 \sin\phi \, d\phi \, d\theta$$

$$= 4^3 c \int_0^{2\pi} (-\cos\phi)\Big|_0^{\frac{\pi}{3}} d\theta = 64\pi c$$

By symmetry, the x and y coordinates of the center of mass are 0.

$$\bar{z} = \frac{1}{M}\iiint_W z\delta(x,y,z)dV = \frac{1}{M}\int_0^{2\pi} \int_0^{\frac{\pi}{3}} \int_0^4 zc\rho^3 \sin\phi \, d\rho \, d\phi \, d\theta$$

$$= \frac{c}{M}\int_0^{2\pi} \int_0^{\frac{\pi}{3}} \int_0^4 \rho\cos\phi\rho^3 \sin\phi \, d\rho \, d\phi \, d\theta$$

$$= \frac{c}{M}\int_0^{2\pi} \int_0^{\frac{\pi}{3}} \left(\frac{\rho^5}{5}\right)\Big|_0^4 \cos\phi \sin\phi \, d\phi \, d\theta$$

$$= \frac{4^5 c}{5M}\int_0^{2\pi} \frac{1}{2}\sin^2\phi \Big|_0^{\frac{\pi}{3}} d\theta$$

$$= \frac{4^5 c}{10(64\pi c)}\left(\sin\frac{\pi}{3}\right)^2 2\pi = \frac{12}{5}$$

Conclusion: $(\bar{x}, \bar{y}, \bar{z}) = \left(0, 0, \dfrac{12}{5}\right)$

CHECK YOUR UNDERSTANDING 11.25

Answer: $\frac{\pi}{8}$

Find the volume of the solid W bounded above by the sphere $x^2 + y^2 + z^2 = z$ and below by the cone $z = \sqrt{x^2 + y^2}$.

	EXERCISES	

CYLINDRICAL COORDINATES

Exercises 1-4. The following points are given in cylindrical coordinates. Express them in rectangular coordinates.

1. $\left(2, \dfrac{5\pi}{6}, 3\right)$ 　　　　2. $\left(3, \dfrac{2\pi}{3}, 1\right)$ 　　　　3. $\left(6, \dfrac{\pi}{4}, 1\right)$ 　　　4. $(0, \pi, -1)$

Exercises 5-8. The following points are given in rectangular coordinates. Express them in cylindrical coordinates.

5. $(-4, -4\sqrt{3}, -8)$ 　6. $\left(-\dfrac{3\sqrt{3}}{2}, -\dfrac{3}{2}, -1\right)$ 　7. $(0, 2, -2)$ 　　8. $(0, \pi, -1)$

Exercises 9-12. Write the equation in cylindrical coordinates.

9. $z = x^2 + y^2$ 　　　　　　　　　　　10. $3x + 2y + z = 1$

11. $x^2 + y^2 = 2y$ 　　　　　　　　　　12. $x^2 + y^2 = 2x$

Exercises 13-16. Evaluate.

13. $\displaystyle\int_0^{2\pi}\int_0^1\int_0^{\sqrt{1-r^2}} r\,dz\,dr\,d\theta$ 　　　　　14. $\displaystyle\int_0^{2\pi}\int_0^3\int_{r^2/3}^{\sqrt{18-r^2}} r\,dz\,dr\,d\theta$

15. $\displaystyle\int_0^1\int_0^{\sqrt{z}}\int_0^{2\pi} (r^2\cos^2\theta + z^2)r\,d\theta\,dr\,dz$ 　　16. $\displaystyle\int_0^{\pi/2}\int_0^{\cos\theta}\int_0^{r^2} r\sin\theta\,dz\,dr\,d\theta$

Exercises 17-18. Evaluate using cylindrical coordinates.

17. $\displaystyle\int_0^1\int_0^{\sqrt{1-x^2}}\int_0^{\sqrt{4-(x^2+y^2)}} dz\,dy\,dx$ 　　18. $\displaystyle\int_{-1}^1\int_0^{\sqrt{1-x^2}}\int_{\sqrt{x^2+y^2}}^1 z^3\,dz\,dy\,dx$

Exercises 19-26. Use cylindrical coordinates to find the volume of the given solid.

19. The solid bounded by the paraboloid $z = x^2 + y^2$ and the plane $z = 9$.

20. The solid bounded by the paraboloid $z = x^2 + y^2$ and the plane $z = x$.

21. The solid lying outside the cone $z = \sqrt{x^2 + y^2}$ and inside the cylinder $x^2 + y^2 = 4$.

22. The solid lying above the cone $z = \sqrt{x^2 + y^2}$ and below the hemisphere
$z = 1 + \sqrt{1 - x^2 - y^2}$.

23. The solid enclosed by the cylinder $x^2 + y^2 = 4$ and the planes $z = -1$ and $x + z = 4$.

24. The solid bounded by the cone $z = \dfrac{h}{a}r$ and the plane $z = h$.

25. The solid cut from the sphere $x^2 + y^2 + z^2 = 4$ by the cylinder $r = 2\sin\theta$.

26. The solid lying above the cone $z = \sqrt{x^2 + y^2}$, below the plane $z = 0$, and bounded on the sides by the cylinder $x^2 + y^2 = 3x$.

Exercises 27-30. Find the mass of the given solid.

27. The ellipsoid $4x^2 + 4y^2 + z^2 = 16$ lying above the xy-plane, with density function $\delta(x, y, z) = kz$.

28. The circular cone of height h and circular base of radius a if the density at each point is proportional to its distance
        (a) from the axis.   (b) from the base.

29. The solid bounded by the paraboloid $z = 1 - (x^2 + y^2)$ and $z = 0$ if the density is proportional to
        (a) its distance from the xy-plane.   (b) the square of the distance from the origin.

30. The solid bounded by the paraboloid $z = x^2 + y^2 - 4$ and $z = 0$ if $\delta(x, y, z) = 1 + x^2 + y^2$.

Exercises 31-36. Find the center of mass of the given solid.

31. The solid is bounded by the paraboloid $z = 4x^2 + 4y^2$ and the plane $z = a$, for $a > 0$, and has constant density k.

32. The solid that lies within the cylinder $x^2 + y^2 = 1$, below the plane $z = 4$, and above the paraboloid $z = 1 - x^2 - y^2$, if $\delta(x, y, z) = z$.

33. The upper hemisphere of the unit sphere, and has constant density k.

34. The cone of constant density with height h and a base of radius b.

35. The solid is bounded by the paraboloids $z = x^2 + y^2$ and $z = 36 - 3x^2 - 3y^2$ and has constant density k.

SPHERICAL COORDINATES

36. Verify that if $P(\rho, \phi, \theta) = P(x, y, z)$, then:

$$\rho^2 = x^2 + y^2 + z^2$$
$$x = \rho \sin\phi \cos\theta \quad y = \rho \sin\phi \sin\theta$$
$$z = \rho \cos\phi$$

Exercises 37-40. The following points are given in spherical coordinates. Express them in rectangular coordinates.

37. $\left(4, \dfrac{\pi}{3}, \dfrac{\pi}{4}\right)$　　　38. $\left(2, -\dfrac{\pi}{2}, \dfrac{\pi}{4}\right)$　　　39. $\left(7, \dfrac{\pi}{2}, \pi\right)$　　　40. $\left(12, \dfrac{5\pi}{6}, \dfrac{2\pi}{3}\right)$

Exercises 41-44. The following points are given in rectangular coordinates. Express them in spherical coordinates.

41. $(1, \sqrt{3}, 2\sqrt{3})$　　　42. $(0, -5, 0)$　　　43. $(0, -1, -1)$　　　44. $(1, 1, 1)$

Exercises 45-48. Write the equation in spherical coordinates.

45. $z^2 = x^2 + y^2$

46. $x + 2y + 3z = 1$

47. $x^2 + z^2 = 9$

48. $x^2 - 2x + y^2 + z^2 = 0$

Exercises 49-52. Evaluate (note the order of integration).

49. $\displaystyle\int_0^{2\pi} \int_0^{\pi} \int_0^1 \rho^3 \sin\phi \, d\rho \, d\phi \, d\theta$

50. $\displaystyle\int_0^{\frac{\pi}{2}} \int_0^{\frac{\pi}{2}} \int_0^{\sin\phi} \rho^2 \sin\phi \cos\phi \, d\rho \, d\theta \, d\phi$

51. $\displaystyle\int_0^{\frac{\pi}{2}} \int_0^{\sin\phi} \int_0^{\frac{\pi}{4}} \rho^2 \sin\phi \, d\theta \, d\rho \, d\phi$

52. $\displaystyle\int_0^{\frac{\pi}{2}} \int_0^{\frac{\pi}{2}} \int_0^{\sin\phi} d\rho \, d\phi \, d\theta$

Exercises 53-56. Evaluate using spherical coordinates.

53. $\displaystyle\int_0^a \int_0^{\sqrt{a^2 - x^2}} \int_0^{\sqrt{a^2 - x^2 - y^2}} (x^2 + y^2 + z^2) \, dz \, dy \, dx$

54. $\displaystyle\int_0^3 \int_0^{\sqrt{9 - y^2}} \int_{\sqrt{x^2 + y^2}}^{\sqrt{18 - x^2 - y^2}} (x^2 + y^2 + z^2) \, dz \, dx \, dy$

55. $\displaystyle\int_0^1 \int_0^{\sqrt{1 - x^2}} \int_{\sqrt{x^2 + y^2}}^{\sqrt{2 - x^2 - y^2}} xy \, dz \, dy \, dx$

56. $\displaystyle\int_0^5 \int_0^{\sqrt{25 - x^2}} \int_{\sqrt{x^2 + y^2}}^{\sqrt{50 - x^2 - y^2}} \sqrt{x^2 + y^2 + z^2} \, dz \, dy \, dx$

Exercises 57-61. Use spherical coordinates to find the volume of the given solid.

57. A sphere of radius a.

58. A spherical shell whose outer radius is 2 and whose inner radius is 1.

59. The solid cut from the cone $\phi = \dfrac{\pi}{4}$ by the sphere $\rho = 2a\cos\phi$.

60. The solid inside both of the spheres $\rho = 1$ and $\rho = 2\cos\phi$

61. The solid bounded below by the cone $z^2 = x^2 + y^2$ lying above the xy-plane and bounded above by the sphere $x^2 + y^2 + z^2 = 9$.

Exercises 62-67. Find the mass of the given solid.

62. A sphere or radius a if the density at a point is proportional to the square of its distance from the center.

63. A sphere of radius a if the density at a point is proportional to its distance from the center.

64. A hemisphere of radius a if the density at a point is proportional to its distance from the base.

65. The solid common to the sphere $x^2 + y^2 + z^2 = 1$ and the cone $z = \sqrt{x^2 + y^2}$ if $\delta(x, y, z) = \sqrt{x^2 + y^2 + z^2}$.

66. The solid enclosed between the spheres $x^2 + y^2 + z^2 = 4$ and $x^2 + y^2 + z^2 = 1$ if $\delta(x, y, z) = (x^2 + y^2 + z^2)^{-1/2}$.

67. The solid ball of radius 1 if the density at each point d units from the center is $\dfrac{1}{1 + d^2}$.

Exercises 68-72. Find the center of mass of the given solid.

68. A hemispherical solid if its density is proportional to the distance from its base.

69. A homogeneous cone of height h and base of radius a, positioned so that its vertex is at $(0, 0, 0)$ and the axis is the positive z-axis. 9

70. A homogeneous solid bounded above by the sphere $\rho = a$ and below by the cone $\phi = \alpha$, where $0 < \phi < \dfrac{\pi}{2}$.

71. The homogeneous solid bounded below by the cone $z^2 = x^2 + y^2$ lying above the xy-plane and bounded above by the sphere $x^2 + y^2 + z^2 = 9$.

72. The homogeneous solid that lies above the cone $z = \sqrt{x^2 + y^2}$ and below the hemisphere $z = 1 + \sqrt{1 - (x^2 + y^2)}$.

	CHAPTER SUMMARY

LIMITS AND CONTINUITY	$$\lim_{(x,y)\to(x_0,y_0)} f(x,y) = L \text{ if:}$$ For any given $\varepsilon > 0$ there exists $\delta > 0$ such that: $$0 < \|(x,y) - (x_0, y_0)\| < \delta \Rightarrow |f(x,y) - L| < \varepsilon$$ The function $f(x,y)$ is **continuous** at (x_0, y_0) if: $$\lim_{(x,y)\to(x_0,y_0)} f(x,y) = f(x_0, y_0)$$ *A* **function that** is continuous at every point in its domain is said to be a **continuous function**.
THEOREMS	If $\displaystyle\lim_{(x,y)\to(x_0,y_0)} f(x,y) = L$ and $\displaystyle\lim_{(x,y)\to(x_0,y_0)} g(x,y) = M$, then: (a) $\displaystyle\lim_{(x,y)\to(x_0,y_0)} [f(x,y) + g(x,y)] = L + M$. (b) $\displaystyle\lim_{(x,y)\to(x_0,y_0)} [f(x,y) - g(x,y)] = L - M$. (c) $\displaystyle\lim_{(x,y)\to(x_0,y_0)} [f(x,y) \cdot g(x,y)] = LM$. (d) $\displaystyle\lim_{(x,y)\to(x_0,y_0)} \frac{f(x,y)}{g(x,y)} = \frac{L}{M}$ if $M \neq 0$. (e) $\displaystyle\lim_{(x,y)\to(x_0,y_0)} [cf(x,y)] = cL$ for any number c. If f and g are continuous at (x_0, y_0) then so are the functions: (a) $f + g$ (b) $f - g$ (c) fg (d) $\dfrac{f}{g}$ [providing $g(x_0, y_0) \neq 0$] (e) cf (for any real number c) If f and g are real-valued continuous functions of one variable, then the following functions are real-valued continuous functions of two variables $$H(x,y) = f(x) + g(y) \text{ and } K(x,y) = f(x)g(y)$$ If f is a real-valued continuous function of one variable, and if g is a real-valued continuous function of two variables, then the function $H(x,y) = f[g(x,y)]$ is a real-valued continuous function of two variables.

DOUBLE INTEGRALS:

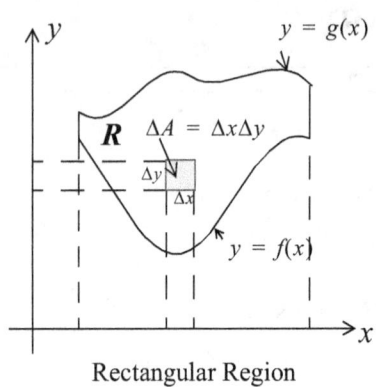

Rectangular Region

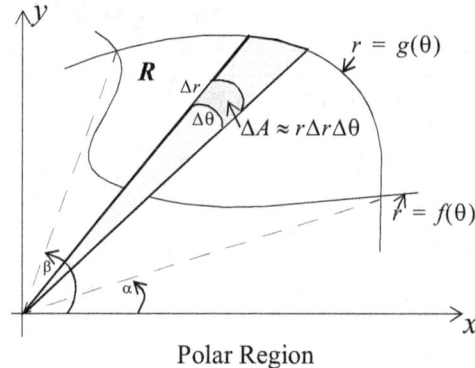

Polar Region

Let $h(x, y)$ be continuous on a region R:

$$\iint\limits_{R} h(x, y)\,dA = \int_{a}^{b} \int_{f(x)}^{g(x)} h(x, y)\,dy\,dx$$

Let $h(x, y)$ be continuous on a region R:

$$\iint\limits_{R} h(r, \theta)\,dA = \int_{\alpha}^{\beta} \int_{f(\theta)}^{g(\theta)} h(r, \theta)\,r\,dr\,d\theta$$

TRIPLE INTEGRALS

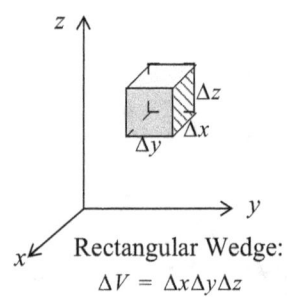

Rectangular Wedge:
$\Delta V = \Delta x \Delta y \Delta z$

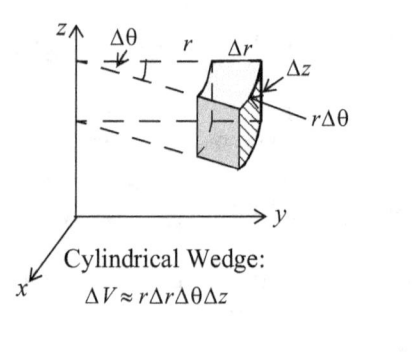

Cylindrical Wedge:
$\Delta V \approx r \Delta r \Delta \theta \Delta z$

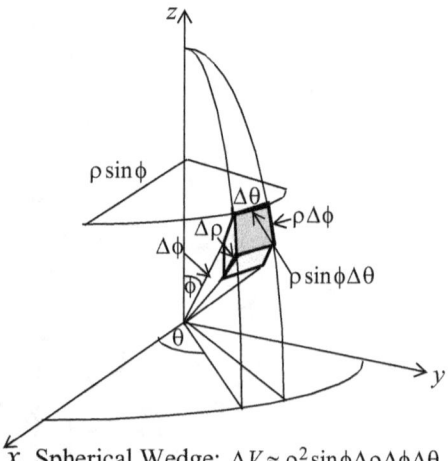

Spherical Wedge: $\Delta V \approx \rho^2 \sin\phi \Delta\rho \Delta\phi \Delta\theta$

RECTANGULAR COORDINATES

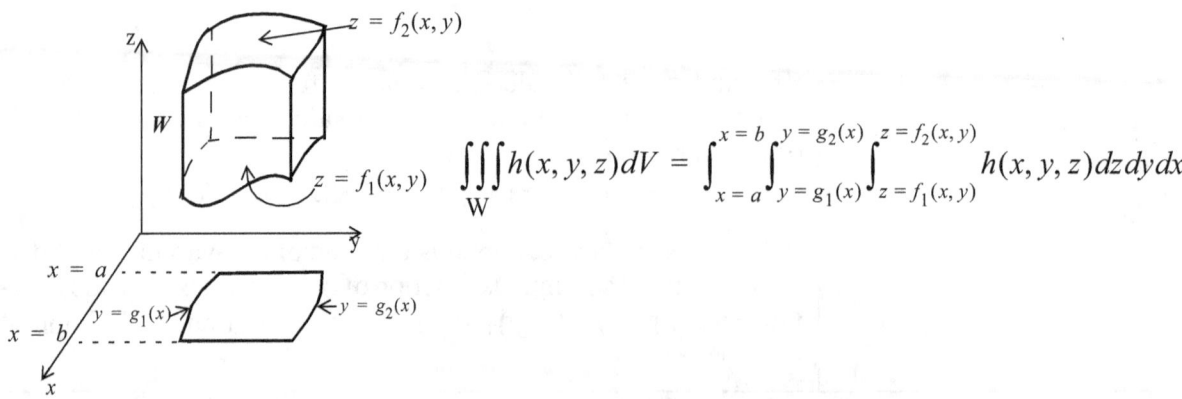

$$\iiint\limits_{W} h(x, y, z)\,dV = \int_{x=a}^{x=b} \int_{y=g_1(x)}^{y=g_2(x)} \int_{z=f_1(x,y)}^{z=f_2(x,y)} h(x, y, z)\,dz\,dy\,dx$$

CYLINDRICAL COORDINATES

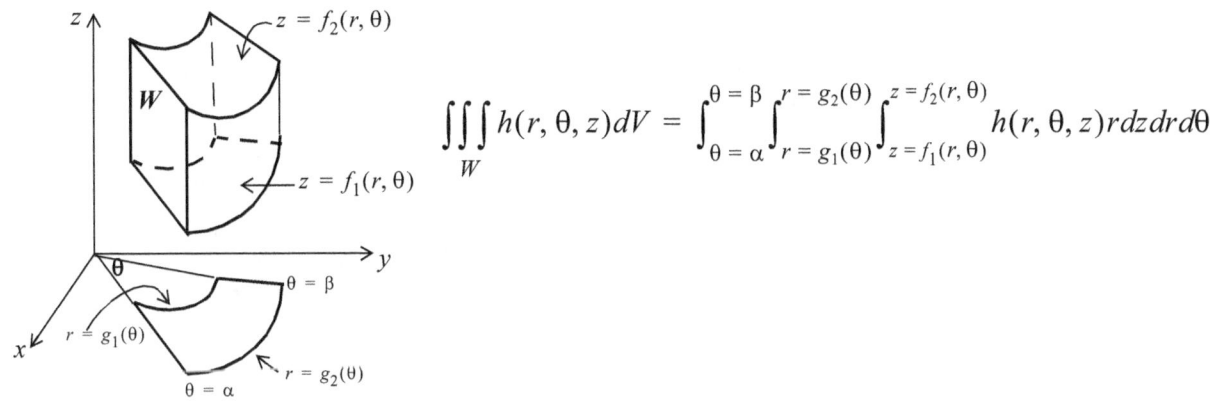

$$\iiint\limits_{W} h(r, \theta, z)dV = \int_{\theta = \alpha}^{\theta = \beta} \int_{r = g_1(\theta)}^{r = g_2(\theta)} \int_{z = f_1(r, \theta)}^{z = f_2(r, \theta)} h(r, \theta, z)r\,dz\,dr\,d\theta$$

VOLUME: SPHERICAL COORDINATES

$$W = \{(\rho, \phi, \theta)\,|\,\rho_1 \le \rho \le \rho_2,\, \phi_1 \le \phi \le \phi_2,\, \alpha \le \theta \le \beta\}$$

$$\iiint\limits_{W} h(\rho, \phi, \theta)dV = \int_{\theta = \alpha}^{\theta = \beta} \int_{\phi = \phi_1(\theta)}^{\phi = \phi_2(\theta)} \int_{\rho = \rho_1(\phi, \theta)}^{\rho = \rho_2(\phi, \theta)} h(\rho, \phi, \theta)\rho^2 \sin\phi\,d\rho\,d\phi\,d\theta$$

CHAPTER 12

Vectors and Vector-Valued Functions

§1. VECTORS IN THE PLANE AND BEYOND

Roughly speaking, a **vector** is a quantity that has both magnitude and direction. A vector in the plane is represented by a directed line segment (or arrow) pointing in the direction of the vector, with length representing its magnitude.

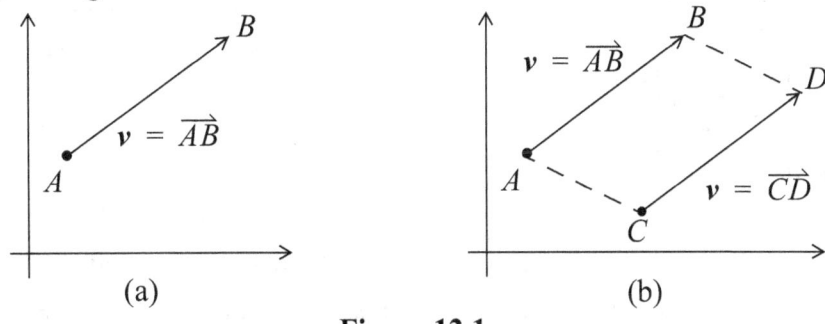

Figure 12.1

Vectors will be denoted by boldface lowercase letters. The vector $v = \overrightarrow{AB}$ in Figure 12.1(a) is said to have **initial point** A, and **terminal point** B.

Since vectors represent magnitude and direction, those possessing the same of both are considered to be equal. If, for example, you pick up the vector v in Figure 12.1(a) and move it in a parallel fashion to the vector in Figure 12.1(b) with initial point C and terminal point D, then you will still have the same vector:

$$v = \overrightarrow{AB} = \overrightarrow{CD}$$

In particular, the vector v in Figure 12.2 with initial point $A = (x_0, y_0)$ and terminal point $B = (x_1, y_1)$ can be moved in a parallel fashion so that its initial point coincides with the origin. When so placed, the vector is said to be in **standard position**.

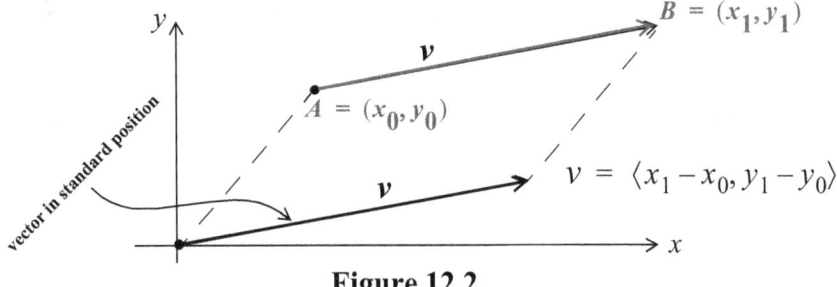

Figure 12.2

<u>**Note:**</u> We will use a "wedged form" to distinguish vectors from points. In particular, the expression $\langle x_1 - x_0, y_1 - y_0 \rangle$ in Figure 12.2 represents the standard position vector, while (x_1, y_1) is a point in the plane.

EXAMPLE 12.1 Sketch the vector with initial point $(-2, 3)$ and terminal point $(4, -1)$. Position that vector in standard position, and identify its terminal point.

SOLUTION: The Figure below tells the whole story.

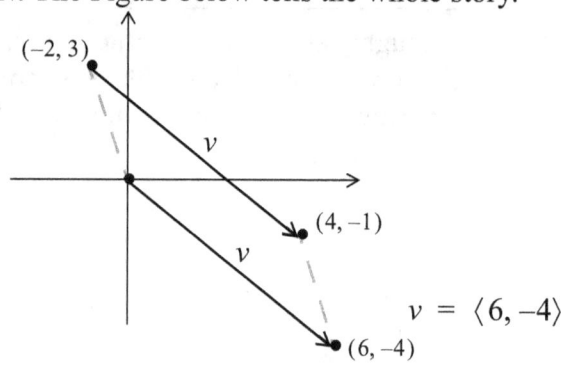

Figure 12.3

The two vectors in the adjacent figure are indeed the "same," since they share a common direction and magnitude:

In either case one can get to the terminal point from the initial point by moving 6 units to the right and 4 units down.

As it is with a standard positioned vector v in the plane [Figure 12.4(a)], a standard position vector v in R^3 can also be expressed in terms of the coordinates of its terminal point, as in Figure 12,4(b). .

The symbol $\Re^n$ will be used to denote the space consisting of n-**tuples**.

The numbers a_i in the n-tuple $(a_1, a_2, ..., a_n)$ are its **components**.

For notational convenience, v_i is often used to represent the components of a vector v in $\Re^n$, as in:

$v = \langle v_1, v_2 \rangle$, $v = \langle v_1, v_2, v_3 \rangle$

and $v = \langle v_1, v_2, ..., v_n \rangle$

Similarly:

$w = \langle w_1, w_2, ..., w_n \rangle$

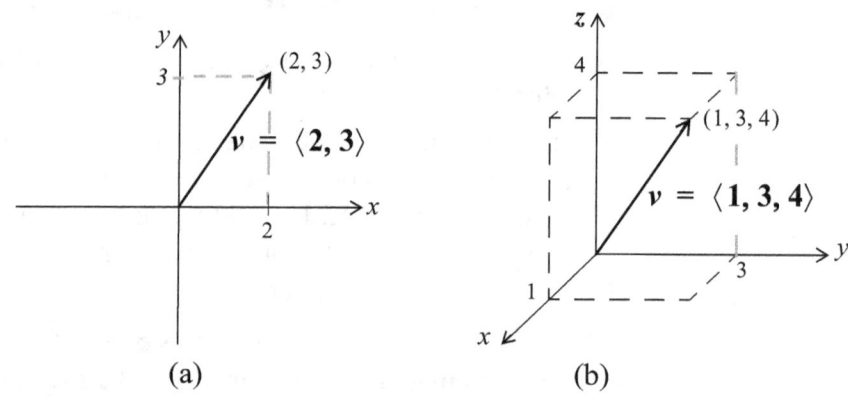

(a)   (b)

Figure 12.4

CHECK YOUR UNDERSTANDING 12.1

Sketch the vector v with initial point $(2, 1)$ and terminal point $(4, -2)$ and determine (a_1, a_2) for which $v = \langle a_1, a_2 \rangle$.

Answers: See page A-11.

SCALAR PRODUCT AND SUMS OF VECTORS

Vectors evolved from the need to adequately represent physical quantities that are characterized by both magnitude and direction. In a way, the quantities themselves tell us how we should define algebraic operations on vectors. Suppose, for example, that the vector $v = \langle 3, 2 \rangle$ of Figure 12.5(a) represents a force. Doubling the magnitude of that force without changing its direction would result in the vector force labeled $2v$ (a vector that is in the same direction as $v = \langle \mathbf{3, 2} \rangle$, and length twice that of v)

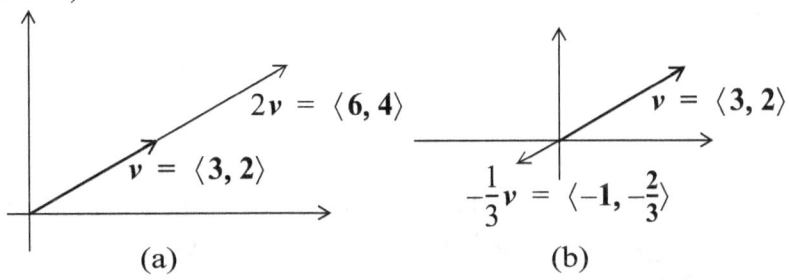

Figure 12.5

Similarly, if a force that is one-third that of $v = \langle \mathbf{3, 2} \rangle$ is applied in the **opposite direction** to v, then the vector representing that new force is the vector $-\frac{1}{3}v$ in Figure 12.5(b).

In general:

Note: Real numbers (as opposed to vectors) are called **scalars.**

DEFINITION 12.1

SCALAR MULTIPLICATION

To any vector $v = \langle v_1, v_2, \ldots, v_n \rangle$ in $\mathfrak{R}^n$, and any $a \in \mathfrak{R}$, we let:

$$av = \langle av_1, av_2, \ldots, av_n \rangle$$

The vector rv is said to be a **scalar multiple of v.**

For example:

$$3 \langle 1, 5 \rangle = \langle 3, 15 \rangle$$

$$-5 \langle 1, 0, -4 \rangle = \langle -5, 0, 20 \rangle$$

$$\sqrt{2} \langle 1, 3, -4, 5 \rangle = \langle \sqrt{2}, 3\sqrt{2}, -4\sqrt{2}, 5\sqrt{2} \rangle$$

VECTOR ADDITION

If two people pull on an object positioned at the origin with forces v and w, then the observed combined effect is the same as that of one individual pulling with force z, where z is the vector coinciding with the diagonal in the parallelogram formed by the vectors v and w [Figure 12.6(a)]. As is revealed in Figure 12.6(b), z can be determined by positioning the initial point of w at the terminal side of v.

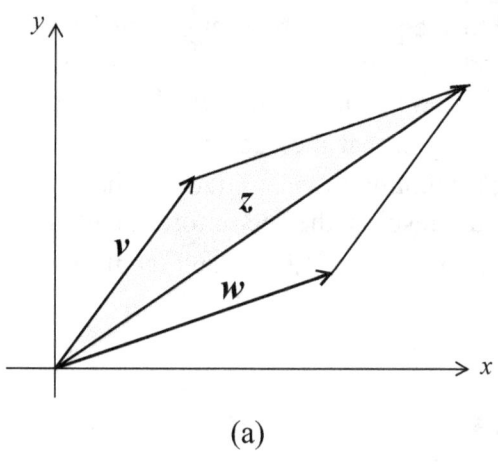

(a)

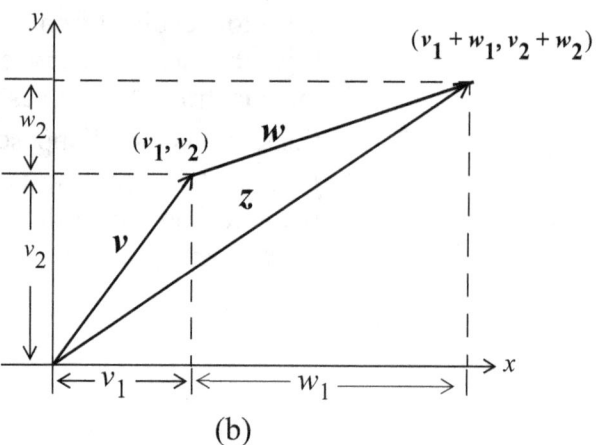

(b)

Figure 12.6

The above vector z is said to be the sum of the vectors v and w, and is denoted by $v + w$. From the figure, we see that:

$$z = v + w = \langle v_1, v_2 \rangle + \langle w_1, w_2 \rangle = \langle v_1 + w_1, v_2 + w_2 \rangle$$

Generalizing, we have:

While identical in form, the "+" in $v + w$ differs in spirit from that in $v_i + w_i$. The latter represents the familiar sum of two numbers, as in $3 + 7$, while the former represents the newly defined sum of two n-tuples, as in:

$(3, -2) + (7, 11) = (10, 9)$

DEFINITION 12.2

VECTOR SUM

The **sum** of the vectors $v = \langle v_1, v_2, \ldots, v_n \rangle$ and $w = \langle w_1, w_2, \ldots, w_n \rangle$ in $\Re^n$, is denoted by $v + w$ and is given by:

$$v + w = \langle v_1 + w_1, v_2 + w_2, \ldots, v_n + w_n \rangle$$

EXAMPLE 12.2   For $v = \langle -2, 3, 1 \rangle$, $w = \langle 1, 5, 0 \rangle$, determine the vector $2v + w$.

SOLUTION: $2 \langle -2, 3, 1 \rangle + \langle 1, 5, 0 \rangle = \langle -4, 6, 2 \rangle + \langle 1, 5, 0 \rangle$

$$= \langle -4 + 1, 6 + 5, 2 + 0 \rangle$$
$$= \langle -3, 11, 2 \rangle$$

CHECK YOUR UNDERSTANDING 12.2

Answers: **(3, 7)**

For $v = \langle 3, 2 \rangle$, $w = \langle -1, 1 \rangle$, determine $2v + 3w$.

The **zero vector** in $\Re^n$, denoted by $\mathbf{0}$, is that vector with each component 0. For example, $\mathbf{0} = \langle 0, 0 \rangle$ is the zero vector in $\Re^2$, and $\mathbf{0} = \langle 0, 0, 0 \rangle$ is the zero vector in $\Re^3$.

No direction is associated with the zero vector. A zero force, for example, is no force at all, and its "direction" would be a moot point.

SUBTRACTION

The **negative** of a vector $v = \langle v_1, v_2, ..., v_n \rangle$, denoted by $-v$, is the vector $-v = (-1)v = \langle -v_1, -v_2, ..., -v_n \rangle$. Note that just as $5 + (-5) = 0$, so then:

$$v + (-v) = \langle v_1, v_2, ..., v_n \rangle + \langle -v_1, -v_2, ..., -v_n \rangle$$
$$= \langle v_1 - v_1, v_2 - v_2, ..., v_n - v_n \rangle = \langle 0, 0, ..., 0 \rangle = \mathbf{0}$$

And just as $7 - 3$ is used to represent the sum $7 + (-3)$, so then for given vectors v and w we define $v - w$ to be the vector $v + (-w)$.

In spite of the above formalities, everything works out just fine. For example, as can be anticipated $\langle 7, -4, 9 \rangle - \langle 2, 8, -1 \rangle = \langle 5, -12, 10 \rangle$:

$$\langle 7, -4, 9 \rangle - \langle 2, 8, -1 \rangle = \langle 7, -4, 9 \rangle + [\langle -2, 8, -1 \rangle]$$
$$= \langle 7, -4, 9 \rangle + \langle -2, -8, 1 \rangle = \langle 5, -12, 10 \rangle$$

For given vectors v and w in the plane, $v - w$ can be represented by the vector with initial point the terminal point of w and with terminal point the terminal point of v:

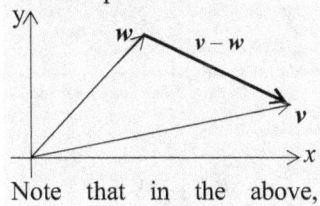

Note that in the above, $w + (v - w)$ is indeed v.

Answers: $\langle -\frac{1}{2}, -\frac{7}{2}, \frac{\sqrt{5}-6}{2} \rangle$

CHECK YOUR UNDERSTANDING 12.3

Determine:

$$\frac{1}{2}[\langle 2, -3, 0 \rangle - \langle 1, 0, -\sqrt{5} \rangle] - \langle 1, 2, 3 \rangle$$

While we are particularly concerned with vectors in the plane and three-dimensional space, it takes little additional effort to lay a foundation in the general setting of n-dimensional space; and so we shall:

THEOREM 12.1 Let u, v, and w be vectors in $\Re^n$, and let a and b be scalars (real numbers). Then:

(a) $u + v = v + u$

(b) $(u + v) + w = u + (v + w)$

(c) $a(u + v) = au + av$

(d) $(a + b)v = av + bv$

(e) $a(bv) = (ab)v$

PROOF: We establish (b) in $\Re^2$, (c) in $\Re^3$, and (e) in $\Re^n$.

To emphasize the important role played by definitions, the symbol $\equiv$ instead of $=$ will temporarily be used to indicate a step in a proof which follows directly from a definition. In addition, the abbreviation "**PofR**" will temporarily be used to denote that a step, such as the additive associative property of the real numbers: $(a + b) + c = a + (b + c)$, that follows directly from a **P**roperty of the **R**eal numbers.

(b): $(\boldsymbol{u} + \boldsymbol{v}) + \boldsymbol{w} = \boldsymbol{u} + (\boldsymbol{v} + \boldsymbol{w})$ (in $\mathfrak{R}^2$).

The associative property eliminates the need for including parenthesis when summing more than two vectors. In particular,

$$\boldsymbol{u} + \boldsymbol{v} + \boldsymbol{w}$$

is perfectly well-defined.

For $\boldsymbol{u} = \langle u_1, u_2 \rangle, \boldsymbol{v} = \langle v_1, v_2 \rangle$, and $\boldsymbol{w} = \langle w_1, w_2 \rangle$:

$$(\boldsymbol{u} + \boldsymbol{v}) + \boldsymbol{w} \equiv [\langle u_1, u_2 \rangle + \langle v_1, v_2 \rangle] + \langle w_1, w_2 \rangle$$

Definition 12.2: $\equiv \langle u_1 + v_1, u_2 + v_2 \rangle + \langle w_1, w_2 \rangle$

Definition 12.2: $\equiv \langle (u_1 + v_1) + w_1, (u_2 + v_2) + w_2 \rangle$

PofR: $= \langle u_1 + (v_1 + w_1), u_2 + (v_2 + w_2) \rangle$

Definition 12.2: $\equiv \langle u_1, u_2 \rangle + \langle (v_1, v_2) + (w_1, w_2) \rangle$

$\equiv \boldsymbol{u} + (\boldsymbol{v} + \boldsymbol{w})$

(c): $a(\boldsymbol{u} + \boldsymbol{v}) = a\boldsymbol{u} + a\boldsymbol{v}$ (in $\mathfrak{R}^3$).

For $\boldsymbol{u} = \langle u_1, u_2, u_3 \rangle$ and $\boldsymbol{v} = \langle v_1, v_2, v_3 \rangle$:

$$a(\boldsymbol{u} + \boldsymbol{v}) \equiv a[\langle u_1, u_2, u_3 \rangle + \langle v_1, v_2, v_3 \rangle]$$

Definition 12.2: $\equiv a \langle u_1 + v_1, u_2 + v_2, u_3 + v_3 \rangle$

Definition 12.1: $\equiv a(u_1 + v_1), a(u_2 + v_2), a(u_3 + v_3)$

PofR: $= (au_1 + av_1, au_2 + av_2, au_3 + av_3)$

Defintion 12.2: $\equiv (au_1, au_2, au_3) + (av_1, av_2, av_3)$

Definition 12.1: $\equiv a(u_1, u_2, u_3) + a(v_1, v_2, v_3)$

$\equiv a\boldsymbol{u} + a\boldsymbol{v}$

(e): $a(b\boldsymbol{v}) = (ab)\boldsymbol{v}$ (in R^n).

For $\boldsymbol{v} = \langle v_1, v_2, \ldots, v_n \rangle$:

$$a(b\boldsymbol{v}) \equiv a[b \langle v_1, v_2, \ldots, v_n \rangle]$$

Definition 12.1: $\equiv a \langle bv_1, bv_2, \ldots, bv_n \rangle$

Definition 12.1: $\equiv \langle a(bv_1), a(bv_2), \ldots, a(bv_n) \rangle$

PofR: $= \langle (ab)v_1, (ab)v_2, \ldots, (ab)v_n \rangle$

Defintion 12.1: $\equiv (ab) \langle v_1, v_2, \ldots, v_n \rangle$

$\equiv (ab)\boldsymbol{v}$

Throughout mathematics:
DEFINITIONS RULE!

Just look at the above proof. It contains but one "logical step," the step labeled **PofR**; all other steps hinge on **DEFINITIONS**.

CHECK YOUR UNDERSTANDING 12.4

Answer: See page A-12.

Establish Theorem 12.1(a) in $\mathfrak{R}^3$, and Theorem 12.1(d) in $\mathfrak{R}^n$.

The length of a vector v, also called the **norm** of v, is denoted by $\|v\|$. Applying the Pythagorean Theorem, we find (see Figure 12.7) that the norm of $v = \langle v_1, v_2 \rangle \in \Re^2$ and $v = \langle v_1, v_2, v_3 \rangle \in \Re^3$ is given by:

$$\|v\| = \sqrt{v_1^2 + v_2^2} \text{ and } \|v\| = \sqrt{v_1^2 + v_2^2 + v_3^2} \text{ (respectively).}$$

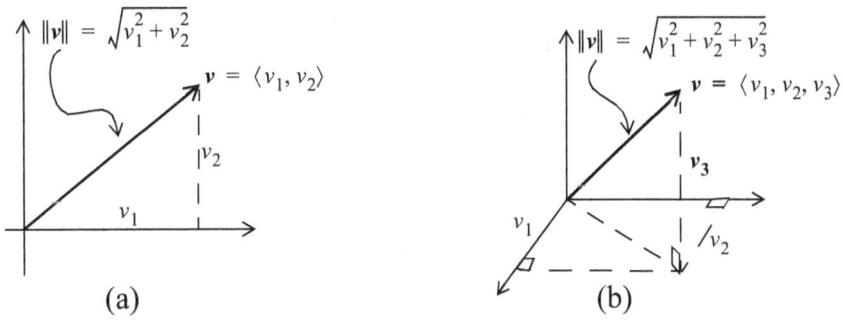

(a)   (b)

Figure 12.7

In general, for $v \in \Re^n$, $\|v\|$ is defined to be the length of v, with:

$$\|v\| = \|\langle v_1, v_2, \ldots, v_n \rangle\| = \sqrt{v_1^2 + v_2^2 + \ldots + v_n^2}$$

Moreover:

$\|u - v\|$ denotes the distance between the endpoints of u and v.

In particular, for $u = \langle u_1, u_2 \rangle$ and $v = \langle v_1, v_2 \rangle$ in $\Re^2$:

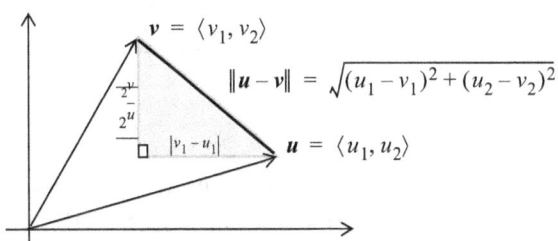

Vectors of norm 1 are said to be **unit vectors**. The vectors $\langle \frac{3}{5}, \frac{4}{5} \rangle$ and $\langle \frac{1}{\sqrt{2}}, 0, \frac{1}{\sqrt{2}} \rangle$, for example, are unit vectors in $\Re^2$ and $\Re^3$, respectively:

$$\left\| \langle \frac{3}{5}, \frac{4}{5} \rangle \right\| = \sqrt{\left(\frac{3}{5}\right)^2 + \left(\frac{4}{5}\right)^2} = \sqrt{\frac{9}{25} + \frac{16}{25}} = \sqrt{\frac{25}{25}} = 1$$

$$\text{and } \left\| \langle \frac{1}{\sqrt{2}}, 0, \frac{1}{\sqrt{2}} \rangle \right\| = \sqrt{\left(\frac{1}{\sqrt{2}}\right)^2 + (0)^2 + \left(\frac{1}{\sqrt{2}}\right)^2} = 1$$

CHECK YOUR UNDERSTANDING 12.5

Answer: (a) $\sqrt{30}$
(b) See page A-12.

(a) Determine $\|3\langle 2, 1, 0 \rangle - \langle 4, -2, 1 \rangle\|$.

(b) Let $v \in \Re^3$ and $c \in \Re$. Prove that $\|cv\| = |c|\|v\|$.

We single out, and label, two vectors in the $\mathfrak{R}^2$ and three in $\mathfrak{R}^3$:

$$\boldsymbol{i} = \langle 1, 0 \rangle, \ \boldsymbol{j} = \langle 0, 1 \rangle$$

and $\ \boldsymbol{i} = \langle 1, 0, 0 \rangle, \ \boldsymbol{j} = \langle 0, 1, 0 \rangle, \ \boldsymbol{k} = \langle 0, 0, 1 \rangle$

The above vectors are called the **unit coordinate vectors** in $\mathfrak{R}^2$ and $\mathfrak{R}^3$, respectively. Each is indeed of length 1, and lies along a positive coordinate axis:

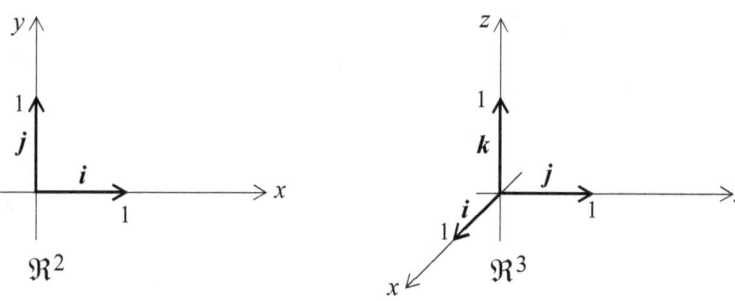

Note that every vector in $\mathfrak{R}^2$ and in $\mathfrak{R}^3$ can be expressed uniquely in terms of the above coordinate vectors. For example:

$$\langle -7, 3 \rangle = -7 \langle 1, 0 \rangle + 3 \langle 0, 1 \rangle = -7\boldsymbol{i} + 3\boldsymbol{j}$$

and $\langle 4, 0, -\frac{1}{2} \rangle = 4 \langle 1, 0, 0 \rangle + 0 \langle 0, 1, 0 \rangle - \frac{1}{2} \langle 0, 0, 1 \rangle$

$$= 4\boldsymbol{i} + 0\boldsymbol{j} - \frac{1}{2}\boldsymbol{k} = 4\boldsymbol{i} - \frac{1}{2}\boldsymbol{k}$$

In general: $a\boldsymbol{i} + b\boldsymbol{j} = \langle a, b \rangle$ and $a\boldsymbol{i} + b\boldsymbol{j} + c\boldsymbol{k} = \langle a, b, c \rangle$.

CHECK YOUR UNDERSTANDING 12.6

Determine:
$$3(2\boldsymbol{i} + 4\boldsymbol{j} - \boldsymbol{k}) - 5(3\boldsymbol{i} - \boldsymbol{k}).$$

Answer: $-9\boldsymbol{i} + 12\boldsymbol{j} + 2\boldsymbol{k}$

As previously noted, force is a vector quantity, subjected to the defined properties of scalar multiplication and vector addition. Consider the following example.

EXAMPLE 12.3   Find the vector (tension) forces, $\boldsymbol{F}_1, \boldsymbol{F}_2$, resulting from the 50 pound force $\boldsymbol{F}$ depicted in the adjacent figure.

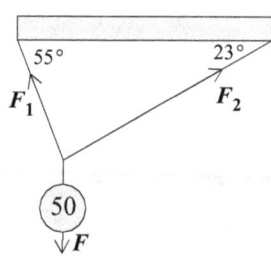

SOLUTION: The two tension forces are broken down into their horizontal and vertical components in the adjacent figure.

Given the stable nature of the system, the sum of all force vectors must equal the zero vector:

Note that the physical length of the attaching cables is of no consequence whatsoever. It is their direction and the magnitude of the force that are relevant (for a vector is determined by it direction and magnitude).

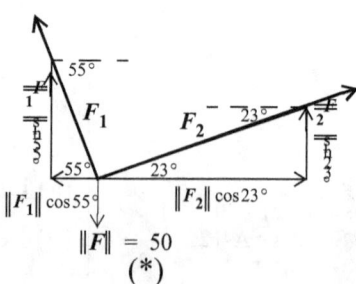

$$(\|F_2\|\cos 23° - \|F_1\|\cos 55°)i + (\|F_1\|\sin 55° + \|F_2\|\sin 23° - 50)j = 0i + 0j$$

Leading us to: (1) $\|F_2\|\cos 23° - \|F_1\|\cos 55° = 0$

(2) $\|F_1\|\sin 55° + \|F_2\|\sin 23° - 50 = 0$

From (1): $\|F_2\| = \dfrac{\|F_1\|\cos 55°}{\cos 23°}$

Substituting in (2): $\|F_1\|\sin 55° + \dfrac{\|F_1\|\cos 55°}{\cos 23°}\sin 23° - 50 = 0$

$$\|F_1\|\sin 55°\cos 23° + \|F_1\|\cos 55°\sin 23° - 50\cos 23° = 0$$

$$\|F_1\| = \dfrac{50\cos 23°}{\sin 55°\cos 23° + \cos 55°\sin 23°} = \dfrac{50\cos 23°}{\sin 78°}$$

see margin

Invoking the sum identity:
$\sin(x+y) = \sin x\cos y + \cos x\sin y$
we have:
$\sin 55°\cos 23° + \cos 55°\sin 23°$
$\qquad = \sin 78°$

Returning to (1): $\|F_2\| = \dfrac{\dfrac{50\cos 23°}{\sin 78°}\cos 55°}{\cos 23°} = \dfrac{50\cos 55°}{\sin 78°}$

Conclusion [see (*) at bottom of previous page]:

Note that
$\quad(-\cos 55°i,\ \sin 55°j)$
and $(\cos 23°i,\ \sin 23°j)$
are unit vectors in the direction
of F_1 and F_2, respectively.

$$F_1 = -\|F_1\|\cos 55°i + \|F_1\|\sin 55°j = \dfrac{50\cos 23°}{\sin 78°}(-\cos 55°i + \sin 55°j)$$

$$F_2 = \|F_2\|\cos 23°i + \|F_2\|\sin 23°j = \dfrac{50\cos 55°}{\sin 78°}(\cos 23°i + \sin 23°j)$$

EXAMPLE 12.4 Find the forces, F_1, F_2, resulting from the 100 pound force F depicted in the adjacent figure.

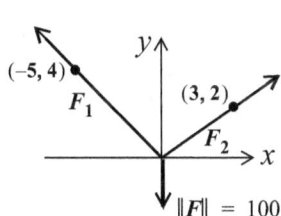

SOLUTION: We know the whole story for the vector F, namely:

$$F = -100j.$$

We can easily find the **unit** vector in the direction of F_1:

$$\dfrac{1}{\sqrt{25+16}}\langle -5, 4\rangle = \dfrac{1}{\sqrt{41}}(-5i + 4j)\,;\ \text{bringing us to:}$$

$$F_1 = \|F_1\|\left[\dfrac{1}{\sqrt{41}}(-5i + 4j)\right] = -\dfrac{5\|F_1\|}{\sqrt{41}}i + \dfrac{4\|F_1\|}{\sqrt{41}}j$$

Similarly:

$$F_2 = \|F_2\|\left[\dfrac{1}{\sqrt{13}}(3i + 2j)\right] = \dfrac{3\|F_2\|}{\sqrt{13}}i + \dfrac{2\|F_2\|}{\sqrt{13}}j$$

Turning to the vector equation: $F_1 + F_2 + F = 0$ we have:

$$\left(-\frac{5\|F_1\|}{\sqrt{41}}i + \frac{4\|F_1\|}{\sqrt{41}}j\right) + \left(\frac{3\|F_2\|}{\sqrt{13}}i + \frac{2\|F_2\|}{\sqrt{13}}j\right) + (-100j) = 0i + 0j$$

Equating components:

$$\left(-\frac{5\|F_1\|}{\sqrt{41}} + \frac{3\|F_2\|}{\sqrt{13}}\right)i = 0i \Rightarrow -5\sqrt{13}\|F_1\| + 3\sqrt{41}\|F_2\| = 0 \tag{1}$$

$$\left(\frac{4\|F_1\|}{\sqrt{41}} + \frac{2\|F_2\|}{\sqrt{13}} - 100\right)j = 0j \Rightarrow 4\sqrt{13}\|F_1\| + 2\sqrt{41}\|F_2\| = 100\sqrt{13}\sqrt{41} \tag{2}$$

From (1): $\|F_1\| = \frac{3\sqrt{41}}{5\sqrt{13}}\|F_2\| \qquad$ (3)

Substituting in (2): $4\sqrt{13}\dfrac{3\sqrt{41}}{5\sqrt{13}}\|F_2\| + 2\sqrt{41}\|F_2\| = 100\sqrt{13}\sqrt{41} \Rightarrow \|F_2\| = \dfrac{500\sqrt{13}}{22}$

Substituting in (3): $\|F_1\| = \dfrac{3\sqrt{41}}{5\sqrt{13}} \cdot \dfrac{500\sqrt{13}}{22} = \dfrac{300\sqrt{41}}{22}$

Conclusion:

$$F_1 = -\frac{5\|F_1\|}{\sqrt{41}}i + \frac{4\|F_1\|}{\sqrt{41}}j = \frac{300}{22}(-5i + 4j) = \frac{150}{11}(-5i + 4j)$$

$$F_2 = \frac{3\|F_2\|}{\sqrt{13}}i + \frac{2\|F_2\|}{\sqrt{13}}j = \frac{500}{22}(3i + 2j) = \frac{250}{11}(3i + 2j)$$

CHECK YOUR UNDERSTANDING 12.7

Find F_2 given that its magnitude is twice that of F_1.

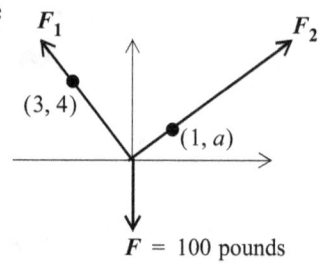

Answer:

$$F_2 = 4(\sqrt{91} - 4)\left(i + \frac{\sqrt{91}}{3}j\right)$$

	EXERCISES	

Exercises 1-6. Sketch the vector with given initial point A and terminal point B. Sketch the same vector in standard position in the plane and identify its terminal point.

1. $A = (-2, -2), B = (0, 1)$ 2. $A = (3, 3), B = (0, -1)$ 3. $A = (1, 1), B = (-2, 3)$

4. $A = (1, 0), B = (0, -1)$ 5. $A = (-2, -1), B = (1, -1)$ 6. $A = (2, 2), B = (1, -2)$

Exercises 7-10. Express, as a 3-tuple, the vector with given initial point A and terminal point B.

7. $A = (1, 2, 3), B = (3, 2, 1)$ 8. $A = (-4, 5, 0), B = (2, -5, 1)$

9. $A = (0, 1, -9), B = (-9, 0, 2)$ 10. $A = (-3, 5, -3), B = (3, -5, 3)$

Exercises 11-18. Perform the indicated vector operations.

11. $5\langle 3, -2\rangle + \langle 0, 1\rangle - \langle 2, -4\rangle$ 12. $\langle 2, 5\rangle + \langle 1, 3\rangle + [-\langle -2, 3\rangle]$

13. $\langle -2, 3, 1\rangle + [-\langle 1, -2, 0\rangle]$ 14. $-[-\langle -1, 2, 3\rangle] + \langle 3, 2, -2\rangle$

15. $4[2i - 4j - (i + 3j)]$ 16. $-2(2i - 4j) - (3i + 3j)$

17. $-2(3i + 2k) + (2j - k) - (i + j + k)$ 18. $5(i - 2j + 3k) - 2[(2i - k) + (i + j + k)]$

Exercises 19-24. Find a unit vector in the direction of the given vector v, and then express v as a product of a scalar times that unit vector.

19. $v = \langle 5, 2\rangle$ 20. $v = 4i - 3j$ 21. $v = 2i - 4j + k$

22. $v = \langle 2, 0, 5\rangle$ 23. $v = \sqrt{2}i - \frac{1}{3}j$ 24. $v = \frac{1}{2}i - \frac{2}{3}j - \sqrt{2}k$

25. For $u = \langle 1, 3\rangle$, $v = \langle 2, 4\rangle$, and $w = \langle 6, -2\rangle$, find scalars a and b such that:

    (a) $au + bv = w$ (b) $-au + bw = v$ (c) $av + (-bw) = u$

26. Find scalars a, b, and c, such that: $a\langle 1, 3, 0\rangle + b\langle 2, 1, 6\rangle + c\langle 1, 4, 6\rangle = \langle 7, 5, 6\rangle$

27. Find scalars a, b, and c, such that: $-a\langle 1, 3, 0\rangle + b\langle 2, 1, 6\rangle + [-c\langle 1, 4, 6\rangle] = \langle 0, 5, 6\rangle$

28. Show that there do not exist scalars a, b, and c, such that
$$r\langle 2, 3, 5\rangle + b\langle 3, 2, 5\rangle + c\langle 1, 2, 3\rangle = \langle 1, 2, 4\rangle$$

29. Find the vector $\langle a, b \rangle \in \Re^2$ of length 5 that has the same direction as the vector with initial point $(1, 3)$ and terminal point $(3, 1)$.

30. Find the vector $\langle a, b \rangle \in \Re^2$ of length 5 that is in the opposite direction to the vector with initial point $(1, 3)$ and terminal point $(3, 1)$.

Exercises 31-34. Determine the force vectors F_1 and F_2 in the given stable situation.

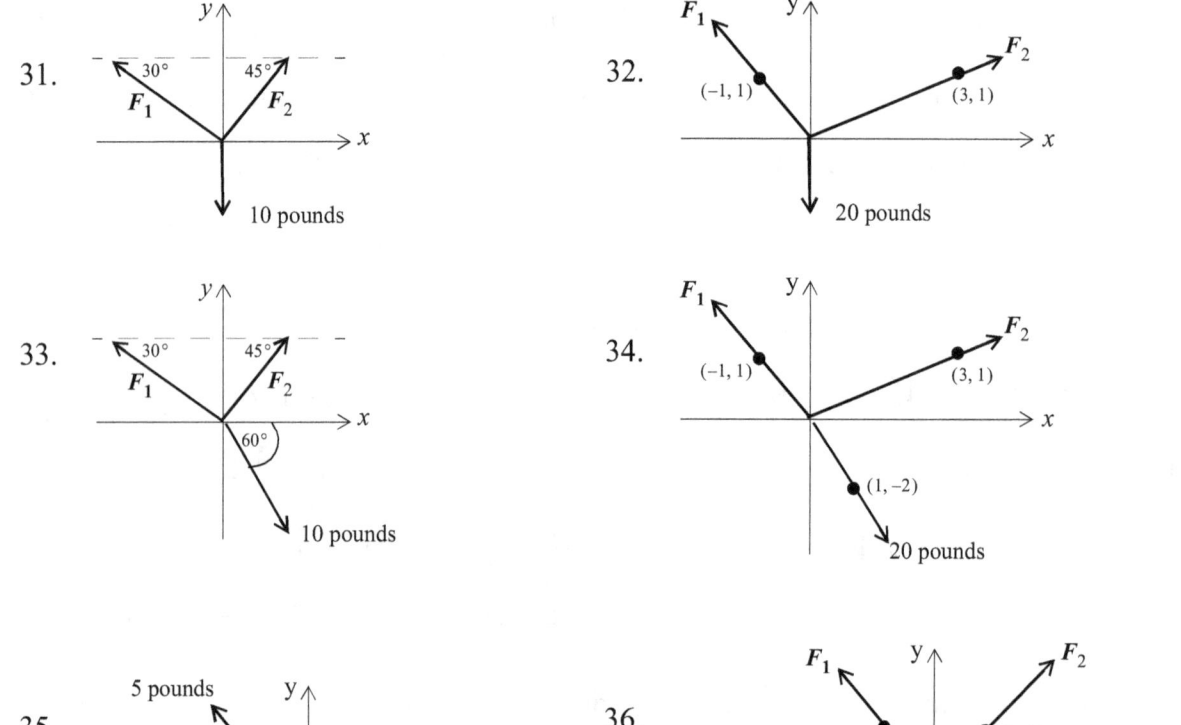

31.

32.

33.

34.

35.

36.

37. A child pulls a sled on a level path with a force of 15 pounds exerted at an angle of $30°$ with respect to the ground. Find the horizontal and vertical components of the force.

38. Ship A is traveling north at a speed of 10 mph, and ship B is heading east at a speed of 15 mph. How fast and in what direction does ship A appear to be moving from the point of view of an individual on ship B?

Exercises 39-42. A **bearing** is used to describe the direction of an object from a given point. Nautical bearing, also used for flights, is specified by an angle measured clockwise from due north (see adjacent figure).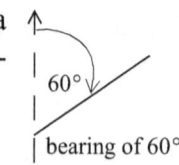

The **airspeed** of a plane is defined to be its speed in still air. The **track** of a plane is the direction resulting from the plane's velocity vector and that of a wind vector, and the **ground speed** of the plane is the magnitude of the sum of those two vectors.

39. An airplane has an airspeed of 400 km/hr in an easterly direction. The wind velocity is 80 km/hr in a southeasterly direction. Determine, to two decimal places, the ground speed of the plane.

40. A plane with airspeed of 250 km/hr is flying at a bearing of $25°$. A 23 km/hr wind is blowing at a bearing of $280°$. Determine, to one decimal place, the track and ground speed of the plane.

41. A plane with airspeed of 300 km/hr is flying at a bearing of $21°$, against a 32 km/hr westerly wind. Determine, to one decimal place, the track and ground speed of the plane.

42. A plane with airspeed of 250 km/hr is flying at a bearing of $25°$. A wind is blowing at a bearing of $15°$. Determine, to one decimal place, the speed of the wind if the ground speed of the plane is 270 km/hr.

43. (a) Referring to the adjacent figure, verify that if $\left\|\overrightarrow{AP}\right\| = \left\|\overrightarrow{PB}\right\|$,
    then: $P = \frac{1}{2}A + \frac{1}{2}B$.

    (b) Use the above result to prove that the line segment joining the midpoints of two sides of a triangle is parallel to, and half the length of the third side.

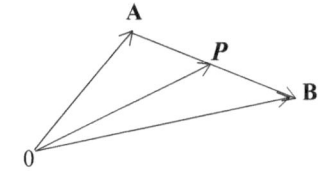

44. Prove that the midpoints of the sides of a quadrilateral ABCD are vertices of a parallelogram PQRS (see adjacent figure).
    Suggestion: Consider Exercise 43.

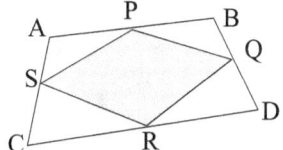

45. Let A and B be two distinct points in the plane and let P be a point on the line segment joining A and B. Show that if $\left\|\overrightarrow{AP}\right\| = r$ and $\left\|\overrightarrow{PB}\right\| = s$, then:

$$P = \left(\frac{s}{r+s}\right)A + \left(\frac{r}{r+s}\right)B$$

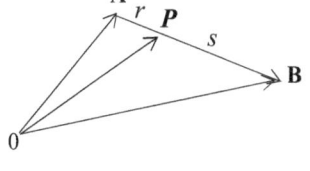

46. A **median** of a triangle is a line segment from a vertex to the midpoint of the opposite side. Prove that the three medians of a triangle have a common point of intersection (see adjacent figure).
    Suggestion: Consider Exercise 45.

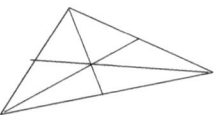

47. Prove that for $v \in \Re^2$, $av = 0$ if and only if $a = 0$ or $v = 0$.

48. Verify that Theorem 12.1(a) holds in $\Re^n$.

49. Verify that Theorem 12.1(b) holds in $\Re^n$.

50. Verify that Theorem 12.1(c) holds in $\Re^n$.

51. Verify that Theorem 12.1(d) holds in $\Re^n$.

§2. DOT AND CROSS PRODUCTS

At this point, we can add vectors (Definition 12.2), and can multiply a vector by a scalar (Definition 12.1). In either of those cases, the end result is again a vector. This operation assigns a real numbers to pairs of vectors in $\mathfrak{R}^n$:

Left to our own devices, we would most likely define the product of two vectors in this manner:
$$\langle 2, 1, 3 \rangle \langle 2, 5, 3 \rangle = \langle 4, 5, 9 \rangle$$
As it turns out, however, the above "natural definition" does not appear to be useful. The same, as you will see, cannot be said for the dot product concept.

DEFINITION 12.3
DOT PRODUCT

The **dot product** of $u = \langle u_1, u_2, ..., u_n \rangle$ and $v = \langle v_1, v_2, ..., v_n \rangle$, denoted by $u \cdot v$, is the real number:
$$u \cdot v = u_1 v_1 + u_2 v_2 + ... + u_n v_n$$

For example:
$$\langle 2, 4, -3, 1 \rangle \cdot \langle 5, 0, 7, -1 \rangle = 2 \cdot 5 + 4 \cdot 0 + (-3) \cdot 7 + 1 \cdot (-1) = -12$$

and: $(2i + 3j - k) \cdot (5i - 7j - 9k) = 10 - 21 + 9 = -2$

Here are four particularly important dot-product properties:

THEOREM 12.2

For $u, v, w \in \mathfrak{R}^n$ and any scalar r:

(a) $v \cdot v \geq 0$, and $v \cdot v = 0$ only if $v = 0$

(b) $u \cdot v = v \cdot u$

(c) $au \cdot v = a(u \cdot v) = u \cdot av$

(d) $(u + v) \cdot w = u \cdot w + v \cdot w$ and
$(u - v) \cdot w = u \cdot w - v \cdot w$

PROOF: We turn to (c) and invite you to verify the rest in the exercises:

$$au \cdot v = a\langle u_1, u_2, ..., u_n \rangle \cdot \langle v_1, v_2, ..., v_n \rangle$$

Defitition 12.1, page 489: $\quad = \langle au_1, au_2, ..., au_n \rangle \cdot \langle v_1, v_2, ..., v_n \rangle$

Definition 12.3: $\quad = (au_1)v_1 + (au_2)v_2 + ... + (au_n)v_n$

$\quad = a(u_1 v_1) + a(u_2 v_2) + ... + a(u_n v_n)$

$\quad = a(u_1 v_1 + u_2 v_2 + ... + u_n v_n)$

Definition 12.3: $\quad = a(u \cdot v)$

As for the other part of (c):

CHECK YOUR UNDERSTANDING 12.8

Verify that for any $u, v \in \mathfrak{R}^n$ and any scalar a:
$$u \cdot av = a(u \cdot v)$$

Answer: See page A-13.

The norm of a vector was introduced on page 493. We now offer a formal definition — one that involves the dot product:

DEFINITION 12.4

NORM IN $\mathfrak{R}^n$

The **norm** of a vector $v = \langle v_1, v_2, \ldots, v_n \rangle$, denoted by $\|v\|$, is given by:

$$\|v\| = \sqrt{v \cdot v}$$

In particular, for $v = \langle v_1, v_2 \rangle \in \mathfrak{R}^2$:

$$\|v\| = \sqrt{v \cdot v} = \sqrt{\langle v_1, v_2 \rangle \cdot \langle v_1, v_2 \rangle} = \sqrt{v_1^2 + v_2^2}$$

$\|v\| = \sqrt{v_1^2 + v_2^2}$

$v = \langle v_1, v_2 \rangle$

v_2

v_1

THEOREM 12.3

For $u, v \in R^n$:

$$\|u - v\|^2 = \|u\|^2 - 2u \cdot v + \|v\|^2$$

[Reminiscent of: $(a - b)^2 = a^2 - 2ab + b^2$]

PROOF:

$$\|u - v\|^2 = (u - v) \cdot (u - v)$$

Theorem 12.2(d): $= u \cdot (u - v) - v \cdot (u - v)$

Theorem 12.2(b) and (d): $= u \cdot u - u \cdot v - v \cdot u + v \cdot v = \|u\|^2 - 2u \cdot v + \|v\|^2$

CHECK YOUR UNDERSTANDING 12.9

(a) Determine $\|3i - 4j + 2k\|$.

(b) Show directly that:

$$\|\langle 5, 1 \rangle - \langle 2, -3 \rangle\|^2 = \|(5, 1)\|^2 - 2\langle 5, 1 \rangle \cdot \langle 2, -3 \rangle + \|\langle 2, -3 \rangle\|^2$$

Answers: (a) $\sqrt{29}$
(b) See page A-13.

ANGLE BETWEEN VECTORS

Applying the Law of Cosines [Figure 12.8(a)] to the nonzero vectors $u, v \in \mathfrak{R}^2$ in Figure 12.8(b), we see that:

$$\|u - v\|^2 = \|u\|^2 + \|v\|^2 - 2\|u\|\|v\| \cos\theta$$

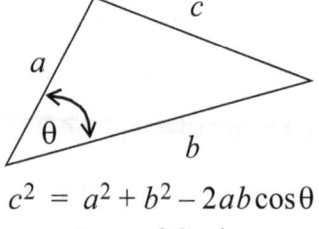

$c^2 = a^2 + b^2 - 2ab\cos\theta$
Law of Cosines

(a)

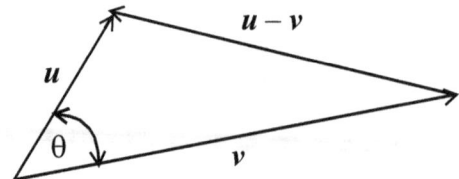

$\|u - v\|^2 = \|u\|^2 + \|v\|^2 - 2\|u\|\|v\| \cos\theta$

(b)

Figure 12.8

We also have (Theorem 12.3):

$$\|u - v\|^2 = \|u\|^2 - 2u \cdot v + \|v\|^2$$

Thus: $\|u\|^2 + \|v\|^2 - 2\|u\|\|v\|\cos\theta = \|u\|^2 - 2u \cdot v + \|v\|^2$

$$-2\|u\|\|v\|\cos\theta = -2u \cdot v$$

$$u \cdot v = \|u\|\|v\|\cos\theta$$

$$\cos\theta = \frac{u \cdot v}{\|u\|\|v\|}$$

see margin: $\theta = \cos^{-1}\left(\frac{u \cdot v}{\|u\|\|v\|}\right)$

For any $-1 \le x \le 1$, $\cos^{-1}x$ is defined to be that angle $0 \le \theta \le \pi$ whose cosine is x.

Formalizing:

DEFINITION 12.5

ANGLE BETWEEN VECTORS

The **angle** θ between two nonzero vectors $u, v \in \Re^n$ is given by:

$$\theta = \cos^{-1}\left(\frac{u \cdot v}{\|u\|\|v\|}\right) \text{ or } \cos\theta = \frac{u \cdot v}{\|u\|\|v\|}$$

And so: $u \cdot v = \|u\|\|v\|\cos\theta$

In Exercise 39 you are asked to verify that

$$\left|\frac{u \cdot v}{\|u\|\|v\|}\right| \le 1$$

Assuring us that:

$\cos^{-1}\left(\dfrac{u \cdot v}{\|u\|\|v\|}\right)$ exists.

EXAMPLE 12.5 Determine the angle between the vectors $u = \langle 1, 2, 0 \rangle$ and $v = \langle -1, 3, 1 \rangle$.

SOLUTION: $\theta = \cos^{-1}\left(\dfrac{u \cdot v}{\|u\|\|v\|}\right) = \cos^{-1}\left(\dfrac{\langle 1, 2, 0\rangle \cdot \langle -1, 3, 1\rangle}{\sqrt{1+4+0}\sqrt{1+9+1}}\right)$

$$= \cos^{-1}\left(\frac{5}{\sqrt{55}}\right) \approx 48°$$

CHECK YOUR UNDERSTANDING 12.10

Answer: $\cos^{-1}\left(\dfrac{1}{\sqrt{2}}\right) = 45°$

Determine the angle between the vectors $i + 2j$ and $-i + 3j$.

ORTHOGONAL VECTORS IN R^n

We remind you that, for any $-1 \le x \le 1$, $\cos^{-1}x$ is that angle $0 \le \theta \le \pi$ such that $\cos\theta = x$.

So, if $\cos^{-1}\left(\dfrac{u \cdot v}{\|u\|\|v\|}\right) = 90°$,

then: $\dfrac{u \cdot v}{\|u\|\|v\|} = 0$

or: $u \cdot v = 0$.

The angle θ between the vectors $u, v \in \Re^2$ depicted in the adjacent figure has a measure of $90°$ ($\frac{\pi}{2}$ radians), and we say that those vectors are perpendicular (or orthogonal). Appealing to Definition 12.5 we see that:

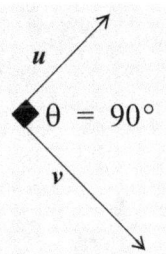

$$\cos^{-1}\left(\frac{u \cdot v}{\|u\|\|v\|}\right) = 90° \text{ or } u \cdot v = 0$$
(see margin)

Bringing us to:

DEFINITION 12.6
ORTHOGONAL VECTORS

Two vectors u and v in $\Re^n$ are **orthogonal** if $u \cdot v = 0$.

Note: The zero vector in $\Re^n$ is orthogonal to every vector in $\Re^n$.

CHECK YOUR UNDERSTANDING 12.11

Which of the following pair of vectors are orthogonal?

(a) $\langle 2, 3 \rangle, \langle 1, -4 \rangle$ (b) $2i + 3j, -3i + 2j$ (c) $\langle 1, 2, 3 \rangle, \langle -1, -1, 1 \rangle$

It is often useful to decompose a vector $v \in \mathfrak{R}^n$ into a sum of two vectors: one parallel to a given nonzero vector u, and the other perpendicular to u. The parallel-vector must be of the form cu for some scalar c. To determine the value of c we note that for $v - cu$ to be orthogonal to u, we must have:

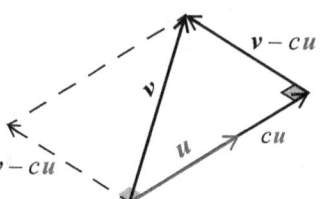

$$(v - cu) \cdot u = 0$$

Theorem 12.2(d): $v \cdot u - (cu) \cdot u = 0$

Theorem 12.2(c): $v \cdot u - c(u \cdot u) = 0$

$$c = \frac{v \cdot u}{u \cdot u} = \frac{v \cdot u}{\|u\|^2} = \frac{u \cdot v}{\|u\|^2}$$

Theorem 12.2(b)

Summarizing:

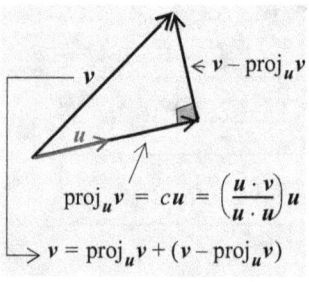

THEOREM 12.4
VECTOR DECOMPOSITION

For given $v \in \mathfrak{R}^n$ and u any nonzero vector in $\mathfrak{R}^n$ (see margin figure):

$$v = \text{proj}_u v + (v - \text{proj}_u v)$$

Where: $\text{proj}_u v = \left(\dfrac{v \cdot u}{u \cdot u} \right) u$

and: $(v - \text{proj}_u v) \cdot \text{proj}_u v = 0$

The vector $\text{proj}_u v$ is said to be the **vector projection of v onto u**, and the vector $v - \text{proj}_u v$ is said to be the **vector component of v orthogonal to u**. Note that:

$$\|\text{proj}_u v\| = \left| \frac{v \cdot u}{u \cdot u} \right| \|u\| = |v \cdot u| \frac{\sqrt{u \cdot u}}{u \cdot u} = \frac{|v \cdot u|}{\sqrt{u \cdot u}}$$

EXAMPLE 12.6   Express the vector $\langle 2, 1, -3 \rangle$ as a sum of a vector parallel to $\langle\langle 1, 4, 0 \rangle\rangle$ and a vector orthogonal to $\langle \mathbf{1, 4, 0} \rangle$.

SOLUTION: For $v = \langle 2, 1, -3 \rangle$ and $u = \langle 1, 4, 0 \rangle$ we have:

$$\text{proj}_u v = \left(\frac{u \cdot v}{u \cdot u} \right) u = \left[\frac{\langle 1, 4, 0 \rangle \cdot \langle 2, 1, -3 \rangle}{\langle 1, 4, 0 \rangle \cdot \langle 1, 4, 0 \rangle} \right] \langle 1, 4, 0 \rangle = \frac{6}{17} \langle 1, 4, 0 \rangle$$

and $v - \text{proj}_u v = \langle 2, 1, -3 \rangle - \langle \frac{6}{17}, \frac{24}{17}, 0 \rangle = \langle \frac{28}{17}, -\frac{7}{17}, -3 \rangle$

Check: $\langle \frac{28}{17}, -\frac{7}{17}, -3 \rangle + \langle \frac{6}{17}, \frac{24}{17}, 0 \rangle = \langle 2, 1, -3 \rangle = v$

and $\langle \frac{28}{17}, -\frac{7}{17}, -3 \rangle \cdot \langle \frac{6}{17}, \frac{24}{17}, 0 \rangle = \left(\frac{28}{17}\right)\left(\frac{6}{17}\right) + \left(-\frac{7}{17}\right)\left(\frac{24}{17}\right) = 0$

Answer:
$\text{proj}_{(0,2)} v = \langle 0, 1 \rangle$
$v - \text{proj}_{(0,2)} v = \langle 3, 0 \rangle$

CHECK YOUR UNDERSTANDING 12.12

Express the vector $v = (3, 1)$ as the sum of a vector parallel to $u = (0, 2)$ and a vector orthogonal to u.

A DETERMINED PAUSE

A **two-by-two matrix** of real numbers is an array consisting of two rows and two columns of numbers, as is the case with A and B below:

$$A = \begin{bmatrix} 2 & 8 \\ -3 & 5 \end{bmatrix}, \quad B = \begin{vmatrix} 0 & 5 \\ 9 & \sqrt{7} \end{vmatrix}$$

The **determinant** of $\begin{bmatrix} a & b \\ c & d \end{bmatrix}$, is the number: $\det \begin{bmatrix} a & b \\ c & d \end{bmatrix} = ad - bc$.

For example: $\det \begin{bmatrix} 2 & 8 \\ -3 & 5 \end{bmatrix} = 2 \cdot 5 - 8(-3) = 34$

CHECK YOUR UNDERSTANDING 12.13

Evaluate:

Answers:
(a) 30   (b) 24   (c) −20

(a) $\det \begin{bmatrix} 5 & -3 \\ 0 & 6 \end{bmatrix}$

(b) $\det \begin{bmatrix} 2 & -3 \\ 4 & 6 \end{bmatrix}$

(c) $\det \begin{bmatrix} 2 & 5 \\ 4 & 0 \end{bmatrix}$

We now define the determinant of a three-by-three array with first row consisting of the unit coordinate vectors i, j, and k as follows:

note negative sign

$$\det \begin{bmatrix} i & j & k \\ a & b & c \\ d & e & f \end{bmatrix} = \det \begin{bmatrix} b & c \\ e & f \end{bmatrix} i - \det \begin{bmatrix} a & c \\ d & f \end{bmatrix} j + \det \begin{bmatrix} a & b \\ d & e \end{bmatrix} k$$

Note:
$\begin{bmatrix} i & j & k \\ a & b & c \\ d & e & f \end{bmatrix}$ $\begin{bmatrix} i & j & k \\ a & b & c \\ d & e & f \end{bmatrix}$ $\begin{bmatrix} i & j & k \\ a & b & c \\ d & e & f \end{bmatrix}$

The reason for such a strange definition will soon surface. For now:

EXAMPLE 12.7   Evaluate:

$$\det \begin{bmatrix} i & j & k \\ 2 & 5 & -3 \\ 4 & 0 & 6 \end{bmatrix}$$

SOLUTION:

$$\det \begin{bmatrix} i & j & k \\ 2 & 5 & -3 \\ 4 & 0 & 6 \end{bmatrix} = \det \begin{bmatrix} 5 & -3 \\ 0 & 6 \end{bmatrix} i - \det \begin{bmatrix} 2 & -3 \\ 4 & 6 \end{bmatrix} j + \det \begin{bmatrix} 2 & 5 \\ 4 & 0 \end{bmatrix} k$$

$$= [(5 \cdot 6) - (-3 \cdot 0)]i - [(2 \cdot 6) - (-3 \cdot 4)]j + [(2 \cdot 0) - (5 \cdot 4)]k$$

$$= 30i - 24j - 20k$$

CHECK YOUR UNDERSTANDING 12.14

Evaluate:

$$\det \begin{bmatrix} i & j & k \\ \dfrac{1}{3} & \dfrac{1}{2} & -1 \\ 0 & -3 & \sqrt{2} \end{bmatrix}$$

Answers:

$$\frac{\sqrt{2}-6}{2}i - \frac{\sqrt{2}}{3}j - k$$

CROSS PRODUCT

We now focus our attention **exclusively** on $\Re^3$, our very own physical three-dimensional space. As you will see, for given $u, v \in \Re^3$, it is often useful to find a particular vector n that is perpendicular to both u and v, and here it is:

DEFINITION 12.7
CROSS PRODUCT

For $u = \langle u_1, u_2, u_3 \rangle$, $v = \langle v_1, v_2, v_3 \rangle$ the **cross product** of u with v, denoted by $u \times v$ and is given by:

$$u \times v = \det \begin{bmatrix} i & j & k \\ u_1 & u_2 & u_3 \\ v_1 & v_2 & v_3 \end{bmatrix}$$

We will soon show that the vector $u \times v$ is indeed perpendicular to both u and v, but first:

EXAMPLE 12.8 Evaluate:
$$(2i + 5j - 3k) \times (4i + 6k)$$

SOLUTION:

$$(2i + 5j - 3k) \times (4i + 6k) = \det \begin{bmatrix} i & j & k \\ 2 & 5 & -3 \\ 4 & 0 & 6 \end{bmatrix} \underset{\underset{\text{Example 12.7}}{\uparrow}}{=} 30i - 24j - 20k$$

CHECK YOUR UNDERSTANDING 12.15

Evaluate:
$$\langle 2, 3, 4 \rangle \times \langle 3, 1, -2 \rangle$$

Answer: $(-10, 16, -7)$

THEOREM 12.5 For any $u, v \in \Re^3$, $u \times v$ is orthogonal to both u and v.

PROOF: Rolling up our sleeves, we simply show that $(u \times v) \cdot u = 0$, and leave it for you to verify that $(u \times v) \cdot v$ is also zero:

$$(u \times v) \cdot u = \det \begin{bmatrix} i & j & k \\ u_1 & u_2 & u_3 \\ v_1 & v_2 & v_3 \end{bmatrix} \cdot (u_1, u_2, u_3) = $$

$$[(u_2 v_3 - u_3 v_2)i - (u_1 v_3 - u_3 v_1)j + (u_1 v_2 - u_2 v_1)k] \cdot (u_1, u_2, u_3)$$

$$= (u_2 v_3 - u_3 v_2)u_1 - (u_1 v_3 - u_3 v_1)u_2 + (u_1 v_2 - u_2 v_1)u_3$$

$$= u_2 v_3 u_1 - u_3 v_2 u_1 - u_1 v_3 u_2 + u_3 v_1 u_2 + u_1 v_2 u_3 - u_2 v_1 u_3 = 0$$

CHECK YOUR UNDERSTANDING 12.16

Verify, directly, that:
$$\langle 2, 3, 4 \rangle \times \langle 3, 1, -2 \rangle \cdot \langle 3, 1, -2 \rangle = 0$$

Answer: See page A-14.

As it turns out, the cross product is not a commutative operator. In particular:

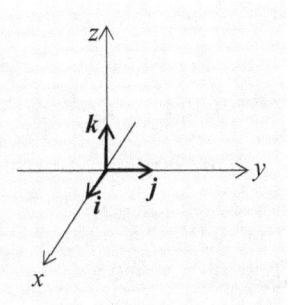

$$i \times j = \det \begin{bmatrix} i & j & k \\ 1 & 0 & 0 \\ 0 & 1 & 0 \end{bmatrix} = k \quad \text{while} \quad j \times i = \det \begin{bmatrix} i & j & k \\ 0 & 1 & 0 \\ 1 & 0 & 0 \end{bmatrix} = -k$$

Since the vectors i and j lie in the xy-plane (margin), and since the cross product of two vectors is orthogonal to both, it comes as no surprise that both of the above cross products must lie on the z-axis.

Note that the direction of the above two cross products follows what is termed the *right-hand-rule*: If the fingers of your right hand curl in the direction of a rotation from i to j, then your thumb points in the k-direction [see Figure 12.9(a)], while the thumb will point in the $-k$-direction if you curl from j to i [see Figure 12.9(b)].

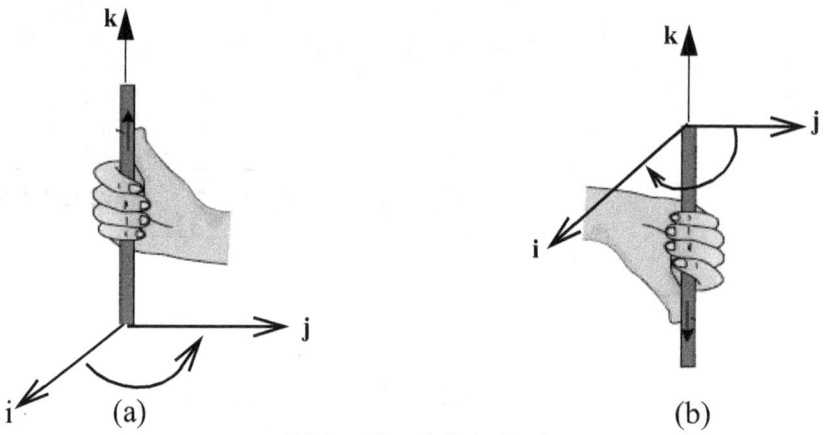

(a)   (b)

Right-Hand Grip Rule

Figure 12.9

Indeed, the *right-hand-rule* holds in general:

For given u and v, if the fingers of your right hand curl in the direction of a rotation from u to v through an angle less than $180°$, then your thumb will point in the direction of $u \times v$.

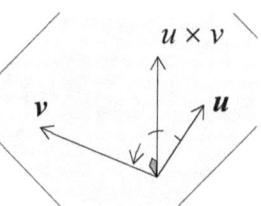

At this point we know the direction of $u \times v$, as for its magnitude:

In particular:
$u \times v = 0$ if and only if
u and v are parallel

THEOREM 12.6   If $0 \le \theta \le \pi$ is the angle between u and v, then:
$$\|u \times v\| = \|u\|\|v\| \sin\theta$$

PROOF: From:
$$u \times v = \det \begin{bmatrix} i & j & k \\ u_1 & u_2 & u_3 \\ v_1 & v_2 & v_3 \end{bmatrix}$$
$$= (u_2 v_3 - u_3 v_2)i - (u_1 v_3 - u_3 v_1)j + (u_1 v_2 - u_2 v_1)k$$

we have:
$$\|u \times v\|^2 = (u_2 v_3 - u_3 v_2)^2 + (u_1 v_3 - u_3 v_1)^2 + (u_1 v_2 - u_2 v_1)^2$$
$$= u_2^2 v_3^2 - 2u_2 v_3 u_3 v_2 + u_3^2 v_2^2 + u_1^2 v_3^2 - 2u_1 v_3 u_3 v_1 + u_3^2 v_1^2$$
$$+ u_1^2 v_2^2 - 2u_1 v_2 u_2 v_1 + u_2^2 v_1^2$$

Which, as you can verify, is but a rearrangement of:
$$(u_1^2 + u_2^2 + u_3^2)(v_1^2 + v_2^2 + v_3^2) - (u_1 v_1 + u_2 v_2 + u_3 v_3)^2$$
$$= \|u\|^2 \|v\|^2 - (u \cdot v)^2$$

At this point we have:
$$\|u \times v\|^2 = \|u\|^2 \|v\|^2 - (u \cdot v)^2 \ (*)$$
We also know that (see Definition 12.5):
$$u \cdot v = \|u\| \|v\| \cos\theta \ (**):$$
Substituting (**) in (*):
$$\|u \times v\|^2 = \|u\|^2 \|v\|^2 - (u \cdot v)^2$$
$$= \|u\|^2 \|v\|^2 - \|u\|^2 \|v\|^2 \cos^2\theta$$
$$= \|u\|^2 \|v\|^2 (1 - \cos^2\theta) = \|u\|^2 \|v\|^2 \sin^2\theta$$
It follows that $\|u \times v\| = \|u\| \|v\| \sin\theta$ (see margin).

Since $\sin\theta \geq 0$ for $0 \leq \theta \leq \pi$:
$$\sqrt{\sin^2\theta} = \sin\theta$$

Answers:
(a) $\begin{array}{l} j \times k = i, k \times j = -i \\ i \times k = -j, k \times i = j \end{array}$

(b) $\begin{array}{l} i \times (i \times j) = -j \\ (i \times i) \times j = 0 \end{array}$

CHECK YOUR UNDERSTANDING 12.17

(a) Determine: $j \times k$, $k \times j$, $i \times k$, and $k \times i$.

(b) Determine: $i \times (i \times j)$ and $(i \times i) \times j$.

The above CYU shows that the cross product is neither a commutative nor associative operator. It does, however, fall under the jurisdiction of some "familiar patterns:"

THEOREM 12.7 For $u, v, w \in \mathfrak{R}^3$ and any scalar c:

(a) $u \times (v + w) = u \times v + u \times w$

(b) $(u + v) \times w = u \times w + v \times w$

(c) $cv \times w = v \times cw = c(v \times w)$

(d) $v \times u = -(u \times v)$

PROOF: We establish (a) and invite you to verify the rest in the exercises.

$$u \times (v + w) = \langle u_1, u_2, u_3 \rangle \times \langle v_1 + w_1, v_2 + w_2, v_3 + w_3 \rangle = \det \begin{bmatrix} i & j & k \\ u_1 & u_2 & u_3 \\ v_1 + w_1 & v_2 + w_2 & v_3 + w_3 \end{bmatrix}$$

$$= [u_2(v_3 + w_3) - u_3(v_2 + w_2)]i - [u_1(v_3 + w_3) - u_3(v_1 + w_1)]j + [u_1(v_2 + w_2) - u_2(v_1 + w_1)]k$$

$$= [(u_2 v_3 - u_3 v_2)i - (u_1 v_3 - u_3 v_1)j + (u_1 v_2 - u_2 v_1)k] + [(u_2 w_3 - u_3 w_2)i - (u_1 w_3 - u_3 w_1)j + (u_1 w_2 - u_2 w_1)k]$$

$$= \det \begin{bmatrix} i & j & k \\ u_1 & u_2 & u_3 \\ v_1 & v_2 & v_3 \end{bmatrix} + \det \begin{bmatrix} i & j & k \\ u_1 & u_2 & u_3 \\ w_1 & w_2 & w_3 \end{bmatrix} = u \times v + u \times w$$

L

CHECK YOUR UNDERSTANDING 12.18

Let u and v be non-zero nonparallel vectors in $\mathfrak{R}^3$. Prove that $\|u \times v\|$ is the area of the parallelogram with u and v as adjacent sides.

Answers: See page A-15.

	EXERCISES	

Exercises 1-6. Evaluate:

1. $2(5i + 2j) \cdot (-3i + j)$

2. $\langle 4, -7 \rangle \cdot \langle 3, 5 \rangle + \langle 1, 5 \rangle \cdot \langle 2, 4 \rangle$

3. $(3, 2, 1) \cdot (-4, 0, 5)$

4. $5(i - 2j + 3k) \cdot (2i - 4j + k)$

5. $4[(2i - 4j) - (i + 3j)] \cdot (i - 2j + 3k)$

6. $[2\langle 3, 1, 0 \rangle + \langle 1, 1, 4 \rangle] \cdot [\langle 0, 2, 1 \rangle - \langle 2, 1, -2 \rangle]$

Exercises 7-18. Determine that angle between the given pair of vectors.

7. $u = -i + 2j, v = 2i + j$

8. $u = \langle 4, 1 \rangle, v = \langle 1, 2 \rangle$

9. $u = \langle 1, 4 \rangle, v = \langle 3, -2 \rangle$

10. $u = 3i - j, v = i + 2j$

11. $u = \langle 2, -1, 1 \rangle, v = \langle 1, 1, -1 \rangle$

12. $u = 2i - 2j + 2k, v = -i + j + 5k$

13. $u = i + j, v = j + k$

14. $u = \langle 2, -2, 2 \rangle, v = \langle -1, 1, 5 \rangle$

15. $u = \langle 3, 0, 4 \rangle, v = \langle 0, \sqrt{7}, -5 \rangle$

16. $u = 4i + 2k, v = 2i - j$

17. $u = 3i - j + 5k, v = -2i + 4j + 3k$

18. $u = \langle 4, -3, 1 \rangle, v = \langle 2, 0, -1 \rangle$

Exercises 19-22. Determine if the given pair of vectors are orthogonal, parallel, or neither.

19. $u = \langle 2, -1, 1 \rangle, v = \langle 3, 1, -5 \rangle$

20. $u = i - \frac{1}{2}j + 3k, v = 4i - 2j + 12k$

21. $u = 3i - j + 5k, v = -2i + 4j + 3k$

22. $u = 0, v = i - 2j + 3k$

Exercises 23-30. Express the vector v as the sum of a vector parallel to u and a vector orthogonal to u.

23. $v = \langle 2, 1 \rangle, u = \langle -3, 2 \rangle$

24. $v = 4i - 3j, u = i + J$

25. $v = 3j + 4k, u = i + j$

26. $v = i + j, u = 3j + 4k$

27. $v = \langle 8, 4, -12 \rangle, u = \langle 1, 2, -1 \rangle$

28. $v = \langle 1, 2, -1 \rangle, u = \langle 8, 4, -12 \rangle$

29. $v = \sqrt{12}i + \sqrt{48}k, u = \frac{1}{\sqrt{3}}i + \frac{1}{\sqrt{3}}j - \frac{1}{\sqrt{3}}k$

30. $v = \langle \frac{1}{2}, 1, -2 \rangle, u = \langle -2, 1, \frac{1}{2} \rangle$

Exercises 31-36. Evaluate:

31. $\langle 2, -1, 6 \rangle \times \langle -3, 4, 1 \rangle$

32. $(3\boldsymbol{i} + 2\boldsymbol{j} - \boldsymbol{k}) \times (\boldsymbol{i} - \boldsymbol{k})$

33. $[(-4\boldsymbol{i} + \boldsymbol{j}) \times (2\boldsymbol{i} - \boldsymbol{j} - 3\boldsymbol{k})] \times (3\boldsymbol{i} - 2\boldsymbol{j} - 3\boldsymbol{k})$

34. $\langle -3, 1, 0 \rangle \times [\langle 2, -1, -3 \rangle \times \langle 3, -2, -3 \rangle]$

35. $[\langle 2, -1, 4 \rangle \times \langle 7, 2, 3 \rangle] \cdot \langle -1, 1, 2 \rangle$

36. $(\boldsymbol{i} - \boldsymbol{k}) \cdot [(2\boldsymbol{i} + \boldsymbol{j} - \boldsymbol{k}) \times (\boldsymbol{i} + 3\boldsymbol{j})]$

37. Prove that $(\boldsymbol{u} - \boldsymbol{v}) \cdot (\boldsymbol{u} - \boldsymbol{v}) = \boldsymbol{u} \cdot (\boldsymbol{u} - \boldsymbol{v}) - \boldsymbol{v} \cdot (\boldsymbol{u} - \boldsymbol{v})$, for:

(a) $\boldsymbol{u}, \boldsymbol{v} \in \Re^2$ (b) $\boldsymbol{u}, \boldsymbol{v} \in R^3$ (c) $\boldsymbol{u}, \boldsymbol{v} \in R^n$

38. Prove that $\boldsymbol{u} \cdot (\boldsymbol{u} - \boldsymbol{v}) - \boldsymbol{v} \cdot (\boldsymbol{u} - \boldsymbol{v}) = \boldsymbol{u} \cdot \boldsymbol{u} - \boldsymbol{u} \cdot \boldsymbol{v} - \boldsymbol{v} \cdot \boldsymbol{u} + \boldsymbol{v} \cdot \boldsymbol{v}$, for:

(a) $\boldsymbol{u}, \boldsymbol{v} \in \Re^2$ (b) $\boldsymbol{u}, \boldsymbol{v} \in R^3$ (c) $\boldsymbol{u}, \boldsymbol{v} \in R^n$

39. (a) Prove, without using the law of cosines, that $\dfrac{|\boldsymbol{u} \cdot \boldsymbol{v}|}{\|\boldsymbol{u}\| \|\boldsymbol{v}\|} \le 1$, for:

(a) $\boldsymbol{u}, \boldsymbol{v} \in \Re^2$ (b) $\boldsymbol{u}, \boldsymbol{v} \in R^3$ (c) $\boldsymbol{u}, \boldsymbol{v} \in R^n$

40. Prove: (a) Theorem 12.2 (a)   (b) Theorem 12.2(b)   (c) Theorem 12.2(d)

41. Prove: (a) Theorem 12.7 (b)   (b) Theorem 12.7(c)   (c) Theorem 12.7(d)

42. Prove that $\boldsymbol{u}$ and $\boldsymbol{v}$ are parallel if and only if $|\boldsymbol{u} \cdot \boldsymbol{v}| = \|\boldsymbol{u}\| \|\boldsymbol{v}\|$, for:

(a) $\boldsymbol{u}, \boldsymbol{v} \in \Re^2$ (b) $\boldsymbol{u}, \boldsymbol{v} \in R^3$ (c) $\boldsymbol{u}, \boldsymbol{v} \in R^n$

43. Prove that a rectangle is a square if and only if its diagonals are perpendicular.

44. Prove that (a) $\|\text{proj}_{c\boldsymbol{u}}(\boldsymbol{v})\| = \|(\text{proj}_{\boldsymbol{u}}\boldsymbol{v})\|$ for any $c \in \Re$.

(b) $\|\text{proj}_{\boldsymbol{u}}(c\boldsymbol{v})\| = |c| \|(\text{proj}_{\boldsymbol{u}}\boldsymbol{v})\|$ for any $c \in \Re$.

45. Establish the Jacoby Identity:
$$(\boldsymbol{u} \times \boldsymbol{v}) \times \boldsymbol{w} + (\boldsymbol{v} \times \boldsymbol{w}) \times \boldsymbol{u} + (\boldsymbol{w} \times \boldsymbol{u}) \times \boldsymbol{v} = 0 \quad (\text{for } \boldsymbol{u}, \boldsymbol{v}, \boldsymbol{w} \in \Re^3)$$

46. Establish the Lagrange Identity:
$$(\boldsymbol{u} \times \boldsymbol{v}) \cdot (\boldsymbol{w} \times \boldsymbol{z}) = (\boldsymbol{u} \cdot \boldsymbol{w})(\boldsymbol{v} \cdot \boldsymbol{z}) - (\boldsymbol{u} \cdot \boldsymbol{z})(\boldsymbol{v} \cdot \boldsymbol{w}) \ (\text{for } \boldsymbol{u}, \boldsymbol{v}, \boldsymbol{w}, \boldsymbol{z} \in \Re^3)$$

Exercises 47-50. (Work) The work done when a constant force F is applied along the line of motion of an object through a distance d, is given by $W = Fd$ (see page 172). More generally, if a force F is applied in a direction that makes an angle θ with the direction of motion, then the work done in moving the body from A to B is defined to be:

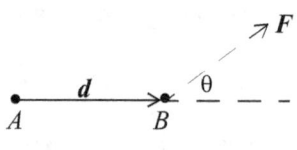

$$W = \|F\| \cos\theta \|d\| = F \cdot d$$

the scalar component of
F in the direction of d

47. Find the work done by a force $F = 20k$ that moves an object along the line from the origin to the point $(1,1,1)$. Assume that the force is measured in pounds and the distance in feet.

48. Find the work done by a force $F = 3i + 2j - 7k$ that moves an object directly from the point $(1, 2, 5)$ to the point $(3, 2, 6)$. Assume that the force is measured in newtons and the distance in meters.

49. How much work does it take to pull a 200 pound railroad car 100 feet along a track by means of a rope that makes a $30°$ angle with the track?

50. Joe pushes on a lawnmower handle with a force of 30 pounds. Determine the angle the handle makes with the ground if 1125 foot-pounds of work is required to mow a line of 75 ft.

Exercises 51-53. (Torque) Let a vector force F be applied to a point P in space and let O be another point is space. The **moment of torque** τ of the force F at the point P about the point O is defined to be:

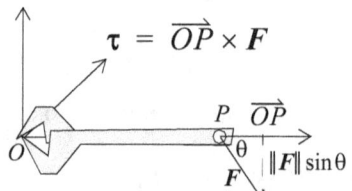

$$\tau = \overrightarrow{OP} \times F$$

In particular, if we apply a force F to a wrench, then the resulting torque acts along the axis of the bolt to drive the bolt forward (see adjacent figure).

51. Prove that $\|\tau\| = \|\overrightarrow{OP}\| \|F\| \sin\theta$.

52. An 8-inch wrench is used to drive a bolt at point O. A force F of magnitude 60 lb is applied at the end of the handle (point P). Determine the magnitude of the torque produced if the angle of applications (see above figure) is:

    (a) $30°$ (b) $90°$ (c) $135°$

53. A 9-inch wrench is used to drive a bolt at point O. A force F of magnitude 30 lb is applied at the end of the handle (point P). Determine the magnitude of the torque produced if the angle of applications (see above figure) is:

    (a) $30°$ (b) $90°$ (c) $135°$

Exercises 54-56. (Triple Scalar Product) The **triple scalar product of** u, v, w in $\Re^3$ is defined to be the number $u \cdot (v \times w)$.

54. Prove that $u \cdot (v \times w) = \det \begin{bmatrix} u_1 & u_2 & u_3 \\ v_1 & v_2 & v_3 \\ w_1 & w_2 & w_3 \end{bmatrix}$, where:

$$\det \begin{bmatrix} u_1 & u_2 & u_3 \\ v_1 & v_2 & v_3 \\ w_1 & w_2 & w_3 \end{bmatrix} = u_1 \det \begin{bmatrix} v_2 & v_3 \\ w_2 & w_3 \end{bmatrix} - u_2 \det \begin{bmatrix} v_1 & v_3 \\ w_1 & w_3 \end{bmatrix} + u_3 \det \begin{bmatrix} v_1 & v_2 \\ w_1 & w_3 \end{bmatrix}$$

55. Prove that $u \cdot (v \times w) = (u \times v) \cdot w$

56. Prove that $u \cdot (v \times w)$ is the volume of the parallelepiped with u, v, w as adjacent sides (see adjacent figure).

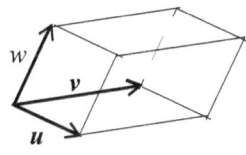

§3. LINES AND PLANES

While the direction (slope) of the line L in Figure 12.10(a), and that of the vector $v = \overrightarrow{PQ} = \langle x_2 - x_1, y_2 - y_1 \rangle$ in Figure 12.10(b) are one and the same, the lines $\bar{L}$ and L, though parallel, are not equal. They will coincide, however, if $\bar{L}$ is moved, in a parallel fashion, so as to contain the point P (any point on L will do just as well). To put it another way:

The line L coincides with the set of **endpoints** of the standard-position vectors $w = (x_1, y_1) + tv$, $-\infty < t < \infty$.

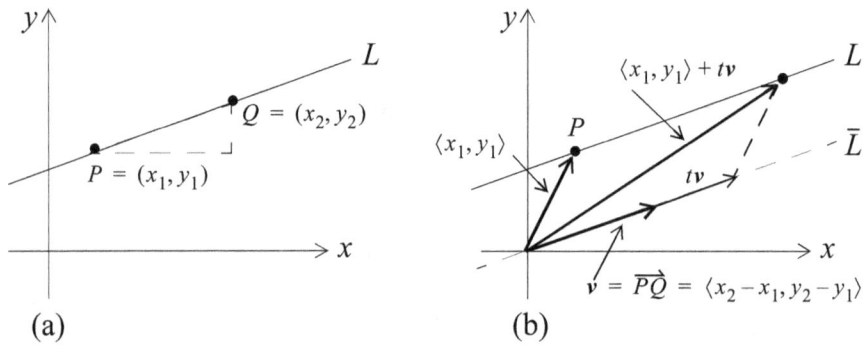

(a)   (b)

Figure 12.10

In general:

Let L be the line passing through two distinct points $P = (x_1, y_1)$ and $Q = (x_2, y_2)$ in the plane. Let $v = \overrightarrow{PQ} = \langle x_2 - x_1, y_2 - y_1 \rangle$ and $u = \langle x_1, y_1 \rangle$, then:

$$w = u + tv, \quad -\infty < t < \infty \quad (*)$$

is said to be a **vector equation** of L.

> The vectors v and u are said to be **direction** and **translation vectors**, respectively, of the line.

EXAMPLE 12.9 Find a vector equation of the line L passing through the points $(1, 5)$ and $(2, 3)$.

SOLUTION: We take $v = \langle 2 - 1, 3 - 5 \rangle = \langle 1, -2 \rangle$ to be our direction vector, and $u = \langle 1, 5 \rangle$ as the translation vector, leading us to the vector equation:

$$w = u + tv = \langle 1, 5 \rangle + t \langle 1, -2 \rangle \qquad .$$

Note that $w = \langle 1, 5 \rangle + t \langle 1, -2 \rangle = \langle 1 + t, 5 - 2t \rangle$ reveals the following **parametric** equations of L:

$$\{(x, y) \mid x = 1 + t, y = 5 - 2t, \ -\infty < t < \infty\}$$

CHECK YOUR UNDERSTANDING 12.19

(a) Find a vector equation and parametric equations of the line L of slope $-\dfrac{2}{3}$ containing the point $(5, 1)$.

(b) Find a vector equation and parametric equations of the vertical line containing the point $(3, 7)$.

The nice thing about vector equations of lines in $\Re^2$ is that they carry over seamlessly to $\Re^3$:

The vectors v and u are again said to be **direction** and **translation vectors**, respectively, of the line.

Let L be the line passing through two distinct points $P = (x_1, y_1, z_1)$ and $Q = (x_2, y_2, z_2)$ in $\Re^3$.

Let $v = \overrightarrow{PQ} = \langle x_2 - x_1, y_2 - y_1, z_2 - z_1 \rangle$ and $u = \langle x_1, y_1, z_1 \rangle$. Then:

$$w = u + tv \qquad (*)$$

is a **vector equation** of L.

EXAMPLE 12.10 Find a vector equation and parametric equations of the line L passing through the points $(2, 0, -3)$ and $(1, 4, 2)$.

SOLUTION: Taking $v = \langle 1 - 2, 4 - 0, 2 - (-3) \rangle = \langle -1, 4, 5 \rangle$ as the direction vector, and $u = \langle 2, 0, -3 \rangle$ as the translation vector we arrive at the vector equation:

$$w = u + tv = \langle 2, 0, -3 \rangle + t \langle -1, 4, 5 \rangle$$

And parametric equations:

$$\{(x, y, z) \mid x = 2 - t, y = 4t, z = -3 + 5t, -\infty < t < \infty\}$$

CHECK YOUR UNDERSTANDING 12.20

Find a vector equation and parametric equations of the line L passing through the points $(1, 2, 9)$ and $(0, 1, -2)$.

EXAMPLE 12.11 Find the distance from the point $P = (3, 1, 3)$ to the line L in $\Re^3$ which passes through the points $(1, 0, 2)$ and $(3, 1, 6)$.

SOLUTION: We first find a direction vector for the given line:

$$u = \langle (3, 1, 6) - (1, 0, 2) \rangle = \langle 2, 1, 4 \rangle$$

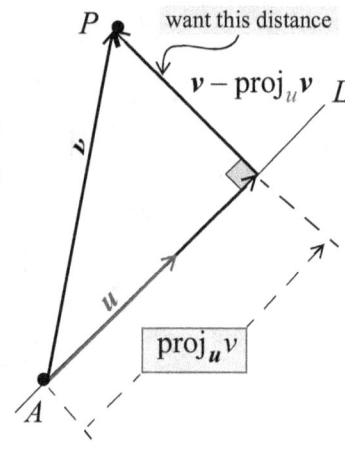

Choosing the point $A = (1, 0, 2)$ on L we determine the vector $\boldsymbol{v}$ from A to P:

$$\boldsymbol{v} = \langle 3, 1, 3 \rangle - \langle 1, 0, 2 \rangle = \langle 2, 1, 1 \rangle$$

Applying Theorem 12.4, page 504, we have:

$$\text{proj}_{\boldsymbol{u}} \boldsymbol{v} = \left(\frac{\boldsymbol{u} \cdot \boldsymbol{v}}{\boldsymbol{u} \cdot \boldsymbol{u}} \right) \boldsymbol{u}$$

$$= \left(\frac{\langle 2, 1, 4 \rangle \cdot \langle 2, 1, 1 \rangle}{\langle 2, 1, 4 \rangle \cdot \langle 2, 1, 4 \rangle} \right) \langle 2, 1, 4 \rangle$$

$$= \frac{9}{21} \langle 2, 1, 4 \rangle = \langle \frac{6}{7}, \frac{3}{7}, \frac{12}{7} \rangle$$

Thus:

$$\left\| \boldsymbol{v} - \text{proj}_{\boldsymbol{u}} \boldsymbol{v} \right\| = \left\| \langle 2, 1, 1 \rangle - \langle \frac{6}{7}, \frac{3}{7}, \frac{12}{7} \rangle \right\|$$

$$= \left\| \langle \frac{8}{7}, \frac{4}{7}, -\frac{5}{7} \rangle \right\| = \frac{1}{7} \| \langle 8, 4, -5 \rangle \| = \frac{1}{7} \sqrt{64 + 16 + 25} = \frac{\sqrt{105}}{7}$$

CHECK YOUR UNDERSTANDING 12.21

Find the distance from the point $P = (2, 5)$ to the line L in $\Re^2$ passing through the points $(1, -2)$ and $(2, 4)$.

(b) Find the distance from the point $P = (1, 0, 1, 3)$ to the line L in $\Re^4$ passing through the points $(1, 2, 0, 1)$ and $(1, 2, 2, 1)$.

Answers: (a) $\dfrac{1}{\sqrt{37}}$ (b) $2\sqrt{2}$

PLANES

Just as a line in $\Re^2$ is determined by a point on the line and its slope, so then is a plane in $\Re^3$ determined by a point on the plane and a nonzero vector orthogonal to the plane (a **normal vector** to the plane). To be more specific, suppose we want the equation of the plane with normal vector $\boldsymbol{n} = \langle a, b, c \rangle$ that contains the point $A_0 = (x_0, y_0, z_0)$. For any point $P = (x, y, z)$ on the plane we have:

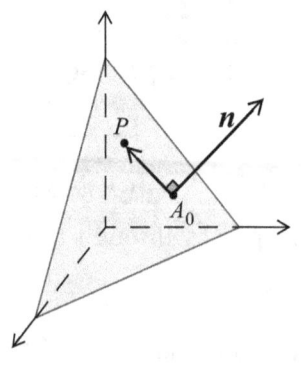

$$\boldsymbol{n} \cdot \overrightarrow{A_0 P} = 0$$

or: $\langle a, b, c \rangle \cdot \langle x - x_0, y - y_0, z - z_0 \rangle = 0$ vector equation

or: $a(x - x_0) + b(y - y_0) + c(z - z_0) = 0$ scalar equation

or: $ax + by + cz = d$, where $d = ax_0 + by_0 + cz_0$ general equation

EXAMPLE 12.12 Find a vector, scalar, and general equation of the plane passing through the point $(1, 3, -2)$ with normal vector $\boldsymbol{n} = \langle 4, -1, 5 \rangle$.

SOLUTION:

$$\text{vector:} \quad \langle 4, -1, 5 \rangle \cdot \langle x-1, y-3, z+2 \rangle = 0$$
$$\text{scalar:} \quad 4(x-1) - 1(y-3) + 5(z+2) = 0$$
$$\text{general:} \qquad\qquad 4x - y + 5z = -9$$

Note that you can easily spot the normal $\boldsymbol{n} = (4, -1, 5)$ to the plane in any of the above equations.

CHECK YOUR UNDERSTANDING 12.22

Find a vector, scalar, and general equation of the plane passing through the point $(1, 3, -2)$ with normal parallel to the line containing the points $(1, 1, 0), (0, 2, 1)$.

Answer: See Page A-16.

EXAMPLE 12.13 Find the general equation of the plane that contains the points $A = (1, 2, -1)$, $B = (2, 3, 1)$, $C = (3, -1, 2)$.

SOLUTION: Noting that the vectors

$$\overrightarrow{AB} = \langle 2, 3, 1 \rangle - \langle 1, 2, -1 \rangle = \langle 1, 1, 2 \rangle$$

and $\overrightarrow{AC} = \langle 3, -1, 2 \rangle - \langle 1, 2, -1 \rangle = \langle 2, -3, 3 \rangle$

are parallel to the plane, we employ Theorem 12.5, page 507, to find a normal to the plane:

$$\boldsymbol{n} = \det \begin{bmatrix} \boldsymbol{i} & \boldsymbol{j} & \boldsymbol{k} \\ 1 & 1 & 2 \\ 2 & -3 & 3 \end{bmatrix} = 9\boldsymbol{i} + \boldsymbol{j} - 5\boldsymbol{k} = (\boldsymbol{9, 1, -5})$$

Choosing the point $A = (1, 2, -1)$ on the plane, we proceed as in Example 12.12 to arrive at the general equation of the plane:

$$\langle 9, 1, -5 \rangle \cdot \langle x-1, y-2, z+1 \rangle = 0$$
$$9(x-1) + (y-2) - 5(z+1) = 0$$
$$9x + y - 5z = 16$$

CHECK YOUR UNDERSTANDING 12.23

Pick three different points A, B, C on the plane $9x + y - 5z = 16$ of Example 12.13 and proceed as in that example to arrive at the very same equation $9x + y - 5z = 16$

No unique approach. One such approach appears on page A-16

EXAMPLE 12.14 Find the (minimal) distance between the point $P = (1, 1, 2)$ and the plane $x + 2y + 3z = 6$.

SOLUTION: From the given equation, we see that $\mathbf{n} = \langle 1, 2, 3 \rangle$ is a normal to the plane, and that $Q = (0, 3, 0)$ is a point on the plane (any other point on the plane would do just as well).

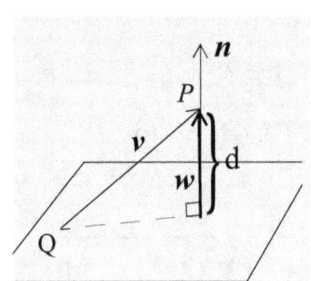

For $\mathbf{v} = \overrightarrow{QP} = \langle 1, 1, 2 \rangle - \langle 0, 3, 0 \rangle = \langle 1, -2, 2 \rangle$, we calculate the length d of the vector $\mathbf{w} = \text{proj}_{\mathbf{n}}\mathbf{v}$ (see margin), as that is the distance between P and the plane:

$$d = \|\text{proj}_{\mathbf{n}}\mathbf{v}\| = \frac{|\mathbf{v} \cdot \mathbf{n}|}{\sqrt{\mathbf{n} \cdot \mathbf{n}}} = \frac{|\langle 1, -2, 2 \rangle \cdot \langle 1, 2, 3 \rangle|}{\sqrt{1 + 4 + 9}} = \frac{3}{\sqrt{14}}$$

Theorem 12.4, page 504

CHECK YOUR UNDERSTANDING 12.24

(a) Find the distance between the point $(2, -3, 4)$ and the plane $x + 2y + 2z = 13$.

(b) Prove that the distance d between a point $P = (x_0, y_0, z_0)$ and the plane $ax + by + cz + d = 0$ is:

$$d = \frac{|ax_0 + by_0 + cz_0 + d|}{\sqrt{a^2 + b^2 + c^2}}$$

Answers: (a) 3
(b) See page A-16.

Two planes in $\Re^3$ are **parallel** if the normal vector of one is a scalar multiple of the normal vector of the other. In particular, the planes
$$3x + 2y - z = 7 \quad \text{and} \quad 6x + 4y - 2z = 99$$
are parallel, since the normal vector $\langle 6, 4, -2 \rangle$ of the latter is twice the normal vector $\langle 3, 2, -1 \rangle$ of the former.

You can see the angle θ between the planes P_1 and P_2 in the adjacent figure, but how is it defined? Like this:

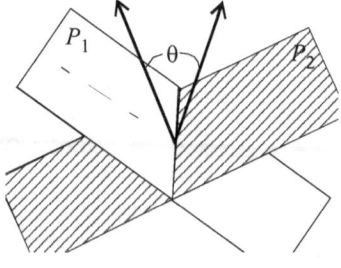

The angle between two nonparallel planes P_1 and P_2, is defined to be the angle $0 \leq \theta < \pi$ between their normals $\mathbf{n_1}$ and

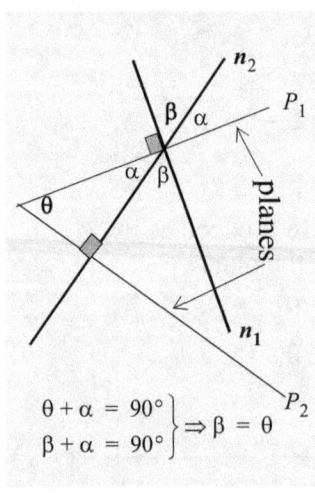

$\theta + \alpha = 90°$
$\beta + \alpha = 90°$ $\Big\} \Rightarrow \beta = \theta$

$\mathbf{n_2}$. Hopefully, the cross-section figure in the margin will convince you that the angle β between the normals is indeed equal to the represented angle θ between the planes.

EXAMPLE 12.15 Find the angle, θ, between the planes $2x + 3y - z = 4$ and $-x + 2y + z = 2$, as well as a vector equation of their line of intersection.

SOLUTION: Here are normal vectors for the two planes:

$$\boldsymbol{n_1} = \langle 2, 3, -1 \rangle, \ \boldsymbol{n_2} = \langle -1, 2, 1 \rangle$$

Appealing to Definition 12.5, page 535 we have:

$$\theta = \cos^{-1}\left(\frac{\boldsymbol{n_1} \cdot \boldsymbol{n_2}}{\|\boldsymbol{n_1}\|\|\boldsymbol{n_2}\|}\right) = \cos^{-1}\left(\frac{\langle 2, 3, -1 \rangle \cdot \langle -1, 2, 1 \rangle}{\|\langle 2, 3, -1 \rangle\|\|\langle -1, 2, 1 \rangle\|}\right)$$

$$= \cos^{-1}\left(\frac{-2 + 6 - 1}{\sqrt{14}\sqrt{6}}\right) = \cos^{-1}\left(\frac{3}{\sqrt{84}}\right) \approx 71°$$

As for their line of intersection:

Since the line lies on both planes, it must be orthogonal to both $\boldsymbol{n_1}$ and $\boldsymbol{n_2}$, and therefore parallel to:

$$\boldsymbol{n_1} \times \boldsymbol{n_2} = \det\begin{bmatrix} \boldsymbol{i} & \boldsymbol{j} & \boldsymbol{k} \\ 2 & 3 & -1 \\ -1 & 2 & 1 \end{bmatrix} = 5\boldsymbol{i} - \boldsymbol{j} + 7\boldsymbol{k} = \langle 5, -1, 7 \rangle$$

We now know that $v = \langle 5, -1, 7 \rangle$ is a direction vector of the line of intersection, but are still missing a translation vector $\boldsymbol{u}$ in the vector equation $w = \boldsymbol{u} + t\langle 5, -1, 7 \rangle$. Any point on the line will direct us to $\boldsymbol{u}$, and any such point must satisfy the equations:

$$2x + 3y - z = 4 \text{ and } -x + 2y + z = 2$$

We chose to find the particular point whose y-coordinate is equal to 0:

$$2x - z = 4$$
$$\underline{-x + z = 2} \quad \xrightarrow{\text{then}} \quad z = 2 + x = 8$$
$$\text{add: } \ x = 6$$

Since $(6, 0, 8)$ is on the line, $\boldsymbol{u} = \langle 6, 0, 8 \rangle$ is a translation vector, bringing us to the vector equation $w = \langle 6, 0, 8 \rangle + t\langle 5, -3, 1 \rangle$.

Answers: 119°
One possible answer:

$w = \left(\frac{1}{4}i - \frac{1}{4}k\right) + t(2i + 4j - 10k)$

CHECK YOUR UNDERSTANDING 12.25

Find the angle, between the planes $x + 2y + z = 0$ and $3x - 4y - z = 1$, as well as a vector equation of their line of intersection.

	EXERCISES	

Exercises 1-4. Find a vector equation and parametric equations of the given line L.

1. L in R^2 has slope 3 and contains the point $(1, 3)$.

2. L in R^2 passes through the points $(2, -5), (1, 0)$.

3. L in R^3 passes through the points $(2, 3, 1), (0, 1, -1)$.

4. L in R^4 passes through the points $(0, 1, 1, 3), (-2, 0, 1, 2)$.

5. With respect to the line L in R^2 passing through the points $(1, 1), (1, 0)$, show that the two vector equations
$$w = \langle 1, 1 \rangle + t(\langle 1, 1 \rangle - \langle 1, 0 \rangle) \quad \text{and} \quad w = \langle 1, 0 \rangle + t\langle \langle 1, 0 \rangle - \langle 1, 1 \rangle \rangle$$
are one and the same by verifying that each vector in either representation is also a vector in the other.

6. With respect to the line L in R^3 passing through the points $(1, 1, 1), (1, 2, 0)$, show that the two vector equations
$$w = \langle 1, 1, 1 \rangle + t(\langle 1, 1, 1 \rangle - \langle 1, 2, 0 \rangle) \quad \text{and} \quad w = \langle 1, 2, 0 \rangle + t(\langle 1, 1, 1 \rangle - \langle 1, 2, 0 \rangle)$$
are one and the same, by verifying that each vector in either representation is also a vector in the other.

7. (a) Find the distance from the point $P = (1, 5)$ to the line L in $\Re^2$ passing through the points $(2, 3)$ and $(0, 4)$.

   (b) Find the distance from the point $P = (2, 0, 3)$ to the line L in $\Re^3$ passing through the points $(1, 2, 2)$ and $(2, 0, 4)$.

   (c) Find the distance from the point $P = (1, 0, 1, 0)$ to the line L in $\Re^4$ passing through the points $(2, 2, 1, 1)$ and $(1, 0, 2, 1)$.

8. (a) Find the distance from the point $P = (5, 3)$ to the line L in $\Re^2$ passing through the points $(2, 2)$ and $(1, 4)$.

   (b) Find the distance from the point $P = (1, 1, -1)$ to the line L in $\Re^3$ passing through the points $(3, 0, 1)$ and $(2, 1, 3)$.

   (c) Find the distance from the point $P = (1, 3, 1, -1)$ to the line L in $\Re^4$ passing through the points $(0, 2, 0, 1)$ and $(1, 3, 1, 1)$.

Exercises 9-12. Find a vector, scalar, and general equation of the plane with given normal vector n that contains the given point Q.

9. $n = \langle 1, 4, 2 \rangle, Q = (2, 0, 3)$

10. $n = 2i + j - 3k, Q = (2, 1, -3)$

11. $n = 3i + 2j - k, Q = (1, 1, 2)$

12. $n = \langle 2, -2, 2 \rangle, Q = (1, -1, 5)$

Exercises 13-16. Find the general equation of the plane containing the given points.

13. $(1, 2, 0)$, $(2, 1, -2)$, $(0, 0, 3)$ 14. $(0, 2, -4)$, $(1, 1, 1)$, $(2, 0, 0)$

15. $(1, 1, 0)$, $(0, -2, 1)$, $(0, 0, 1)$ 16. $(1, 1, 0)$, $(2, 1, 2)$, $(0, 0, 0)$

17. Find the distance between the point $(1, 2, 2)$ and the plane $x + y + 2z = 4$.

18. Find the distance between the point $(3, 0, 1)$ and the plane $-2x + 2z = 3$.

19. Find the distance between the origin and the plane with normal $\boldsymbol{n} = \langle 1, 4, 2 \rangle$ that contains the point $(2, 1, 3)$.

20. Find the distance between the point $(3, -3, 2)$ and the plane containing the points $(1, 1, 0)$, $(0, -2, 1)$, $(0, 0, 1)$.

21. Find the distance between the point $(2, -3, 3)$ and the plane containing the point $(1, 2, 2)$ that is parallel to the plane $3x - 2y + 3z = 1$.

22. Find the values of x for which $(x, 1, 1)$ is 10 units from the plane $3x - 2y + 3z = 1$.

23. Find the values of a for which the point $(3, 0, 1)$ is 1 unit from the plane $ax + 2y + z = 3$.

24. Find an equation of the line that is common to the planes $3x - 2y + 3z = 1$ and $ax + 2y - z = 3$.

25. Find two planes that intersect in the line $y = 3x + 2$ in the x, y-plane.

26. Find the angle between the planes $2x - 3y + 6z = 5$ and $3x + y + 2z = 6$.

27. Find the angle between the planes $x + y + z = 1$ and $x - y + z = 1$.

28. Find an equation of the plane consisting of all point that are equidistant from the points $(3, 4, 0)$ and $(1, 0, -2)$.

29. Find an equation of the line containing the point $(0, 1, 2)$ that is parallel to the plane $x + y + z = 2$ and perpendicular to the line $\boldsymbol{w} = \langle 1, 1, 0 \rangle + t\langle 1, -1, 2 \rangle$.

30. Verify that the plane $5x - 3y - z - 6 = 0$ contains the line $\boldsymbol{w} = \langle 1, -1, 2 \rangle + t\langle 2, 3, 1 \rangle$.

31. Find an equation of the plane that contains the point $(2, 1, -3)$ and the line of intersection of the planes $3x + y - z = 2$ and $2x + y + 4z = 1$.

32. Find an equation of the plane containing the point $(4, 1, -6)$ that is perpendicular to the line passing through the points $(-1, 6, 2)$ and $(-8, 10, -2)$.

33. Find an equation of the plane that is perpendicular to each of the planes $2x + y - z = -2$ and $x - y + 3z = 1$ and contains the point $(1, 3, -2)$.

34. Show that the line $x = -1 + t, \ y = 3 + 2t, \ z = -t$ is parallel to the plane $-2x + 2y + 2z = 3$ and find the distance between them.

35. Find the minimal distant between the line $x = 1 + 4t, \ y = 5 - 4t, \ z = -1 + 5t$ and the line $x = 2 + 8t, \ y = 4 - 3t, \ z = 5 + t$.

§4. VECTOR-VALUED FUNCTIONS

A vector-valued functions of a real variable is a functions of the form:

$$r(t) = f(t)i + g(t)j + h(t)k$$

Or: $r(t) = (f(t), g(t), h(t))$
And, in a more general setting:
$r(t) = (f_1(t), f_2(t), ..., f_n(t))$

Such functions can effectively be used to describe curves in $\Re^3$.

Assume, for example, that the coordinates of a particle moving in space during a time interval I are given by:

$$x = f(t), \; y = g(t), z = h(t), \text{ for } t \in I.$$

The point $(x, y, z) = (f(t), g(t), h(t))$ on the particle's path can also be described by means of the **position vector** function:

$$r(t) = f(t)i + g(t)j + h(t)k$$

EXAMPLE 12.16
A HELIX

Describe the curve traced out by a particle with position vector:

$$r(t) = (\cos t)i + (\sin t)j + tk \text{ for } t \geq 0.$$

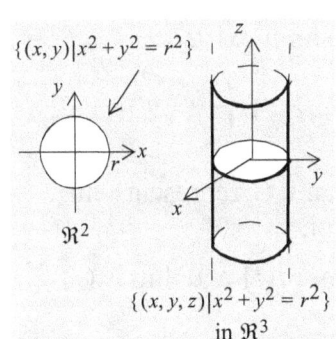

$\{(x, y) | x^2 + y^2 = r^2\}$

$\Re^2$

$\{(x, y, z) | x^2 + y^2 = r^2\}$
in $\Re^3$

SOLUTION: We begin by noting that while the equation $x^2 + y^2 = r^2$ represents a circle in the plane, it denotes a cylinder in $\Re^3$ (see margin). In particular, since $x = \cos t$ and $y = \sin t$ satisfy the equation $x^2 + y^2 = 1$, the terminal point of the position vector:

$$r(t) = (\cos t)i + (\sin t)j + tk$$
$$= \langle \cos t, \sin t, t \rangle$$

lies on the cylinder:

$$\{(x, y, z) | x^2 + y^2 = 1\}$$

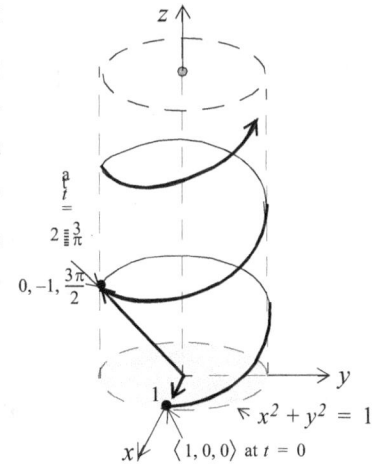

At time $t = 0$, for example, the particle has position vector
$$r(0) = (\cos 0)i + (\sin 0)j + 0k = \langle 1, 0, 0 \rangle$$
with terminal point $(1, 0, 0)$, while at $t = \dfrac{3\pi}{2}$:

$$r\left(\frac{3\pi}{2}\right) = (\cos\frac{3\pi}{2})i + (\sin\frac{3\pi}{2})j + \frac{3\pi}{2}k = \langle 0, -1, \frac{3\pi}{2} \rangle$$

CHECK YOUR UNDERSTANDING 12.26

Describe the curve traced out by the position vector:
$$r(t) = ti + t^2j + 2tk \qquad \text{for } 1 \leq t \leq 3.$$

Indicate the coordinates of the initial position of the particle (when $t = 1$), and its terminal position (when $t = 3$).

Answer: See page A-18.

Some previously encountered concepts naturally extend to vector-valued functions:

Compare with Definition 2.2, page 53.

DEFINITION 12.8

LIMIT

For given $r(t) = f(t)i + g(t)j + h(t)k$ and vector $L = ai + bj + ck$, we say that r has **limit L** as t approaches t_0 and write:

$$\lim_{t \to t_0} r(t) = L$$

if for any given $\varepsilon > 0$ there exists $\delta > 0$ such that:

$$\text{if } 0 < |t - t_0| < \delta \text{ then } \|r(t) - L\| < \varepsilon$$

Compare with Definition on page 57.

CONTINUOUS

$r(t)$ is **continuous** at t_0 if:

$$\lim_{t \to t_0} r(t) = r(t_0)$$

Compare with Definition 3.1, page 66.

DERIVATIVE

The **derivative** of $r(t)$ at t_0 is the vector $r'(t_0)$ given by:

$$r'(t_0) = \lim_{h \to 0} \frac{1}{h}[r(t_0 + h) - r(t_0)]$$

(providing the limit exists)

Compare with Definition 5.2, page 167.

INDEFINITE INTEGRAL

$$\int r(t)dt = R(t) + C$$

where $R'(t) = r(t)$ and C represents an arbitrary constant vector.

In the exercises you are invited to show that if $\lim_{t \to t_0} r_1(t)$ and $\lim_{t \to t_0} r_2(t)$ exist, then:

$$\lim_{t \to t_0} [r_1(t) + r_2(t)] = \lim_{t \to t_0} r_1(t) + \lim_{t \to t_0} r_2(t)$$

(the limit of a sum is the sum of the limits)

That being the case, by looking at $r(t) = f(t)i + g(t)j + h(t)k$ as the sum of the three vector-valued functions $f(t)i$, $g(t)j$, and $h(t)k$, we see that:

$$\lim_{t \to t_0} r(t) = \left[\lim_{t \to t_0} f(t)\right]i + \left[\lim_{t \to t_0} g(t)\right]j + \left[\lim_{t \to t_0} h(t)\right]k$$

(providing the component-limits exist)

Moreover (see CYU 12.27 below):

$$r'(t) = f'(t)i + g'(t)j + h'(t)k$$

(providing the component-derivatives exist)

Also (see Exercise 31):

$$\int r(t)dt = [\int f(t)dt]i + [\int g(t)dt]j + [\int h(t)dt]k + C$$

(providing the component-integrals exist)

So: You can arrive at the limit, or derivative, or integral of a vector-valued functions by performing those operations on the real-valued components of the function.

EXAMPLE 12.17 Let $r(t) = (t^3 + 1)i + (\sin t)j + 9k$. Determine:

(a) $\lim\limits_{t \to 2} r(t)$ (b) $r'(t)$ (c) $\int r(t)dt$

SOLUTION:

(a) $\lim\limits_{t \to 2} r(t) = \lim\limits_{t \to 2}(t^3 + 1)i + \lim\limits_{t \to 2}(\sin t)j + \lim\limits_{t \to 2}(9)k$

$= 9i + (\sin 2)j + 9k$

(b) $r'(t) = (t^3 + 1)'i + (\sin t)'j + (9)'k = (3t^2)i + (\cos t)j$

(c) $\int r(t)dt = [\int(t^3 + 1)dt]i + [\int \sin t\,dt]j + [\int 9\,dt]k$

$= \left(\frac{t^4}{4} + t\right)i - (\cos t)j + (9t)k + C$

CHECK YOUR UNDERSTANDING 12.27

Answers: (i) $ei + \frac{1}{2}j + \sin(1)k$

(ii) $(e^t)i + \frac{-t^2 + 1}{(t^2 + 1)^2}j + (\cos t)k$

(iii) $e^t i + \left[\frac{1}{2}\ln(t^2 + 1)\right]j$

$\qquad - (\cos t)k + C$

(b) See page A18.

(a) For $r(t) = (e^t)i + \left(\dfrac{t}{t^2 + 1}\right)j + (\sin t)k$, determine:

(i) $\lim\limits_{t \to 1} r(t)$ (ii) $r'(t)$ (iii) $\int r(t)dt$

(b) Show that if the real-valued functions f, g, and h, are differentiable, them so is the function $r(t) = f(t)i + g(t)j + h(t)k$, and:

$$r'(t) = f'(t)i + g'(t)j + h'(t)k$$

Given that the derivative of vector-valued functions may be performed by differentiating their real-valued components, one might expect that the derivative formulas for such functions mimic those of real-valued functions, and so they do:

As it is with real-valued functions, a vector-valued function $r(t)$ is said to be differentiable if $r'(t)$ exists throughout its domain.

THEOREM 12.8 For given differentiable vector-valued functions $u(t)$, $v(t)$ in R^3, and any scalar c, and any real-valued function $f(t)$:

Theorems (a) through (f) hold in any R^n.

(a) $[cu(t)]' = cu'(t)$

Since in a scalar times a vector expression such as cv, the scalar c appears to the left of the vector v, the chain rule is generally not expressed in what might be considered to be the more reminiscent form: $(u[f(t)])' = u'[f(t)]f'(t)$.

(b) $[f(t)u(t)]' = f(t)u'(t) + f'(t)u(t)$

(c) $[u(f(t))]' = f'(t)u'[f(t)]$ (see margin)

(d) $[u(t) + v(t)]' = u'(t) + v'(t)$

(e) $[u(t) - v(t)]' = u'(t) - v'(t)$

(f) $[u(t) \cdot v(t)]' = u(t) \cdot v'(t) + u'(t) \cdot v(t)$

Why is there no quotient rule?

(g) $[u(t) \times v(t)]' = u(t) \times v'(t) + u'(t) \times v(t)$

PROOF: We establish (b) and (f). You are asked to verify (d) in CYU 12.28, and are invited to prove the rest in the exercises.

(b) $[f(t)\boldsymbol{u}(t)]' = [f(t)(u_1(t)\boldsymbol{i} + u_2(t)\boldsymbol{j} + u_3(t)\boldsymbol{k})]'$

$= [f(t)u_1(t)]'\boldsymbol{i} + [f(t)u_2(t)]'\boldsymbol{j} + [f(t)u_3(t)]'\boldsymbol{k}$

$= [f(t)u_1'(t) + u_1(t)f'(t)]\boldsymbol{i} + [f(t)u_2'(t) + u_2(t)f'(t)]\boldsymbol{j} + [f(t)u_3'(t) + u_3(t)f'(t)]\boldsymbol{k}$

$= f(t)[u_1'(t)\boldsymbol{i} + u_2'(t)\boldsymbol{j} + u_3'(t)\boldsymbol{k}] + f'(t)[u_1(t)\boldsymbol{i} + u_2(t)\boldsymbol{j} + u_3(t)\boldsymbol{k}]$

$= f(t)\boldsymbol{u}'(t) + f'(t)\boldsymbol{u}(t)$

(f) $[\boldsymbol{u}(t) \cdot \boldsymbol{v}(t)]' = [(u_1(t)\boldsymbol{i} + u_2(t)\boldsymbol{j} + u_3(t)\boldsymbol{k}) \cdot (v_1(t)\boldsymbol{i} + v_2(t)\boldsymbol{j} + v_3(t)\boldsymbol{k})]'$

$= [u_1(t)v_1(t) + u_2(t)v_2(t) + u_3(t)v_3(t)]'$

$= [u_1(t)v_1'(t) + v_1(t)u_1'(t)] + [u_2(t)v_2'(t) + v_2(t)u_2'(t)] + [u_3(t)v_3'(t) + v_3(t)u_3'(t)]$

$= [u_1(t)\boldsymbol{i} + u_2(t)\boldsymbol{j} + u_3(t)\boldsymbol{k}] \cdot [v_1'(t)\boldsymbol{i} + v_2'(t)\boldsymbol{j} + v_3'(t)\boldsymbol{k}]$

$\hookrightarrow + [v_1(t)\boldsymbol{i} + v_2(t)\boldsymbol{j} + v_3(t)\boldsymbol{k}] \cdot [u_1'(t)\boldsymbol{i} + u_2'(t)\boldsymbol{j} + u_3'(t)\boldsymbol{k}]$

$= \boldsymbol{u}(t) \cdot \boldsymbol{v}'(t) + \boldsymbol{v}(t) \cdot \boldsymbol{u}'(t)$

In the event that a vector-valued function has constant magnitude, we have:

THEOREM 12.9 Let $\boldsymbol{r}(t)$ be differentiable. If $\|\boldsymbol{r}(t)\| = c$, then:

$$\boldsymbol{r}(t) \cdot \boldsymbol{r}'(t) = 0 \ (\text{i.e: } \boldsymbol{r}(t) \perp \boldsymbol{r}'(t))$$

PROOF: From $\boldsymbol{r}(t) \cdot \boldsymbol{r}(t) = \|\boldsymbol{r}(t)\|^2 = c^2$, we have:

$$[\boldsymbol{r}(t) \cdot \boldsymbol{r}(t)]' = (c^2)'$$

Theorem 12.8(f): $\boldsymbol{r}(t) \cdot \boldsymbol{r}'(t) + \boldsymbol{r}'(t) \cdot \boldsymbol{r}(t) = 0$

$$2[\boldsymbol{r}(t) \cdot \boldsymbol{r}'(t)] = 0$$

$$\boldsymbol{r}(t) \cdot \boldsymbol{r}'(t) = 0$$

CHECK YOUR UNDERSTANDING 12.28

Prove Theorem 12.8(d):
$$[\boldsymbol{u}(t) + \boldsymbol{v}(t)]' = \boldsymbol{u}'(t) + \boldsymbol{v}'(t)$$

Answer: See page A-18.

We adopt the form:
$$\lim_{h \to 0} \frac{\boldsymbol{r}(t_0 + h) - \boldsymbol{r}(t_0)}{h}$$
to conform with Definition 3.2, page 67. Technically speaking, we really should write:
$$\lim_{h \to 0} \frac{1}{h}[\boldsymbol{r}(t_0 + h) - \boldsymbol{r}(t_0)]$$
(See Definition 12.1, page 520)

Let's focus on the geometrical significance of the derivative expression:

$$\boldsymbol{r}'(t_0) = \lim_{h \to 0} \frac{\boldsymbol{r}(t_0 + h) - \boldsymbol{r}(t_0)}{h}$$

Multiplying the vector $\boldsymbol{r}(t_0 + h) - \boldsymbol{r}(t_0)$ in the adjacent figure by $\frac{1}{h}$ will result in a vector parallel to $\boldsymbol{r}(t_0 + h) - \boldsymbol{r}(t_0)$. Moreover, as h gets smaller, the corresponding vector $\boldsymbol{r}(t_0 + h) - \boldsymbol{r}(t_0)$ pivots to what we will now call the **tangent vector** to the curve at $\boldsymbol{r}(t_0)$. More formally:

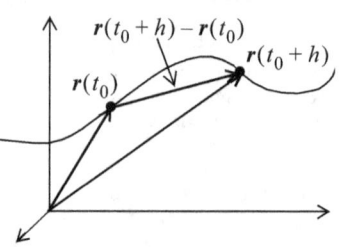

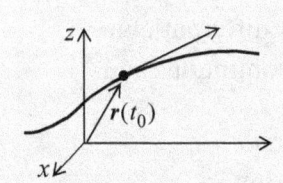

In general, $r(t)$ is said to be be a **smooth curve** if $r'(t)$ exists throughout its domain.

DEFINITION 12.9 The **tangent vector** to the curve $r(t) = f(t)\mathbf{i} + g(t)\mathbf{j} + h(t)\mathbf{k}$, at $t = t_0$ is the vector $r'(t_0)$, providing the derivative exists and is not $\mathbf{0}$.

Why is $r'(t_0) = \mathbf{0}$ ostracized in the above definition? Because the tangent lines at $(f(t_0), g(t_0), h(t_0))$ is the line containing that point that is parallel to the tangent vector $r'(t_0)$. But no line can be parallel to $r'(t_0)$ if $r'(t_0) = \mathbf{0}$, as the zero vector has no direction. It is because of its lack of direction that we do not associate a tangent vector to a curve at $t = t_0$ if $r'(t_0) = \mathbf{0}$.

EXAMPLE 12.18 Determine the tangent line to the curve traced out by the vector valued function

$$r(t) = t^2\mathbf{i} + 3t\mathbf{j} + \mathbf{k} \quad \text{when } t = \frac{1}{2}.$$

SOLUTION: The tangent vector of $r(t) = t^2\mathbf{i} + 3t\mathbf{j} + \mathbf{k} = \langle t^2, 3t, 1 \rangle$ at $t = \frac{1}{2}$ is $\langle \left(\frac{1}{2}\right)^2, 3\left(\frac{1}{2}\right), 1 \rangle = \langle \frac{1}{4}, \frac{3}{2}, 1 \rangle$. This gives us the direction of the tangent line at the point $\left(\frac{1}{4}, \frac{3}{2}, 1\right)$ on the curve. Any vector parallel to $\langle \frac{1}{4}, \frac{3}{2}, 1 \rangle$ can be used as the direction vector. That being the case, we elect to go with $4\langle \frac{1}{4}, \frac{3}{2}, 1 \rangle = \langle 1, 6, 4 \rangle$.

Proceeding as in Example 12.9, page 514, we determine a parametric equation of the tangent line L at $\left(\frac{1}{4}, \frac{3}{2}, 1\right)$:

$$L = \left\{ (x, y, z) \mid x(t) = \frac{1}{4} + t, y = \frac{3}{2} + 6(t), z = 1 + 4t \right\}$$

In the event that $s(t) = x(t)\mathbf{i} + y(t)\mathbf{j} + z(t)\mathbf{k}$ represents a position vector for a particle as a function of time, then the tangent vector points in the direction of motion of the particle, and represents the rate of change of its position with respect to time. That being the case:

Let $s(t) = x(t)i + y(t)j + z(t)k$ for $t \in I$ be a differentiable position vector for a particle for which $s'(t)$ is continuous and never $\mathbf{0}$. Then:

$v(t) = s'(t)$ is the particle's **velocity** and

$a(t) = s''(t)$ is the particle's **acceleration**.

(providing the second derivative exists and is distinct from $\mathbf{0}$)

Moreover: $\|v(t_0)\|$ denotes the particle's speed at $t = t_0$.

Note that velocity and acceleration are **vectors** indicating both magnitude and direction. Recall that even in the one-dimensional consideration of free falling objects on page 171, both acceleration and velocity were linked to a direction — a positive or negative direction, for up or down, respectively.

EXAMPLE 12.19 A particle moves along a curve so that its coordinates at time t are:

$$x = 2t, \quad y = \frac{1}{2}t^2, \quad z = e^t$$

Find the velocity, speed, and acceleration of the particle at $t = 3$.

SOLUTION: Focusing on the position vector:

$$s(t) = 2ti + \frac{1}{2}t^2 j + e^t k$$

we have: $v(t) = r'(t) = 2i + tj + e^t k$

and: $a(t) = v'(t) = j + e^t k$

In particular, at $t = 3$:

$$v(3) = 2i + 3j + e^3 k \quad \text{and} \quad a(3) = j + e^3 k$$

And: Speed at $t = 3$: $\|v(3)\| = \sqrt{2^2 + 3^2 + (e^3)^2} = \sqrt{13 + e^6}$.

Answer:
$v(t) = (-\sin t)i + (\cos t)j + k$
$a(t) = -(\cos t)i - (\sin t)j$
$\|v(t)\| = \sqrt{2}$

CHECK YOUR UNDERSTANDING 12.29

Find the velocity, acceleration, and speed of the particle moving along the helix traced by $s(t) = (\cos t)i + (\sin t)j + tk$.

EXAMPLE 12.20 Find $v(t)$ and $s(t)$ of a particle with initial velocity $v(0) = i + 2j$ and initial position $s(0) = i + j + k$ given that:

$$a(t) = 4ti + 2tj - 3t^2 k \quad \text{for } t \geq 0.$$

SOLUTION: Since $a(t) = \dfrac{dv}{dt}$:

Letting $t = 0$ in
$v(t) = 2t^2i + t^2j - t^3k + C$
we find that:
$\qquad C = v(0) = v_0$
(as in Theorem 5.5, page 171)

$$v(t) = \int a(t)dt = \int(4ti + 2tj - 3t^2k)dt = 2t^2i + t^2j - t^3k + C$$
$$= 2t^2i + t^2j - t^3k + v_0$$
$$= 2t^2i + t^2j - t^3k + (i + 2j)$$
$$= (2t^2 + 1)i + (t^2 + 2)j - t^3k$$

Since $v(t) = \dfrac{dr}{dt}$:

At $t = 0$: $s(0) = s_0$.

$$s(t) = \int v(t)dt = \int[(2t^2 + 1)i + (t^2 + 2)j - t^3k]dt$$
$$= \left(\frac{2t^3}{3} + t\right)i + \left(\frac{t^3}{3} + 2t\right)j - \frac{t^4}{4}k + s_0$$
$$= \left(\frac{2t^3}{3} + t\right)i + \left(\frac{t^3}{3} + 2t\right)j - \frac{t^4}{4}k + (i + j + k)$$
$$= \left(\frac{2t^3}{3} + t + 1\right)i + \left(\frac{t^3}{3} + 2t + 1\right)j - \left(\frac{t^4}{4} - 1\right)k$$

CHECK YOUR UNDERSTANDING 12.30

Answer:
$5\sqrt{2}$ and $\sqrt{146}$ ft/sec.

Determine the minimum and maximum speeds of a particle with initial velocity $v_0 = i - 12j + k$ (feet per second) if:
$$a(t) = i + 2j - k \text{ for } 0 \le t \le 5.$$

EXAMPLE 12.21 A projectile is fired at a speed of 1200 feet per second from ground level, at a 60° angle of elevation. Determine its maximum height and range.

SOLUTION: We establish our axes so that the projectile's plane of trajectory lies within the yz-plane, and so that it is positioned at the origin at $t = 0$.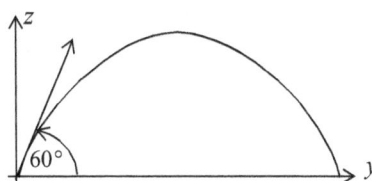

Due to the force of gravity, the projectile will be subjected to a downward acceleration of 32 ft/sec^2; which is to say: $a(t) = -32k$. Consequently:

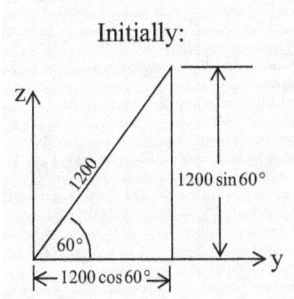
Initially:

$$v(t) = \int(-32k)dt = (-32t)k + v(0)$$
$$= (-32t)k + (1200\cos60°j + 1200\sin60°k)$$
$$= (-32t)k + (600j + 600\sqrt{3}k)$$
$$= 600j + (-32t + 600\sqrt{3})k \quad (*)$$

and:

$$s(t) = \int [600\boldsymbol{j} + (-32t + 600\sqrt{3})\boldsymbol{k}]dt$$

$$= (600t)\boldsymbol{j} + (-16t^2 + 600\sqrt{3}t)\boldsymbol{k} + s(\boldsymbol{0})$$

$$s(0) = 0: \quad = (600t)\boldsymbol{j} + (-16t^2 + 600\sqrt{3}t)\boldsymbol{k} \quad (**)$$

Now, when the projectile reaches its maximum height, its velocity's **vertical component** must be zero. Turning to (*) we find the value of t for that occurrence:

$$-32t + 600\sqrt{3} = 0, \text{ or: } t = \frac{600\sqrt{3}}{32} = \frac{75\sqrt{3}}{4} \text{ sec.}$$

The maximum height of the projectile is the vertical component of $s(t)$ at that point in time:

$$\text{maximum height} = -16\left(\frac{75\sqrt{3}}{4}\right)^2 + 600\sqrt{3}\left(\frac{75\sqrt{3}}{4}\right) = 16,875 \text{ ft}$$

Another approach: Since it takes as long to reach the ground as it does to reach the maximum height:

$$t = 2\left(\frac{75\sqrt{3}}{4}\right) = \frac{75\sqrt{3}}{2}$$

To find the range of the projectile, we first determine the value of t for which the vertical component in (**) is 0:

$$-16t^2 + 600\sqrt{3}t = 0 \Rightarrow t = \frac{600\sqrt{3}}{16} = \frac{75\sqrt{3}}{2} \text{ sec.}$$

$$\uparrow$$
(see margin)

We then determine the horizontal component of (**) at that instant of time:

$$\text{range} = 600\left(\frac{75\sqrt{3}}{2}\right) \approx 38,971 \text{ ft.}$$

CHECK YOUR UNDERSTANDING 12.31

Answer: 3131 m from the base of the perch, at 175 m/sec.

A projectile is fired at a speed of 175 meters per second at a 45° angle of elevation from a perch that is 10 meters above ground level. Where does the projectile hit the ground, and at what speed?

Note: $\boldsymbol{a}(t) = -9.8\boldsymbol{k}$ m/sec^2

	EXERCISES	

Exercises 1-6. Sketch, in R^3, the graph of the curve with given position vector.

1. $r(t) = ti$, $0 \leq t \leq 1$

2. $r(t) = ti + 2tj$, $0 \leq t \leq 1$

3. $r(t) = ti + tj + 4k$, $1 \leq t \leq 2$

4. $r(t) = (\cos t)i - (\sin t)j + k$, $0 \leq t \leq \frac{\pi}{2}$

5. $r(t) = 3i + tj + t^2k$, $-1 \leq t \leq 2$

6. $r(t) = ti + tj + (\sin t)k$, $0 \leq t \leq 2\pi$

Exercises 7-16. Determine $\lim\limits_{t \to 1} r(t)$, $r'(t)$, and $\int r(t)dt$ for the given vector function.

7. $r(t) = 2ti + 5t^2j$

8. $r(t) = 2ti + 5t^2j + k$

9. $r(t) = (\sin t)i - t^2j$

10. $r(t) = (\sin t)i - t^2j + e^tk$

11. $r(t) = ti + j + k$

12. $r(t) = (e^{t^2})i - \sin(2t)j + \frac{t-1}{t}k$

13. $r(t) = ti - 3j + \left(\frac{\ln t}{t}\right)k$

14. $r(t) = \frac{t}{t+1}i + \frac{t+1}{t^3}j - k$

15. $r(t) = ti - t^2j + (\tan t)k$

16. $r(t) = (t\sin t^2)i + tj - (t\cos t)k$

Exercises 17-20. Find parametric equations for the tangent line at the specified point on the curve with given position vector.

17. $r(t) = 2ti + 5t^2j$, $t = 1$.

18. $r(t) = ti + t^2j + t^3k$, $t = 0$

19. $r(t) = (\sin t)i - (2\cos t)j + k$, $t = 0$

20. $r(t) = (\sin t)i - t^2j + e^tk$, $t = \pi$

Exercises 21-24. Find the velocity, acceleration, and speed of the particle moving along the given path at the indicated time.

21. $r(t) = ti + 2tj$, $t = 1$

22. $r(t) = ti + j + t^2k$, $t = 0$

23. $r(t) = (\sin t)i + tj + (\cos t)k$, $t = 0$

24. $r(t) = \sqrt{t}\,i - tj + e^{2t}k$, $t = 4$

25. Determine the minimum and maximum speed of a particle with initial velocity $v_0 = i - 12j + k$ (feet per second) and $a(t) = 2i + j - k$, within the interval $0 \leq t \leq 6$.

26. Determine the minimal and maximum speed of a particle with initial velocity $v_0 = i - k$ (feet per second) and $a(t) = (\sin t)i + (\cos t)j - tk$, within the interval for $0 \le t \le \frac{\pi}{2}$.

Exercises 27-30. Prove:

27. Theorem 12.8(a) and (e)   28. Theorem 12.8(g)   29. Theorem 12.8(c)

30. Show that if $\lim_{t \to t_0} r_1(t)$ and $\lim_{t \to t_0} r_2(t)$ exist, then:
$$\lim_{t \to t_0} [r_1(t) + r_2(t)] = \lim_{t \to t_0} r_1(t) + \lim_{t \to t_0} r_2(t)$$

31. Show that for $r(t) = f(t)i + g(t)j + h(t)k$:
$$\int r(t)dt = [\int f(t)dt]i + [\int g(t)dt]j + [\int h(t)dt]k + C$$
(providing the component-integrals exist)

32. Prove that any two antiderivatives of a vector-valued function $r(t)$ can only differ by a constant vector C.

33. A projectile is fired at a speed of 500 meters per second at a $45°$ angle of elevation, from a point that is 30 meters above ground level. Find its speed when it hits the ground.

34. A projectile is fired at a speed of 800 feet per second from ground level, at a $45°$ angle of elevation. Determine its maximum height, range, and speed when it hits the ground.

35. A projectile is fired from ground level with initial velocity $v(0) = 100j + 100k$ feet per second. Determine its maximum height, range, and speed when it hits the ground.

36. Repeat Exercise 32 given a constant wind velocity of:
   (a) $2j$ feet per second.   (b) $2j + k$ feet per second.

37. A stone is thrown downward from the top of a 168 foot building, at an angle of depression of $60°$ at a speed of 80 ft/sec. How far from the base of the building will the stone land?

38. A golf ball is hit at a speed of 90 feet per second at a $30°$ angle of elevation. Will it clear the top of a 35 foot tree that is 135 feet away?

39. Referring to Exercise 36, determine the minimum speed for which the ball will clear the tree (maintaining the $30°$ angle of elevation).

40. Referring to Exercise 36, can the angle of elevation be adjusted so as to clear the tree (maintaining the 90 feet per second initial speed)?

41. At what speed must a stone be thrown horizontally from a point that is 25 feet above ground level if it is to hit a bottle sitting on a 4 foot pole 45 feet away?

42. At what speed must a stone be thrown at an angle of elevation of $30°$ so as to achieve a maximum height of 25 feet?

43. At what speed must a stone be thrown at an angle of elevation of 30° so as to achieve a range of 50 feet?

44. Find two angles of elevation that will enable a projectile fired from ground level at a speed of 800 feet per second to hit a ground-level target 10,000 feet away.

45. Show that in order to achieve maximum range, a projectile must be fired at a 45° angle of elevation.

46. Show that a projectile fired at an angle α, with $0 < \alpha < 90°$, has the same range as one fired at the same speed at the angle $90° - \alpha$.

47. Show that doubling a projectile's initial speed while maintaining its launching angle will quadruple its range.

48. Show that a projectile attains three-quarters of its maximum height in half the time it takes for it to reach its maximum height.

49. Prove that if the acceleration vector of a particle is always **0**, then the path of the particle is a line.

50. Assume that you take true aim at a bottle that is sitting on a y foot pole that is x feet down the road from you. Show that if the bottle starts to fall just as you pull the trigger, then the bottle will be hit, regardless of the muzzle speed of the gun.

51. Prove that any non-vertical trajectory of a projectile subjected solely to the force of gravity is parabolic.

§5. ARC LENGTH AND CURVATURE

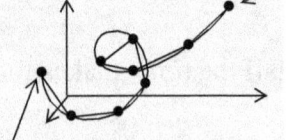

Requiring that the length L of a parametrized curve

$$x = x(t), \quad y = y(t) \quad \text{for} \quad a \le t \le b$$

be the limit of the lengths of inscribed pogonal segments (see margin), led us to Definition 10.1 of page 401:

$$L = \int_a^b \sqrt{\left(\frac{dx}{dt}\right)^2 + \left(\frac{dy}{dt}\right)^2}\, dt$$

Taking two-dimensional paths and whisking them into three-dimensional space (see margin) brings us to:

DEFINITION 12.10
ARC LENGTH

The **arc length**, L, of a smooth curve that is traced out exactly once by

$$r(t) = x(t)\boldsymbol{i} + y(t)\boldsymbol{j} + z(t)\boldsymbol{k}$$

as t increases from a to b is given by:

$$L = \int_a^b \sqrt{\left(\frac{dx}{dt}\right)^2 + \left(\frac{dy}{dt}\right)^2 + \left(\frac{dz}{dt}\right)^2}\, dt$$

$$\text{Or:}\ L = \int_a^b \|r'(t)\|\, dt$$

EXAMPLE 12.22   Find the length of one turn of the helix of Example 12.16, page 523:

$$r(t) = (\cos t)\boldsymbol{i} + (\sin t)\boldsymbol{j} + t\boldsymbol{k} \quad \text{for } 0 \le t \le 2\pi$$

SOLUTION: $L = \displaystyle\int_0^{2\pi} \|r'(t)\|\, dt = \int_0^{2\pi} \sqrt{(-\sin t)^2 + (\cos t)^2 + (1)^2}\, dt$

$$= \int_0^{2\pi} \sqrt{2}\, dt = \sqrt{2}\, t \Big|_0^{2\pi} = 2\pi\sqrt{2}$$

CHECK YOUR UNDERSTANDING 12.32

Answer: $\dfrac{13}{3}$

Find the arc length of $r(t) = \left\langle 2t, \dfrac{t^3}{3}, t^2 \right\rangle$ for $1 \le t \le 2$.

Among all of the possible parametrizations for a given curve C, one stands out above the rest, for it is based on an intrinsic property of the curve itself. To be more specific:

A curve traced out by $r(t)$ for which $r'(t)$ is continuous and never $\boldsymbol{0}$ is said to be a **smooth** curve.

Consider a smooth curve C that is traced out exactly once by $r(t) = f(t)\boldsymbol{i} + g(t)\boldsymbol{j} + h(t)\boldsymbol{k}$, for t contained in some interval I. If we choose a base point $P(t_0) = (f(t_0), g(t_0), h(t_0))$ on C, we can then assign a directed distance $s(t)$ to each $t \in I$ as follows:

$$s(t) = \int_{t_0}^{t} \|r'(u)\| \, du \quad (*)$$

can't use t, as it is being used for the upper limit of integration

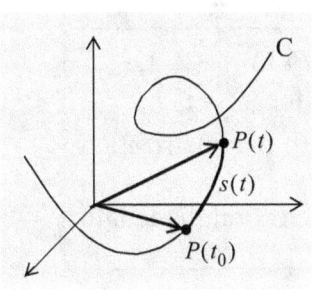

If f is continuous on $[a, b]$, then:

$$\frac{d}{dx}\left[\int_a^x f(t)dt\right] = f(x)$$

(Theorem 5.7, page 178)

Observe that if $t \geq t_0$, then $s(t)$ is the distance along the curve between the points $P(t_0)$ and $P(t) = (f(t), g(t), h(t))$ (see margin). If $t < t_0$, then $s(t)$ is the negative of the distance between the points. In any event, $s(t)$ is called the **arc length parameter** of C with base point $P(t_0)$, and $r(s)$ is said to be an **arc length parametrization** of the curve.

Applying the Principal Theorem of Calculus (see margin) to (*), we have:

THEOREM 12.10
$$\frac{ds}{dt} = \|r'(t)\|$$

Note that $r(0) = (1, 0, 0)$

EXAMPLE 12.23 Find the arc length parameter with base point $P(0) = (1, 0, 0)$ of the helix traced out by:
$$r(t) = (\cos t)i + (\sin t)j + tk.$$

SOLUTION: Turning directly to the above development we have:

$$s(t) = \int_0^t \|r'(u)\| \, du = \int_0^t \sqrt{(-\sin u)^2 + (\cos u)^2 + (1)^2} \, du$$

$$= \int_0^t \sqrt{2} \, du = \sqrt{2} u \Big|_0^t = \sqrt{2} t$$

In particular, note that $s(2\pi)$ yields the answer in Example 12.22.

EXAMPLE 12.24 Find the arc length parameterization of one turn of the helix traced out by:
$$r(t) = (\cos t)i + (\sin t)j + tk \text{ for } 0 \leq t \leq 2\pi$$
using the arc length parameter of Example 12.23.

SOLUTION: From Example 12.23:
$$s = s(t) = \int_0^t \|r'(u)\| \, du = \sqrt{2} t$$

From $s = \sqrt{2} t$, we have $t = \dfrac{s}{\sqrt{2}}$. Replacing t with $\dfrac{s}{\sqrt{2}}$ in
$$r(t) = (\cos t)i + (\sin t)j + tk \text{ for } 0 \leq t \leq 2\pi$$
yields the arc length parametrization:
$$r(s) = \cos\left(\frac{s}{\sqrt{2}}\right)i + \sin\left(\frac{s}{\sqrt{2}}\right)j + \left(\frac{s}{\sqrt{2}}\right)k \text{ for } 0 \leq s \leq 2\sqrt{2}\pi.$$

CHECK YOUR UNDERSTANDING 12.33

(a) Find the arc length parameter with base point $P(1) = (2, 1, -2)$ for the curve traced out by $r(t) = \langle 2t, t, -2t \rangle$ for $0 \le t \le 3$.

(b) Use the parameter in (a) to obtain an arc length parametrization of the curve.

(c) Use both the parametrizations in (a) and in (b) to find the length of the curve,

Answer: See page A-20.

TANGENT, NORMAL AND BINORMAL VECTORS

The tangent vector $r'(t)$ to a curve $r(t)$ was defined on page 527. We now define:

$$T(t) = \frac{r'(t)}{\|r'(t)\|} \text{ to be the } \textbf{unit tangent} \text{ vector,}$$

$$N(t) = \frac{T'(t)}{\|T'(t)\|} \text{ to be the } \textbf{unit normal} \text{ vector, and}$$

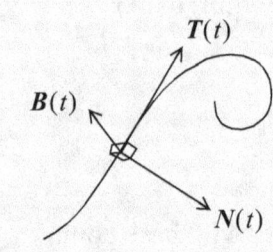

$$B(t) = T(t) \times N(t) \text{ to be the } \textbf{unit binormal} \text{ vector to the curve.}$$

Clearly both $T(t)$ and $N(t)$ are unit vectors. In particular, since $\|T(t)\|$ is constant we have (see Theorem 12.9, page 526:

$$T(t) \cdot T'(t) = 0$$

It follows that $T(t)$ and $N(t)$ are orthogonal:

$$T(t) \cdot N(t) = T(t) \cdot \frac{T'(t)}{\|T'(t)\|} \underset{\uparrow}{=} \frac{1}{\|T'(t)\|}(T(t) \cdot T'(t)) = 0$$

Theorem 12.2(c), page 501

Since $B(t)$ is the cross product of $T(t)$ and $N(t)$, it is orthogonal to both. And it is indeed a unit vector:

$$\|B(t)\| = \|T(t) \times N(t)\| \underset{\uparrow}{=} \|T(t)\|\|N(t)\|\sin 90° = 1 \cdot 1 \cdot 1 = 1$$

Theorem 12.6, page 540 $T(t) \perp N(t)$

EXAMPLE 12.25 Find the unit tangent, unit normal, and unit binormal vectors of the helix

$$r(t) = (\cos t)i + (\sin t)j + tk$$

SOLUTION:

$$T(t) = \frac{r'(t)}{\|r'(t)\|} = \frac{(-\sin t)i + (\cos t)j + k}{\sqrt{\sin^2 t + \cos^2 t + 1}} = \frac{1}{\sqrt{2}}[(-\sin t)i + (\cos t)j + k]$$

$$N(t) = \frac{T'(t)}{\|T'(t)\|} = \frac{\frac{1}{\sqrt{2}}[(-\cos t)i - (\sin t)j]}{\frac{1}{\sqrt{2}}\sqrt{\cos^2 t + \sin^2 t}} = (-\cos t)i - (\sin t)j$$

$$B(t) = T(t) \times N(t) = \det \begin{bmatrix} i & j & k \\ -\dfrac{\sin t}{\sqrt{2}} & \dfrac{\cos t}{\sqrt{2}} & \dfrac{1}{\sqrt{2}} \\ -\cos t & -\sin t & 0 \end{bmatrix} = \dfrac{1}{\sqrt{2}}[(\sin t)i - (\cos t)j + k]$$

EXAMPLE 12.26 Find the unit tangent, unit normal, and unit binormal vectors of the curve:

$$r(t) = \langle 0, t, t^2 \rangle \text{ at } t = 0$$

SOLUTION: The curve $r(t) = (0, t, t^2)$ is easy to sketch (and to see), for it is the parabola $z = y^2$, residing in the y-z plane (see margin). As for the unit vectors:

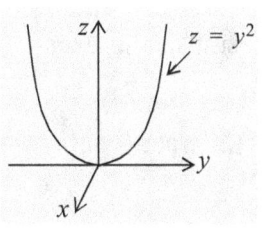

$$T(t) = \frac{r'(t)}{\|r'(t)\|} = \frac{\langle 0, 1, 2t \rangle}{\sqrt{1 + 4t^2}} \Rightarrow T(0) = \langle 0, 1, 0 \rangle = j$$

$$N(t) = \frac{T'(t)}{\|T'(t)\|} = \frac{\left\langle 0, \dfrac{1}{\sqrt{1+4t^2}}, \dfrac{2t}{\sqrt{1+4t^2}} \right\rangle'}{\left\| \left\langle 0, \dfrac{1}{\sqrt{1+4t^2}}, \dfrac{2t}{\sqrt{1+4t^2}} \right\rangle' \right\|}$$

$$= \frac{\left\langle 0, -\dfrac{4t}{(4t^2+1)^{3/2}}, \dfrac{2}{(4t^2+1)^{3/2}} \right\rangle}{\left\| \left\langle 0, -\dfrac{4t}{(4t^2+1)^{3/2}}, \dfrac{2}{(4t^2+1)^{3/2}} \right\rangle \right\|} \Rightarrow N(0) = \frac{\langle 0, 0, 2 \rangle}{2} = k$$

$$B(0) = T(0) \times N(0) = \det \begin{bmatrix} i & j & k \\ 0 & 1 & 0 \\ 0 & 0 & 1 \end{bmatrix} = i$$

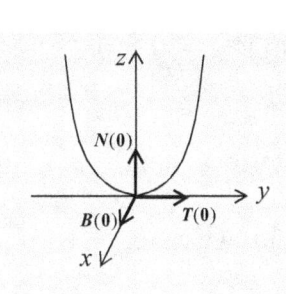

The pairwise orthogonality of the above three unit vectors is visually apparent in the margin figure. Indeed, they even coincide with the unit coordinate vectors i, j, and k.

In a general setting, though always at right angles to each other, the directions of the unit vectors $T(t)$, $N(t)$, and $B(t)$ will vary as t varies. For a body moving along the curve, these three vectors provide information on how the body's path twists and turns.

In general, the plane determined by the normal and binormal vectors is called the **normal plane** to the curve. It consists of all vectors orthogonal to the tangent vector. In particular, the normal plane in the margin figure is the x-z plane.

The plane determined by the normal and tangent vectors to a point on a curve is called the **osculating plane**, derived from the Latin *osculum*, for: "kiss." It is a passionate kiss in that it contains more of the curve than any other plane at that point. The osculating plane in the above margin figure is the *y-z* plane which contains the entire curve.

CHECK YOUR UNDERSTANDING 12.34

Find the unit tangent, unit normal, and unit binormal vectors of the curve $r(t) = \langle 2t, t^2, -t \rangle$ at $t = 1$, and then verify, directly that each of the vectors is orthogonal to the remaining two.
Determine the normal and osculating planes at $t = 1$.

Answer: See page A-20.

You are invited to establish the following useful result in the exercises:

THEOREM 12.11 If $r(t) = f(t)\boldsymbol{i} + g(t)\boldsymbol{j} + h(t)\boldsymbol{k}$ traces out a smooth curve, then the vector

$$\overline{\boldsymbol{B}(t)} = \det \begin{bmatrix} \boldsymbol{i} & \boldsymbol{j} & \boldsymbol{k} \\ f'(t) & g'(t) & h'(t) \\ f''(t) & g''(t) & h''(t) \end{bmatrix}$$

has the same direction as the unit binormal

$$\boldsymbol{B}(t) \text{ i.e: } \boldsymbol{B}(t) = \frac{\overline{\boldsymbol{B}(t)}}{\|\overline{\boldsymbol{B}(t)}\|}$$

EXAMPLE 12.27 Find the unit tangent, unit normal, and unit binormal vectors of the curve $r(t) = \langle 2t, t^2, -t \rangle$ at $t = 1$

SOLUTION (Compare with solution of CYU 12.34):

$$T(t) = \frac{\langle 2t, t^2, -t \rangle'}{\|\langle 2t, t^2, -t \rangle'\|} = \frac{\langle 2, 2t, -1 \rangle}{\sqrt{5 + 4t^2}} \Rightarrow T(1) = \frac{1}{3}\langle 2, 2, -1 \rangle$$

$$\overline{\boldsymbol{B}(1)} = \det \begin{bmatrix} \boldsymbol{i} & \boldsymbol{j} & \boldsymbol{k} \\ f'(1) & g'(1) & h'(1) \\ f''(1) & g''(1) & h''(1) \end{bmatrix} = \det \begin{bmatrix} \boldsymbol{i} & \boldsymbol{j} & \boldsymbol{k} \\ 2 & 2 & -1 \\ 0 & 2 & 0 \end{bmatrix} = \langle 2, 0, 4 \rangle \Rightarrow \boldsymbol{B}(1) = \frac{1}{\sqrt{5}}\langle 1, 0, 2 \rangle$$

$$\langle f(t), g(t), h(t) \rangle = \langle 2t, t^2, -t \rangle \Rightarrow \langle f'(t), g'(t), h'(t) \rangle = \langle 2, 2t, -1 \rangle \Rightarrow \langle f''(t), g''(t), h''(t) \rangle = \langle 0, 2, 0 \rangle$$

Using the fact that $N(t) = B(t) \times T(t)$ (Exercise 30) we have:

$$N(1) = B(1) \times T(1) = \det \begin{bmatrix} \boldsymbol{i} & \boldsymbol{j} & \boldsymbol{k} \\ \dfrac{1}{\sqrt{5}} & 0 & \dfrac{2}{\sqrt{5}} \\ \dfrac{2}{3} & \dfrac{2}{3} & -\dfrac{1}{3} \end{bmatrix} = \frac{1}{3\sqrt{5}}\langle -4, 5, 2 \rangle$$

CURVATURE

The curve in the adjacent figure appears to be "curving" less at a than at b, and less at b than at c. Roughly speaking, the curvature at a point is a measure of how rapidly the direction of the curve is changing at that point. To be more precise, we define the curvature to be the magnitude of the rate of change of the unit tangent vector T with respect to **arc length**:

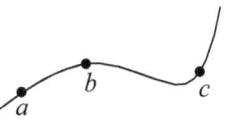

Arc length, which is a natural parameter for any curve, is chosen for consistency.

DEFINITION 12.11
CURVATURE

The **curvature** of a smooth curve is:

$$\kappa(s) = \left\| \frac{dT(s)}{ds} \right\|$$

Greek letter κ : kappa

Our first objective is to show that the above definition does do what it is supposed to do, at least for lines and circles. Specifically, since a line is aways heading in one direction, we expect that its curvature is zero, and it is [Example 12.28(a) below]. Also, since a circle of radius 1 "bends quicker" then one of radius 2, we expect that the curvature of the smaller circle is greater than that of the larger circle, and it is [Example 12.28(b)].

EXAMPLE 12.28   (a) Show that the curvature of a line is zero.

(b) Show that the curvature of a circle of radius a is $\frac{1}{a}$.

SOLUTION: (a) The unit tangent vector T always points in the direction of L. Being a constant vector:

$$\kappa(s) = \left\| \frac{dT(s)}{ds} \right\| = \|0\| = 0$$

(b) We restrict our attention to the circle $x^2 + y^2 = a^2$ with vector equation:

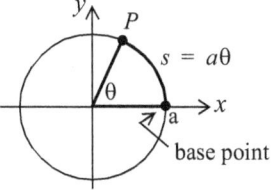

$$r(\theta) = (a\cos\theta)i + (a\sin\theta)j, \ 0 \le \theta < 2\pi \ (*)$$

Note that $(a\cos\theta)^2 + (a\sin\theta)^2 = a^2$

The arc length to any point P on the circle is given by $s = a\theta$ (see margin). Replacing θ with $\frac{s}{a}$ in (*) we arrive at the arc length parametrization:

$$\frac{s}{2\pi a} = \frac{\theta}{2\pi}$$

s is to the circumference of the circle as θ is to a complete revolution

$$r(s) = \left(a\cos\frac{s}{a}\right)i + \left(a\sin\frac{s}{a}\right)j, \ 0 \le s < 2\pi a$$

With unit vector (see margin):

$$T(t) = \frac{r'(t)}{\|r'(t)\|}$$

$$T(s) = \frac{dr(s)}{ds} = \left(-\sin\frac{s}{a}\right)i + \left(\cos\frac{s}{a}\right)j$$

Theorem 12.10 $= \dfrac{r'(t)}{\frac{ds}{dt}}$

and:

$$= \frac{dr}{dt}\frac{dt}{ds} = \frac{dr}{ds}$$

$$\left\| \frac{dT(s)}{ds} \right\| = \left\| \left(-\frac{1}{a}\cos\frac{s}{a}\right)i - \left(\frac{1}{a}\sin\frac{s}{a}\right)j \right\| = \sqrt{\frac{1}{a^2}\left[\cos^2\left(\frac{s}{a}\right) + \sin^2\left(\frac{s}{a}\right)\right]} = \frac{1}{a}$$

The following curvature formula involves an arbitrary parameter t:

THEOREM 12.12 If $r(t) = f(t)i + g(t)j + h(t)k$ traces out a smooth curve, then:

$$\kappa = \frac{\|T'(t)\|}{\|r'(t)\|}$$

Since the derivative of a vector-valued function is defined in terms of the derivatives of its components, the Chain Rule Theorem extends to vector-valued functions.

PROOF $\kappa = \left\|\dfrac{dT(s)}{ds}\right\| \underset{\text{Chain Rule}}{\uparrow} = \dfrac{\left\|\dfrac{dT}{dt}\right\|}{\dfrac{ds}{dt}} = \dfrac{\|T'(t)\|}{\dfrac{ds}{dt}} \underset{\text{Theorem 12.10}}{\uparrow} = \dfrac{\|T'(t)\|}{\|r'(t)\|}$

EXAMPLE 12.29 Use the above theorem to address Example 12.28(b).

SOLUTION: Using the parametrization:

$$r(t) = (a\cos t)i + (a\sin t)j$$

we have:

$$\kappa = \frac{\|T'(t)\|}{\|r'(t)\|} = \frac{\left\|\left[\dfrac{r'(t)}{\|r'(t)\|}\right]'\right\|}{\|r'(t)\|} = \frac{\left\|\left[\dfrac{(-a\sin t)i + (a\cos t)j}{\sqrt{(-a\sin t)^2 + (a\cos t)^2}}\right]'\right\|}{\sqrt{(-a\sin t)^2 + (a\cos t)^2}}$$

$$= \frac{\left\|\left[\dfrac{(-a\sin t)i + (a\cos t)j}{a}\right]'\right\|}{a}$$

$$= \frac{\|[(-\sin t)i + (\cos t)j]'\|}{a} = \frac{\|[(-\cos t)i - (\sin t)j]\|}{a}$$

$$= \frac{\sqrt{(\cos t)^2 + (\sin t)^2}}{a} = \frac{1}{a}$$

While more intimidating than the expression in Theorem 12.12, the following representation for curvature is often the more practical to utilize:

THEOREM 12.13 Let C be a smooth curve traced out by

$$r(t) = f(t)i + g(t)j + h(t)k$$

The curvature at the point $r(t)$ on C is given by:

$$\kappa = \frac{\|r'(t) \times r''(t)\|}{\|r'(t)\|^3}$$

(providing $r''(t)$ exists)

PROOF: From $T(t) = \dfrac{r'(t)}{\|r'(t)\|}$ we have:

$$r'(t) = \|r'(t)\| T(t) \underset{\text{Theorem 12.10}}{\uparrow} = \frac{ds}{dt} T(t)$$

So: $r''(t) = \left(\dfrac{ds}{dt} T(t)\right)' \underset{\text{Theorem 12.8(b), page 525}}{\uparrow} = \dfrac{ds}{dt} T'(t) + \dfrac{d^2 s}{dt^2} T(t)$

Thus: $r'(t) \times r''(t) = \dfrac{ds}{dt}T(t) \times \left(\dfrac{ds}{dt}T'(t) + \dfrac{d^2s}{dt^2}T(t)\right)$

Theorem 12.7(a) and (c): $= \left(\dfrac{ds}{dt}\right)^2 (T(t) \times T'(t)) + \left(\dfrac{ds}{dt}\right)\left(\dfrac{d^2s}{dt^2}\right)(T(t) \times T(t))$
(page 509)

Theorem 12.6, page 508: $= \left(\dfrac{ds}{dt}\right)^2 (T(t) \times T'(t)) + \mathbf{0}$
(see margin)

$\|u \times v\| = \|u\|\|v\|\sin\theta$
$\Rightarrow$
$\|T \times T\| = \|T\|\|T\|\sin 0 = 0$
$\Rightarrow$
$T \times T = \mathbf{0}$

Consequently: $\|r'(t) \times r''(t)\| = \left(\dfrac{ds}{dt}\right)^2 \|T(t) \times T'(t)\|$

Theorem 12.6, page 508
and Theorem 12.9, page 526: $= \left(\dfrac{ds}{dt}\right)^2 \underset{1}{\|T(t)\|}\|T'(t)\| = \left(\dfrac{ds}{dt}\right)^2 \|T'(t)\|$

Hence: $\|T'\| = \dfrac{\|r'(t) \times r''(t)\|}{\left(\dfrac{ds}{dt}\right)^2} = \dfrac{\|r'(t) \times r''(t)\|}{\|r'\|^2}$

And finally: $\kappa = \dfrac{\|T'(t)\|}{\|r'(t)\|} = \dfrac{\dfrac{\|r'(t) \times r''(t)\|}{\|r'(t)\|^2}}{\|r'(t)\|} = \dfrac{\|r'(t) \times r''(t)\|}{(\|r'(t)\|)^3}$

Theorem 12.12

EXAMPLE 12.30   Find the curvature of the helix
$$r(t) = (\cos t)i + (\sin t)j + tk$$

SOLUTION: For $r(t) = (\cos t)i + (\sin t)j + tk$ we have:

$r'(t) = (-\sin t)i + (\cos t)j + k$

$r''(t) = (-\cos t)i + (-\sin t)j$

$r'(t) \times r''(t) = \det\begin{bmatrix} i & j & k \\ -\sin t & \cos t & 1 \\ -\cos t & -\sin t & 0 \end{bmatrix} = (\sin t)i + (-\cos t)j + k$

Thus: $\kappa = \dfrac{\|r'(t) \times r''(t)\|}{\|r'(t)\|^3} = \dfrac{\sqrt{\sin^2 t + \cos^2 t + 1}}{\left(\sqrt{\sin^2 t + \cos^2 t + 1}\right)^3} = \dfrac{\sqrt{2}}{(\sqrt{2})^3} = \dfrac{1}{2}$

CHECK YOUR UNDERSTANDING 12.35

Find the curvature of the curve traced out by:
$$r(t) = ti + t^2j + t^3k$$
at the points $(1, 1, 1)$ and $(2, 4, 8)$.

Answer: $\dfrac{2\sqrt{19}}{14^{3/2}}$ and $\dfrac{2\sqrt{181}}{161^{3/2}}$.

You can think of curvature as an extension of the concavity concept introduced on page 131. Unlike concavity, which only tells us if a curve bends up or down, curvature yields a measure of how "fast" a curve is bending. But in what direction is it bending? This should help:

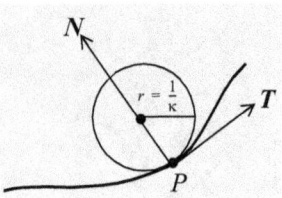

> The **circle of curvature** (or **osculating circle**) at a point P on a smooth curve C where $\kappa \neq 0$, is the circle of radius $r = 1/\kappa$ that is tangent to the curve at P (shares the tangent line to C at P) and lies on the concave side of C (in the direction of the normal to C at P).
>
> As such, the circle of curvature is contained in the osculating plane to the curve at P, and shares T, N, and κ with C at P (recall that the curvature of a circle of radius a is $1/a$).

The graph of the function $f(x) = x^2$ is concave up everywhere but appears to be bending more rapidly at the origin than it does at the point $(1, 1)$. We will turn to the circle of curvature at those two points in a bit; but first:

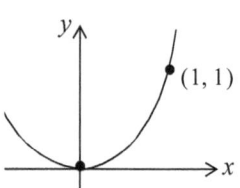

CHECK YOUR UNDERSTANDING 12.36

Prove that the curvature of the graph of $y = f(x)$ at $(x, f(x))$ is:

$$\kappa = \frac{|f''(x)|}{[1 + (f'(x))^2]^{3/2}}$$

Answer: See page A-22.

Returning to the parabola $f(x) = x^2$ we have:

$$\kappa = \frac{|f''(x)|}{[1 + (f'(x))^2]^{3/2}} = \frac{2}{(1 + 4x^2)^{3/2}}$$

We find that, as expected, the curvature at $(1, 1)$, namely $\dfrac{2}{5^{3/2}}$, is smaller than the curvature at $(0, 0)$: 2. To put it another way: the radius of the circle of curvature at $(1, 1)$, namely $r = \dfrac{5^{3/2}}{2} \approx 5.6$, is larger than it is at the origin: $r = \dfrac{1}{2}$. As for the rest:

CHECK YOUR UNDERSTANDING 12.37

Answer: $x^2 + \left(y - \dfrac{1}{2}\right)^2 = \dfrac{1}{4}$ and

$(x + 4)^2 + \left(y - \dfrac{7}{2}\right)^2 = \dfrac{125}{4}$

Find an equation for the circle of curvature to the graph of $f(x) = x^2$ at $(0, 0)$ and at $(1, 1)$.

Hint: The center of the circle is located at the endpoint of the normal vector whose length is the radius of the circle and whose initial point is the given point.

	EXERCISES	

Exercises 1-8. Find the length of the given curve.

1. $r(t) = (2\sin t)i + 5tj + (2\cos t)k, \quad -10 \le t \le 10$

2. $r(t) = \langle 2t, \dfrac{t^3}{3}, t^2 \rangle, \quad 1 \le t \le 2$

3. $r(t) = a(1 - \sin t)i + a(1 - \cos t)j, \quad 0 \le t \le 2\pi$

4. $r(t) = \sin(t^2)i - \cos(t^2)j + t^2 k, \quad 0 \le t \le \pi$

5. $r(t) = (a\cos t)i + (a\sin t)j + (bt)k, \quad 0 \le t \le 2\pi$

6. $r(t) = \langle 2\cos t, 2\sin t, t^2 \rangle, \quad 0 \le t \le 1$

7. $r(t) = \langle t, \ln(\sec t), 3 \rangle, \quad 0 \le t \le \dfrac{\pi}{4}$

8. $r(t) = (2t)i + (t^2 - 2)j + (1 - t^2)k, \quad 0 \le t \le 2$

Exercises 9-14. Find parametric equations for the given curve using the arc length parameter with base point at $t = 0$.

9. $r(t) = (3t - 2)i + (4t + 3)j$

10. $r(t) = (3 + \cos t, 2 + \sin t)$

11. $r(t) = \langle \dfrac{1}{3}t^3, \dfrac{1}{2}t^2 \rangle$

12. $r(t) = (e^t \cos t)i + (e^t \sin t)j$

13. $r(t) = (\sin t)i + (\cos t)j + tk$

14. $r(t) = (2\cos t)i + (2\sin t)j + 2tk$

Exercises 15-19. Find the unit tangent, unit normal, unit binormal vectors and curvature of the given curve.

15. $r(t) = \langle 3\sin t, 3\cos t, 4t \rangle$

16. $r(t) = (\sin t)i + (\cos t)j + \dfrac{t^2}{2}k$

17. $r(t) = (e^t \cos t)i + (e^t \sin t)j + 2k$

18. $r(t) = (2\sin t)i + (5t)j + (2\cos t)k$

19. $r(t) = \langle \dfrac{t^3}{3}, \dfrac{t^2}{2}, 0 \rangle$ for $t > 0$

Exercises 20-23. Determine the normal and osculating planes at the indicated point.

20. $r(t) = \langle \cos t, \sin t, t \rangle;\ (1, 0, 0)$

21. $r(t) = (2\sin 3t)i + tj + (2\cos 3t)k;\ t = \pi$

22. $r(t) = ti + t^2 j + t^3 k;\ (1, 1, 1)$

23. $r(t) = \langle \sin t, \cos t, t \rangle;\ \left(\dfrac{1}{\sqrt{2}}, \dfrac{1}{\sqrt{2}}, \dfrac{\pi}{4} \right)$

Exercises 24-25. Find the equation for the circle of curvature at the indicated point.

24. $r(t) = \langle 2\sin t, 5t \rangle$ at $(0, 5\pi)$

25. $r(t) = ti + (\sin 2t)j$ at $\left(\dfrac{\pi}{4}, 1 \right)$

26. Prove that if the curvature of a curve is 0, then the curve is a line.

27. Show that the parabola $y = ax^2$ achieves its maximum curvature at its vertex.

28. Find the point on the graph of the function $y = e^x$ at which the curvature is maximum.

29. Prove Theorem 12.11.

30. Show that for a smooth curve $r(t)$: $T(t) = N(t) \times B(t)$ and $N(t) = B(t) \times T(t)$.

CHAPTER SUMMARY	
SCALAR PRODUCT AND VECTOR ADDITION	For $v = \langle v_1, v_2, ..., v_n \rangle$ and $r \in \Re$: $$rv = \langle rv_1, rv_2, ..., rv_n \rangle$$ For $v = \{v_1, v_2, ..., v_n\}$ and $w = \langle w_1, w_2, ..., w_n \rangle$: $$v + w = (v_1 + w_1, v_2 + w_2, ..., v_n + w_n)$$
THEOREMS	Let u, v, and w be vectors, and let r and s be scalars (real numbers). Then: (a) $u + v = v + u$ (b) $(u + v) + w = u + (v + w)$ (c) $r(u + v) = ru + rv$ (d) $(r + s)v = rv + sv$ (e) $r(sv) = (rs)v$
DOT PRODUCT	For $u = \langle u_1, u_2, ..., u_n \rangle$ and $v = \langle v_1, v_2, ..., v_n \rangle$: $$u \cdot v = u_1v_1 + u_2v_2 + ... + u_nv_n$$
THEOREMS	Let u, v, and w be vectors and let r be a scalar. Then: (a) $v \cdot v \geq 0$, and $v \cdot v = 0$ only if $v = \mathbf{0}$ (b) $u \cdot v = v \cdot u$ (c) $ru \cdot v = r(u \cdot v) = u \cdot rv$ (d) $(u + v) \cdot w = u \cdot w + v \cdot w$ and $(u - v) \cdot w = u \cdot w - v \cdot w$
NORM	The **norm** of a vector $v = \langle v_1, v_2, ..., v_n \rangle$, denoted by $\|v\|$, is given by $\|v\| = \sqrt{v \cdot v}$.
ANGLE BETWEEN VECTORS	The **angle** θ between two nonzero vectors $u, v \in \Re^n$ is given by: $$\theta = \cos^{-1}\left(\frac{u \cdot v}{\|u\|\|v\|}\right) \quad \text{or} \quad u \cdot v = \|u\|\|v\|\cos\theta$$
ORTHOGONAL VECTORS	Two vectors u and v are **orthogonal** if $u \cdot v = 0$
VECTOR DECOMPOSITION	For given v and u any nonzero vector: $$v = \text{proj}_u v + (v - \text{proj}_u v) \text{ where:}$$ $$\text{proj}_u v = \left(\frac{v \cdot u}{u \cdot u}\right)u \quad \text{and: } (v - \text{proj}_u v) \cdot \text{proj}_u v = 0$$

CROSS PRODUCT	For $u = \langle u_1, u_2, u_3 \rangle$ and $v = \langle v_1, v_2, v_3 \rangle$: $$u \times v = \det \begin{bmatrix} i & j & k \\ u_1 & u_2 & u_3 \\ v_1 & v_2 & v_3 \end{bmatrix}$$
THEOREMS	For any $u, v \in \Re^3$, $u \times v$ is orthogonal to both u and v. If $0 \le \theta \le \pi$ is the angle between u and v, then: $$\|u \times v\| = \|u\|\|v\|\sin\theta$$ For $u, v, w \in \Re^3$ and $c \in \Re$: (a) $u \times v = -(v \times u)$ (b) $u \times (v + w) = u \times v + u \times w$ (c) $(u + v) \times w = u \times w + v \times w$ (d) $cv \times w = v \times cw = c(v \times w)$
LINES AND PLANES	Let L be the line passing through two distinct points $P = (x_1, y_1)$ and $Q = (x_2, y_2)$ in the plane. Let $v = \overrightarrow{PQ} = \langle x_2 - x_1, y_2 - y_1 \rangle$ and let $u = (x_0, y_0)$ be such that its endpoint lies on L Then: $$w = u + tv$$ is said to be a **vector equation** for L. If $n = \langle a, b, c \rangle$ is a normal vector to a plane containing the point $A_0 = (x_0, y_0, z_0)$, then for any point $P = (x, y, z)$ on the plane we have: $$n \cdot \overrightarrow{A_0 P} = 0$$ **or:** $\langle a, b, c \rangle \cdot \langle x - x_0, y - y_0, z - z_0 \rangle = 0$ **or:** $a(x - x_0) + b(y - y_0) + c(z - z_0) = 0$ **or:** $ax + by + cz = d$, where $d = ax_0 + by_0 + cz_0$
LIMITS AND DERIVATIVE OF VECTOR VALUED FUNCTIONS	For $r(t) = f(t)i + g(t)j + h(t)k$: $$\lim_{t \to t_0} r(t) = \left[\lim_{t \to t_0} f(t)\right]i + \left[\lim_{t \to t_0} g(t)\right]j + \left[\lim_{t \to t_0} h(t)\right]k$$ and: $\quad r'(t) = f'(t)i + g'(t)j + h'(t)k$

THEOREMS	For given differentiable vector-valued functions $u(t)$, $v(t)$ in R^3, and any scalar c, and any real-valued function $f(t)$: (a) $[cu(t)]' = cu'(t)$ (b) $[f(t)u(t)]' = f(t)u'(t) + f'(t)u(t)$ (c) $(u[f(t)])' = f'(t)u'[f(t)]$ (d) $[u(t) + v(t)]' = u'(t) + v'(t)$ (e) $[u(t) - v(t)]' = u'(t) - v'(t)$ (f) $[u(t) \cdot v(t)]' = u(t) \cdot v'(t) + u'(t) \cdot v(t)$ (g) $[u(t) \times v(t)]' = u \times v'(t) + u'(t) \times v(t)$ Let $r(t)$ be differentiable. If $\|r(t)\| = c$, then $r(t)$ and $r'(t)$ are orthogonal.
ARC LENGTH	The **arc length**, L, of a smooth curve that is traced out exactly once by $$r(t) = f(t)\mathbf{i} + g(t)\mathbf{j} + h(t)\mathbf{k}$$ as t increases from a to b is given by: $$L = \int_a^b \sqrt{\left(\frac{dx}{dt}\right)^2 + \left(\frac{dy}{dt}\right)^2 + \left(\frac{dz}{dt}\right)^2}\, dt$$ where: $x = f(t), y = g(t)$, and $z = h(t)$ Or: $L = \int_a^b \|r'(t)\|\, dt$
UNIT TANGENT, NORMAL AND BINORMAL VECTORS **NORMAL AND OSCULATING PLANES**	Unit tangent vector: $T(t) = \dfrac{r'(t)}{\|r'(t)\|}$ Unit normal vector: $N(t) = \dfrac{T'(t)}{\|T'(t)\|}$ } mutually orthogonal Unit binormal vector: $B(t) = T(t) \times N(t)$ Normal Plane: Plane containing $N(t)$ and $B(t)$ (with normal $T(t)$) Osculating Plane: Plane containing $N(t)$ and $T(t)$ (with normal $B(t)$)
CURVATURE	$$\kappa(s) = \left\|\frac{dT}{ds}\right\| = \frac{\|T'(t)\|}{\|r'(t)\|} = \frac{\|r'(t) \times r''(t)\|}{\|r'(t)\|^3}$$
CIRCLE OF CURVATURE	The **circle of curvature** (or **osculating circle**) at a point P on a smooth curve C where $\kappa \neq 0$, is the circle of radius $r = 1/\kappa$ that is tangent to the curve at P (shares the tangent line to C at P) and lies on the concave side of C (in the direction of the normal to C at P).

CHAPTER 13
Differentiating Functions of Several Variables

§1. PARTIAL DERIVATIVES AND DIFFERENTIABILITY

The following derivative concept, involving a function of two (or more) variables, is not far removed from that of a single-variable function:

Alternate notation for the partial derivative of $z = f(x, y)$ with respect to x and with respect to y:

$$\frac{\partial z}{\partial x}, \frac{\partial f}{\partial x}, z_x, f_x$$

$$\frac{\partial z}{\partial y}, \frac{\partial f}{\partial y}, z_y, f_y$$

Moreover:

$$f_x(x_0, y_0) \text{ and } \frac{\partial f}{\partial x}\Big|_{(x_0, y_0)}$$

denote the value of the partial derivative (with respect to x) at the point (x_0, y_0).

DEFINITION 13.1

PARTIAL DERIVATIVES

If $z = f(x, y)$, then the **partial derivative of f with respect to x** at (x, y) is

$$\frac{\partial z}{\partial x} = \lim_{h \to 0} \frac{f(x+h, y) - f(x, y)}{h}$$

while $\dfrac{\partial z}{\partial y} = \lim\limits_{h \to 0} \dfrac{f(x, y+h) - f(x, y)}{h}$

(Providing the limits exist)

Note that, in both cases, only one of the variables is allowed to vary, while the other is held fixed — as if it were a **constant**! The partial derivative is therefore obtained by taking the "regular" derivative with respect to the "varying" variable. For example:

For $z = x^3 + y^2 + xy$

y is constant

$$\frac{\partial z}{\partial x} = 3x^2 + y$$
$$\uparrow$$
$$\frac{\partial}{\partial x}(x^3 + y^2 + xy)$$

$\rightarrow (x^3 + 5^2 + x \cdot 5)' = 3x^2 + 5$

x is constant

$$\frac{\partial z}{\partial y} = 2y + x$$
$$\uparrow$$
$$\frac{\partial}{\partial y}(x^3 + y^2 + xy)$$

$(5^3 + y^2 + 5 \cdot y)' = 2y + 5 \leftarrow$

Geometrically speaking, for given (x_0, y_0, z_0) on the surface of $z = f(x, y)$, $\dfrac{\partial z}{\partial x}\Big|_{(x_0, y_0)}$ is the slope of the tangent line to the curve of intersection of the surface with the plane $y = y_0$ [see Figure 13.1(a)], while $\dfrac{\partial z}{\partial y}\Big|_{(x_0, y_0)}$ is the slope of the tangent line to the curve of intersection with the plane $x = x_0$ [see Figure 13.1(b)].

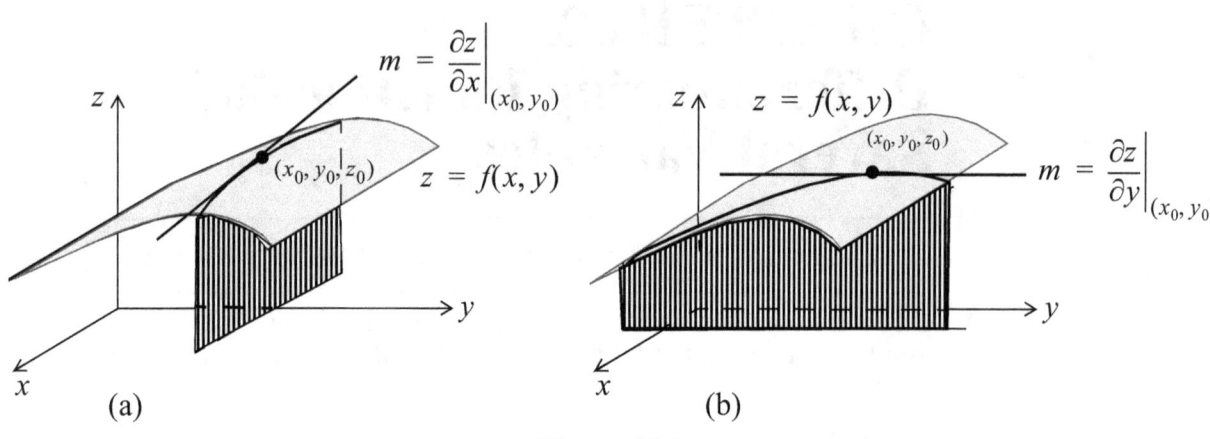

Figure 13.1

EXAMPLE 13.1 For $z = f(x, y) = 5x^2 - y^3 + \sin(xy)$ find:

(a) $\dfrac{\partial z}{\partial x}$ (b) $\dfrac{\partial z}{\partial y}$ (c) $f_x\left(1, \dfrac{\pi}{4}\right)$ (d) $\dfrac{\partial z}{\partial y}\bigg|_{\left(1, \frac{\pi}{4}\right)}$

SOLUTION: (a) When determining $\dfrac{\partial z}{\partial x}$, remember that **y is fixed** and must be treated as if it were a **constant**:

$$\frac{\partial}{\partial x}(5x^2 - \overbrace{y^3}^{\text{constant}} + \sin xy) = 10x + (\cos xy)\frac{\partial}{\partial x}(xy)$$
$$= 10x + (\cos xy)(y) \quad \text{just like: } (5x)' = 5$$

(b) When it comes to $\dfrac{\partial z}{\partial y}$, **x is fixed**:

$$\frac{\partial}{\partial y}(5\overbrace{x^2}^{\text{constant}} - y^3 + \sin xy) = -3y^2 + (\cos xy)x$$

(c) Since $f_x(x, y) = 10x + y\cos xy$ [see (a)]:

$$f_x\left(1, \frac{\pi}{4}\right) = 10 \cdot 1 + \frac{\pi}{4}\cos\left(1 \cdot \frac{\pi}{4}\right) = 10 + \frac{\pi}{4}\left(\frac{1}{\sqrt{2}}\right) = 10 + \frac{\pi}{4\sqrt{2}}$$

(d) Since $\dfrac{\partial z}{\partial y} = -3y^2 + x\cos(xy)$ [see (b)]:

$$\frac{\partial z}{\partial y}\bigg|_{\left(1, \frac{\pi}{4}\right)} = -3\left(\frac{\pi}{4}\right)^2 + \cos\frac{\pi}{4} = -\frac{3\pi^2}{16} + \frac{1}{\sqrt{2}}$$

> Stating the obvious:
> Since partial derivatives are "regular derivatives" with all but one variable assuming the role of a constant, the familiar derivative rules still hold. For example, for given $f(x, y)$ and $g(x, y)$:
>
> $$\frac{\partial}{\partial x}(f + g) = \frac{\partial f}{\partial x} + \frac{\partial g}{\partial x}, \quad \frac{\partial}{\partial y}(fg) = f\frac{\partial g}{\partial y} + g\frac{\partial f}{\partial y}, \text{ etc.}$$

CHECK YOUR UNDERSTANDING 13.1

Answers: (a) $2xy + e^{x+3y}$
(b) $x^2 + 3e^{x+3y}$
(c) e^2 (d) $4 + 3e^2$

For $z = f(x, y) = x^2y + e^{x+3y}$ determine:

(a) $\dfrac{\partial z}{\partial x}$ (b) $\dfrac{\partial z}{\partial y}$ (c) $f_x(2, 0)$ (d) $\dfrac{\partial z}{\partial y}\bigg|_{(2, 0)}$

HIGHER ORDER PARTIAL DERIVATIVES

One may be able to take partial derivatives of partial derivatives of a function $z = f(x, y)$, and here is the "∂-notation" for that activity:

$$(*) \quad \frac{\partial^2 z}{\partial x^2} = \frac{\partial}{\partial x}\left(\frac{\partial z}{\partial x}\right) \qquad \frac{\partial^2 z}{\partial y^2} = \frac{\partial}{\partial y}\left(\frac{\partial z}{\partial y}\right)$$

$$(**) \quad \frac{\partial^2 z}{\partial x \partial y} = \frac{\partial}{\partial x}\left(\frac{\partial z}{\partial y}\right) \qquad \frac{\partial^2 z}{\partial y \partial x} = \frac{\partial}{\partial y}\left(\frac{\partial z}{\partial x}\right)$$

While the translation of (*) to the subscript notation is quite safe:

$$\frac{\partial^2 z}{\partial x^2} = \frac{\partial}{\partial x}\left(\frac{\partial z}{\partial x}\right) = z_{xx} \qquad \frac{\partial^2 z}{\partial y^2} = \frac{\partial}{\partial y}\left(\frac{\partial z}{\partial y}\right) = z_{yy}$$

the same cannot be said about (**):

Note: For $z = f(x, y)$, one can replace z with f throughout; as in f_{xy} for z_{xy}.

$$\frac{\partial^2 z}{\partial x \partial y} = \frac{\partial}{\partial x}\left(\frac{\partial z}{\partial y}\right) \overset{\text{subscript notation}}{=} z_{yx} \qquad \frac{\partial^2 z}{\partial y \partial x} = \frac{\partial}{\partial y}\left(\frac{\partial z}{\partial x}\right) \overset{\text{subscript notation}}{=} z_{xy}$$

first y then x first x then y

[Actually the danger is not so great (see Theorem 13.1 on the next page)]

EXAMPLE 13.2   For $z = f(x, y) = e^{xy} + x\ln y$ find:

(a) $\dfrac{\partial^2 z}{\partial x^2}$ (b) z_{yy}

(c) f_{yx} (d) f_{xy}

SOLUTION:

keep in mind that y is held fixed — treat it as if it were a constant [see (1) in margin]

(a) $\dfrac{\partial^2 z}{\partial x^2} = \dfrac{\partial}{\partial x}\left[\dfrac{\partial}{\partial x}(e^{xy} + x\ln y)\right] = \dfrac{\partial}{\partial x}[ye^{xy} + \ln y] = y^2 e^{xy}$

y is held fixed [see (2) in margin]

x is held fixed

(b) $z_{yy} = \dfrac{\partial}{\partial y}\left[\dfrac{\partial}{\partial y}(e^{xy} + x\ln y)\right] = \dfrac{\partial}{\partial y}\left[xe^{xy} + \dfrac{x}{y}\right] = x^2 e^{xy} - \dfrac{x}{y^2}$

x is held fixed   now y is held fixed

(c) $f_{yx} = \dfrac{\partial}{\partial x}\left[\dfrac{\partial}{\partial y}(e^{xy} + x\ln y)\right] = \dfrac{\partial}{\partial x}\left[xe^{xy} + \dfrac{x}{y}\right] = xye^{xy} + e^{xy} + \dfrac{1}{y}$

first do y

(d) $f_{xy} = \dfrac{\partial}{\partial y}\left[\dfrac{\partial}{\partial x}(e^{xy} + x\ln y)\right] = \dfrac{\partial}{\partial y}[ye^{xy} + \ln y] = xye^{xy} + e^{xy} + \dfrac{1}{y}$

first do x

CHECK YOUR UNDERSTANDING 13.2

For $z = f(x, y) = \dfrac{x^2 + y}{5 + x - y}$, determine:

(a) $\dfrac{\partial^2 z}{\partial y^2}$ (b) $\dfrac{\partial^2 z}{\partial x^2}\bigg|_{(0,4)}$ (c) f_{yx} (d) $f_{yx}(2, 3)$

In Example 13.2 we observed that for $z = e^{xy} + x\ln y$:

$$f_{xy} = f_{yx}$$

Was this a fluke? Nearly not, for one has to look pretty hard to find a function $z = f(x, y)$ with existing mixed second order partial derivatives that differ. We offer the following useful result, without proof:

You need look no further than Exercise 60.

S is an open region containing (x_0, y_0) if there exists $r > 0$ such that the circle of radius r centered at (x_0, y_0) is contained in S:

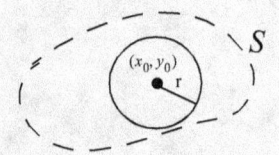

THEOREM 13.1   If, for given $z = f(x, y)$, the partial derivatives f_x, f_y, f_{xx}, f_{yy}, f_{xy}, f_{yx} are defined and continuous in an open region containing (x_0, y_0) (see margin), then:

$$f_{xy}(x_0, y_0) = f_{yx}(x_0, y_0)$$

Both the notion and the notation of partial derivatives naturally extend to accommodate higher order partial derivatives and to include functions involving more that two variables. Nudging nature along:

CHECK YOUR UNDERSTANDING 13.3

(a) For $z = f(x, y) = x^4 y^3$, determine:

(i) $\dfrac{\partial^3 z}{\partial x^3}$ (ii) z_{xyx} (iii) z_{xxy}

(b) For $w = f(x, y, z) = xy^2 + z^3 y - x^3 yz$, determine:

(i) $\dfrac{\partial w}{\partial x}$ (ii) $\dfrac{\partial^2 w}{\partial y^2}$ (iii) w_{xxx} (iv) w_{yzx}

DIFFERENTIABILITY OF A FUNCTION OF TWO VARIALBES.

Starting with Definition 3.2, page 67:

$$f'(x_0) = \lim_{h \to 0} \frac{f(x_0 + h) - f(x_0)}{h}$$

we replace h with Δx and the resulting $f(x_0 + \Delta x) - f(x_0)$ with Δy to arrive at the form:

$$f'(x_0) = \lim_{\Delta x \to 0} \frac{\Delta y}{\Delta x}$$

To put it another way, the function is differentiable at x_0 if and only if the difference $\varepsilon = f'(x_0) - \dfrac{\Delta y}{\Delta x}$ goes to zero as $\Delta x \to 0$; or, equivalently, if and only if:

$$\Delta y = f'(x_0)\Delta x + \varepsilon \Delta x \text{ where } \varepsilon \to 0 \text{ as } \Delta x \to 0.$$

Generalizing the above to accommodate a function of two variables $z = f(x, y)$ brings us to:

DEFINITION 13.2
DIFFERENTIABLE

A function $z = f(x, y)$ is **differentiable** at (x_0, y_0) if $\Delta z = f(x_0 + \Delta x, y_0 + \Delta y) - f(x_0, y_0)$ can be expressed in the form:

$$\Delta z = f_x(x_0, y_0)\Delta x + f_y(x_0, y_0)\Delta y + \varepsilon_1 \Delta x + \varepsilon_2 \Delta y$$

where $\varepsilon_1, \varepsilon_2 \to 0$ as $\Delta x, \Delta y \to 0$.

A function f is said to be **differentiable** if it is differentiable at each point in its domain.

Built into the above definition is the requirement that both partial derivatives of f need to exist for f to be differentiable at a given point. That, as it turns out, is not quite enough. However:

THEOREM 13.2 If the partial derivatives f_x and f_y exist and are **continuous** in an open region about (x_0, y_0), then f is differentiable at (x_0, y_0).

PROOF: Appendix B, page B-2.

EXAMPLE 13.3 Verify that $z = f(x, y) = y^2 e^{3x}$ is a differentiable function.

SOLUTION: Since both of the partial derivatives: $f_x = 3y^2 e^{3x}$ and $f_y = 2y e^{3x}$ are defined and continuous everywhere, $f(x, y) = y^2 e^{3x}$ is differentiable.

CHECK YOUR UNDERSTANDING 13.4

Answer: See page A-24.

Verify that $f(x, y) = x^2 \sin y$ is a differentiable function.

As it is with functions of one variable:

THEOREM 13.3 If $z = f(x, y)$ is differentiable at (x_0, y_0), then f is continuous at (x_0, y_0).

In the exercises you are invited to show that though both $f_x(0, 0)$ and $f_y(0, 0)$ exist for the function:

$$f(x, y) = \begin{cases} \dfrac{xy}{x^2 + y^2} & \text{if } (x, y) \neq (0, 0) \\ 0 & \text{if } (x, y) = (0, 0) \end{cases}$$

the function fails to be continuous at $(0, 0)$, and is therefore not differentiable at that point.

PROOF: We show that $\displaystyle\lim_{(x, y) \to (x_0, y_0)} [f(x, y) - f(x_0, y_0)] = 0$:

From Definition 13.2):
$$\Delta z = f_x(x_0, y_0)\Delta x + f_y(x_0, y_0)\Delta y + \varepsilon_1 \Delta x + \varepsilon_2 \Delta y \text{ where:}$$

$$\Delta z = f(x_0 + \Delta x, y_0 + \Delta y) - f(x_0, y_0) \text{ and } \varepsilon_1, \varepsilon_2 \to 0 \text{ as } \Delta x, \Delta y \to 0.$$

Letting $\Delta x = x - x_0$ and $\Delta y = y - y_0$ we have:
$$f(x, y) - f(x_0, y_0) = f(x_0 + \Delta x, y_0 + \Delta y) - f(x_0, y_0) = \Delta z.$$

So:
$$\lim_{(x, y) \to (x_0, y_0)} [f(x, y) - f(x_0, y_0)] = \lim_{(x, y) \to (x_0, y_0)} \Delta z$$

$$= \lim_{(x, y) \to (x_0, y_0)} f_x(x_0, y_0)\Delta x + f_y(x_0, y_0)\Delta y + \varepsilon_1 \Delta x + \varepsilon_2 \Delta y = 0$$

If $y = f(x)$ and $z = g(y)$ are differentiable functions, then the composite function $z = g[f(x)]$ is also differentiable, and:
$$\frac{dz}{dx} = \frac{dz}{dy}\frac{dy}{dx}$$

CHAIN RULE

Here are some partial derivative variations of the single-variable chain rule of page 94 (see margin):

THEOREM 13.4   (a) If $z = f(x, y)$ is differentiable, and if $x = g(t)$ and $y = h(t)$ are differentiable functions, then the single variable function $z = f[g(t), h(t)]$ is differentiable, and:

$$\frac{dz}{dt} = \frac{\partial z}{\partial x}\frac{dx}{dt} + \frac{\partial z}{\partial y}\frac{dy}{dt}$$

(b) If $z = f(x, y)$ is differentiable, and if $x = g(u, v)$ and $y = h(u, y)$ are differentiable functions, then the two variable function $z = f[g(u, v), h(u, v)]$ is differentiable, and:

$$\frac{\partial z}{\partial u} = \frac{\partial z}{\partial x}\frac{\partial x}{\partial u} + \frac{\partial z}{\partial y}\frac{\partial y}{\partial u} \qquad \frac{\partial z}{\partial v} = \frac{\partial z}{\partial x}\frac{\partial x}{\partial v} + \frac{\partial z}{\partial y}\frac{\partial y}{\partial v}$$

PROOF: (a)

$$\frac{dz}{dt} = \lim_{\Delta t \to 0}\frac{\Delta z}{\Delta t} = \lim_{\Delta t \to 0}\frac{\dfrac{\partial z}{\partial x}\Delta x + \dfrac{\partial z}{\partial y}\Delta y + \varepsilon_1 \Delta x + \varepsilon_2 \Delta y}{\Delta t}$$

$$= \frac{\partial z}{\partial x}\lim_{\Delta t \to 0}\frac{\Delta x}{\Delta t} + \frac{\partial z}{\partial y}\lim_{\Delta t \to 0}\frac{\Delta y}{\Delta t} + \lim_{\Delta t \to 0}\varepsilon_1 \lim_{\Delta t \to 0}\frac{\Delta x}{\Delta t} + \lim_{\Delta t \to 0}\varepsilon_2 \lim_{\Delta t \to 0}\frac{\Delta y}{\Delta t}$$

$$= \frac{\partial z}{\partial x}\frac{dx}{dt} + \frac{\partial z}{\partial y}\frac{dy}{dt} \quad \leftarrow \text{as } \Delta t \to 0: \varepsilon_1 \to 0 \text{ and } \varepsilon_2 \to 0$$

(b) If v is held fixed, then $x = g(u, v)$ and $y = h(u, y)$ become functions of u alone. Applying (a) with u in place of t, and if we use δ instead if d to indicate that the variable v is fixed, we obtain:

$$\frac{\partial z}{\partial u} = \frac{\partial z}{\partial x}\frac{\partial x}{\partial u} + \frac{\partial z}{\partial y}\frac{\partial y}{\partial u}$$

A similar argument can be used to establish the formula for $\dfrac{\partial z}{\partial v}$:

EXAMPLE 13.4   (a) Determine $\dfrac{dz}{dt}$ if:

$$z = xy^2, \quad x = t^2, \quad y = \sqrt{t}$$

(b) Determine $\dfrac{\partial z}{\partial s}$ if:

$$z = xy^2, x = s + t^2, y = s\cos t$$

SOLUTION: (a) Using the chain rule:

$$\frac{dz}{dt} = \frac{\partial z}{\partial x}\frac{dx}{dt} + \frac{\partial z}{\partial y}\frac{dy}{dt} = y^2(2t) + 2yx\left(\frac{1}{2t^{1/2}}\right)$$

$$= (\sqrt{t})^2(2t) + 2\sqrt{t}(t^2)\left(\frac{1}{2t^{1/2}}\right)$$

$$= 2t^2 + t^2 = 3t^2$$

Alternatively, we can express z explicitly as a function of t:

$$z = xy^2 = t^2(\sqrt{t})^2 = t^3$$

and then differentiate: $\dfrac{dz}{dt} = \dfrac{d}{dt}t^3 = 3t^2$.

(b) Let $z = xy^2$, $x = s + t^2$, $y = s\cos t$. Using the chain rule:

$$\frac{\partial z}{\partial s} = \frac{\partial z}{\partial x}\frac{\partial x}{\partial s} + \frac{\partial z}{\partial y}\frac{\partial y}{\partial s} = y^2(1) + 2xy(\cos t)$$

$$= s^2\cos^2 t + 2(s + t^2)(s\cos t)(\cos t)$$

$$= 3s^2\cos^2 t + 2st^2\cos^2 t$$

Without the chain rule:

$$z = xy^2 = (s + t^2)(s^2\cos^2 t) = s^3\cos^2 t + t^2 s^2\cos^2 t$$

And again we have:

$$\frac{\partial z}{\partial s} = \frac{\partial}{\partial s}(s^3\cos^2 t + t^2 s^2\cos^2 t) = 3s^2\cos^2 t + 2st^2\cos^2 t$$

CHECK YOUR UNDERSTANDING 13.5

(a) Let: $z = (2x + e^y)^4$, $x = \sin t$, $y = t^2$. Use the chain rule to find $\dfrac{dz}{dt}$ and check your result by expressing z explicitly as a function of t and differentiating directly.

(b) Let: $z = e^{xy}$, $x = s + 3t$, $y = st^2$. Use the chain rule to find $\dfrac{\partial z}{\partial t}$ and check your result by expressing z explicitly as a function of s and t and applying $\dfrac{\partial}{\partial t}$ to that expression.

Answers:

(a) $8(2\sin t + e^{t^2})^3(\cos t + te^{t^2})$

(b) $ste^{(s^2t^2 + 3st^3)}(9t + 2s)$

We conclude this section by noting that both the notion and the notation of partial derivatives extend to higher order partial derivatives, and can also include functions involving more than two variables. That being the case (your turn):

CHECK YOUR UNDERSTANDING 13.6

(a) For $z = f(x, y) = e^{xy} + x\ln y$, find: (i) $\dfrac{\partial^3 z}{\partial x^3}$ and (ii) z_{xyx}

(b) For $w = f(x, y, z) = xy^2 + z^3 y - xyz$, find:

(i) $\dfrac{\partial w}{\partial x}$ (ii) $\dfrac{\partial^2 w}{\partial y^2}$ (iii) w_{xyz}

(c) Show that $w = f(x, y, z) = zy^2 e^{3x}$ is a differentiable function.

(d) Use the chain rule to determine $\dfrac{dw}{dt}$ if:

$$w = xy^2 z^3, \quad x = t^2, \quad y = \sin t, \quad z = e^t$$

(e) Use the chain rule to determine $\dfrac{\partial w}{\partial t}$ if:

$$w = xe^{yz}, \quad x = rst^2, \quad y = \sin t, \quad z = \cos r^2$$

Answers: (a-i) $y^3 e^{xy}$
(a-ii) $ye^{xy}(xy + 2)$
(b-i) $y^2 - yz$ (b-ii) $2x$
(b-iii) -1
(c) See page A-25
(d) $te^{3t}(\sin t)(2\sin t + 2t\cos t + 3t\sin t)$
(e) $rste^{\sin t\cos r^2}(2 + t\cos r^2\cos t)$

	EXERCISES	

Exercises 1-12. Find $\dfrac{\partial z}{\partial x}$ and $\dfrac{\partial z}{\partial y}$.

1. $z = x^2 + y^2$ 2. $z = 3x^2 - xy + y$ 3. $z = (xy - 1)^2$

4. $z = \sqrt{x^3 + y^2}$ 5. $z = 4e^{x^2}y^3$ 6. $z = \dfrac{1}{x+y}$

7. $z = \dfrac{x+y}{xy-1}$ 8. $z = \sin^2(x - 3y)$ 9. $z = e^{x\sin y}$

10. $z = \ln(x^2 + y^2)$ 11. $z = e^{x\cos y}\ln x$ 12. $z = \sin x \cos y$

Exercises 13-21. Find f_{xx}, f_{yy}, f_{xy}, and f_{yx}, and verify that $f_{xy} = f_{yx}$.

13. $f(x,y) = 6x^2 - 8xy + 9y^2$ 14. $f(x,y) = \ln\dfrac{x}{y}$ 15. $f(x,y) = e^{2x+3y}$

16. $f(x,y) = \sin^2(x^2 - y)$ 17. $f(x,y) = \sqrt{x^2 + y^2}$ 18. $f(x,y) = \sqrt{\dfrac{x}{x+y}}$

19. $f(x,y) = \dfrac{\sin x}{\cos y}$ 20. $f(x,y) = \ln(ye^x)$ 21. $f(x,y) = \ln(xy - y^2)$

Exercises 22-27. Find f_{xz}, f_{zx}, f_{yyy}, f_{zzy}, f_{zyz}, f_{xyz}, and f_{zyx}.

22. $f(x,y,z) = xyz$ 23. $f(x,y,z) = x^3y^2z$ 24. $f(x,y,z) = xy^2\sin z$

25. $f(x,y,z) = e^{xyz}$ 26. $f(x,y,z) = \ln(x+y+z)$ 27. $f(x,y,z) = z^2 + y\cos x$

Exercise 28-36. Verify that the given function is differentiable.

28. $f(x,y) = x^3y$ 29. $f(x,y) = x^2y - 9$ 30. $f(x,y) = 3x^2 + y^2$

31. $f(x,y) = e^{2x+3y}$ 32. $f(x,y) = \dfrac{\sin x}{\cos y}$ 33. $f(x,y) = \ln(xy - y^2)$

34. $f(x,y) = \dfrac{y^2}{x}$ 35. $f(x,y,z) = x^3y^2z^4$ 36. $f(x,y,z) = e^{2z}\cos xy$

Exercises 37-39. (Implicit Partial Differentiation) Find $\dfrac{\partial z}{\partial x}$ and $\dfrac{\partial z}{\partial y}$.

Note: $\dfrac{\partial}{\partial x}(xy^2 + \boxed{x^2 z^3}) = y^2 + \boxed{x^2 \left[3z^2 \dfrac{\partial z}{\partial x}\right] + 2xz^3}$

37. $x^2 z + 2y^2 z^2 + y = 0$ 38. $e^{xyz} - z^2 = xy$ 39. $x^3 + y^3 + z^3 + 6xyz = 0$

Exercises 40-42. Find a function $f(x, y)$ such that:

40. $f_x = x$ and $f_y = y$ 41. $f_x = y^2$ and $f_y = 2xy$ 42. $f_x = e^x \sin y$ and $f_y = e^x \cos y$

Exercises 43-44. (Cauchy-Riemann Equations) Show that the functions $f(x, y)$, $g(x, y)$ satisfies the following Cauchy-Riemann equations: $f_x = g_y$ and $f_y = -g_x$:

43. $f(x, y) = x^2 - y^2$, $g(x, y) = 2xy$ 44. $f(x, y) = e^x \cos y$, $g(x, y) = e^x \sin y$

45. Let $z = \tan^{-1}\dfrac{x}{y}$. Show that $z_{xx} + z_{yy} = 0$.

46. Let $f(x, y) = e^{cy} \cos cy$. Show that $f_{xx} + f_{yy} = 0$.

47. Show that if $f(x, y) = g(x) + h(y)$, then $f_{xy} = 0$.

48. Show that if $f(x, y, z) = g(x, y) + h(y, z) + k(z, x)$, then $f_{xyz} = 0$.

49. Show that the function $z = f(x, y) = x^4 + 2x^2 y^2 + y^4$ satisfies the partial differential equation $x\dfrac{\partial z}{\partial x} + y\dfrac{\partial z}{\partial y} = 4z$.

50. Show that the function $z = f(x, y) = \dfrac{x^2 y^2}{x + y}$ satisfies the partial differential equation

$x\dfrac{\partial z}{\partial x} + y\dfrac{\partial z}{\partial y} = 3z$

51. Show that the function $w = f(x, y, z) = x^2 y + y^2 z + z^2 x$ satisfies the partial differential equation $\dfrac{\partial w}{\partial x} + \dfrac{\partial w}{\partial y} + \dfrac{\partial w}{\partial z} = (x + y + z)^2$.

52. Show that the function $w = f(x, y, z) = \dfrac{xz + y^2}{yz}$ satisfies the partial differential equation

$x\dfrac{\partial w}{\partial x} + y\dfrac{\partial w}{\partial y} + z\dfrac{\partial w}{\partial z} = 0$.

53. Show that the function $z = \cos(x + y) + \cos(x - y)$ satisfies the second order partial differential equation $z_{xx} - z_{yy} = 0$.

54. Show that the function $z = \sin(x + y) + \ln(x - y)$ satisfies the second order partial differential equation: $z_{xx} - z_{yy} = 0$.

55. The pressure P of a gas confined in a container of volume V and temperature T is related by an equation of the form $PV = kT$, where k is a positive constant. Verify that:

$$V\frac{\partial P}{\partial V} = -P, \quad V\frac{\partial P}{\partial V} + T\frac{\partial P}{\partial T} = 0, \quad \text{and} \quad \left(\frac{\partial V}{\partial T}\right)\left(\frac{\partial T}{\partial P}\right)\left(\frac{\partial P}{\partial V}\right) = -1$$

56. The kinetic energy of a body of mass m moving at a velocity v is given by $K = \frac{1}{2}mv^2$. Verify that $\frac{\partial K}{\partial m} \cdot \frac{\partial^2 K}{\partial v^2} = K$.

57. When two resisters of resistance R_1 and R_2 ohms are connected in parallel, their combined resistance R in ohms is given by $R = \frac{R_1 R_2}{R_1 + R_2}$. Verify that $\frac{\partial^2 R}{\partial R_1^2} \cdot \frac{\partial^2 R}{\partial R_2^2} = \frac{4R^2}{(R_1 + R_2)^4}$.

58. Find the slope of the tangent line to the curve of intersection of the curve $z = x^2 + 4y^2$ and the plane $x = -1$ if $y = 1$.

59. Find the slope of the tangent line to the curve of intersection of the sphere $x^2 + y^2 + z^2 = 9$ and the plane $x = 2$ at $(2, 1, 2)$.

60. Let $f(x, y) = \begin{cases} \dfrac{x^3 y - xy^3}{x^2 + y^2} & \text{if } (x, y) \neq (0, 0) \\ 0 & \text{if } (x, y) = (0, 0) \end{cases}$. Verify that $f_{xy}(0, 0) = -1$ and $f_{yx}(0, 0) = 1$.

Exercises 61-62. Find $\frac{dz}{dt}$, both with and without using the chain rule.

61. $z = xy$, $x = \cos t$, $y = \sin t$

62. $z = x^3 y + e^{xy}$, $x = t^2$, $y = \sin x$

Exercises 63-64. Find $\frac{dw}{dt}$, both with and without using the chain rule.

63. $w = xy + z$, $x = \cos t$, $y = \sin t$, $z = t$

64. $w = z\sin(xy^2)$, $x = t^2$, $y = e^t$, $z = xy$

Exercises 65-66. Find $\frac{\partial z}{\partial t}$ and $\frac{\partial z}{\partial s}$, both with and without using the chain rule.

65. $z = x\sin y$, $x = s + 3t$, $y = t^2$

66. $z = xe^{xy}$, $x = \sin st$, $y = st^2$

Exercises 67-68. Find $\frac{\partial w}{\partial t}$ and $\frac{\partial w}{\partial s}$, both with and without using the chain rule.

67. $w = x + 2y - 3z$, $x = st$, $y = e^{st}$, $z = \sin t$

68. $w = xe^{yz}$, $x = st$, $y = e^{st}$, $z = t$

§2. DIRECTIONAL DERIVATIVES, GRADIENT VECTORS, AND TANGENT PLANES

The partial derivatives, $f_x(x, y)$ and $f_y(x, y)$, represent the rates of change of the function $z = f(x, y)$ in directions parallel to the x- and y-axes, respectively [see Definition 13.1 and Figure 13.1 (page 549)].

Here is a natural extension of that definition:

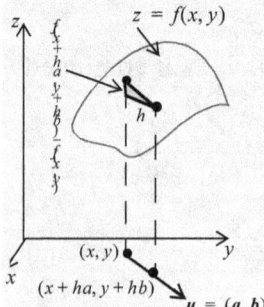

DEFINITION 13.3
DIRECTIONAL DERIVATIVE

The **directional derivative** of f in the direction of the <u>unit</u> vector $\boldsymbol{u} = \langle a, b \rangle = a\boldsymbol{i} + b\boldsymbol{j}$ is given by:

$$D_{\boldsymbol{u}}f(x, y) = \lim_{h \to 0} \frac{f(x + ha, y + hb) - f(x, y)}{h}$$

(providing the limit exists)

Observe that if $\boldsymbol{u}$ is the unit vector $\boldsymbol{i} = \langle 1, 0 \rangle$, then $D_{\boldsymbol{u}}f = D_{\boldsymbol{i}}f = f_x$. Similarly: $D_{\boldsymbol{j}}f = f_y$.

EXAMPLE 13.5

Find the directional derivative of $f(x, y) = x^2 + y^2$ in the direction $\boldsymbol{u}$ of the unit vector making an angle $30°$ with the positive x-axis. Calculate $D_{\boldsymbol{u}}f(\sqrt{3}, 5)$, $D_{\boldsymbol{u}}f(\sqrt{3}, -5)$, and indicate what those numbers represent.

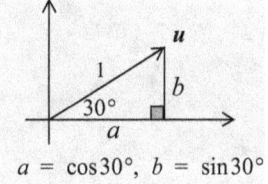

$a = \cos 30°, \ b = \sin 30°$

SOLUTION: Since $\boldsymbol{u} = \cos 30° \boldsymbol{i} + \sin 30° \boldsymbol{j}$ (see margin):

$$D_{\boldsymbol{u}}f(x, y) = \lim_{h \to 0} \frac{f(x + h\cos 30°, y + h\sin 30°) - f(x, y)}{h}$$

$$= \lim_{h \to 0} \frac{f\left(x + \frac{\sqrt{3}}{2}h, y + \frac{1}{2}h\right) - (x^2 + y^2)}{h}$$

$$= \lim_{h \to 0} \frac{\left(x + \frac{\sqrt{3}}{2}h\right)^2 + \left(y + \frac{1}{2}h\right)^2 - (x^2 + y^2)}{h}$$

$$= \lim_{h \to 0} \frac{x^2 + \sqrt{3}hx + \frac{3h^2}{4} + y^2 + hy + \frac{h^2}{4} - x^2 - y^2}{h}$$

$$= \lim_{h \to 0} (\sqrt{3}x + y + h) = \sqrt{3}x + y$$

In particular:

$$D_{\boldsymbol{u}}f(\sqrt{3}, 5) = \sqrt{3}\sqrt{3} + 5 = 8 \text{ and } D_{\boldsymbol{u}}f(\sqrt{3}, -5) = \sqrt{3}\sqrt{3} - 5 = -2$$

Interpretation: If you move a bit away from the point $(\sqrt{3}, 5)$, in the direction of $u = \frac{\sqrt{3}}{2}i + \frac{1}{2}j$, then the function value will increase by approximately 8 times that bit. On the other hand, if you move a bit away from the point $(\sqrt{3}, -5)$, in the direction of $u = \frac{\sqrt{3}}{2}i + \frac{1}{2}j$, then the function value will decrease by approximately 2 times that bit.

One seldom resorts to Definition 13.3 directly to compute the directional derivative of a function. Here is the preferred choice:

THEOREM 13.5 If $z = f(x, y)$ is differentiable, then for any **unit** vector $u = \langle a, b \rangle = ai + bj$:
$$D_u f(x, y) = af_x(x, y) + bf_y(x, y)$$

PROOF: Appendix B, page B-3.

EXAMPLE 13.6 (a) Use Theorem 13.5 to solve Example 13.5.
(b) Find the directional derivative of $f(x, y) = x^2y + 7xy$ at the point $(2, -3)$ in the direction of the vector $3i - 4j$.

SOLUTION: (a) Applying Theorem 13.5 with $f(x, y) = x^2 + y^2$ and $u = \cos 30°i + \sin 30°j = \frac{\sqrt{3}}{2}i + \frac{1}{2}j$, we have:
$$D_u f(x, y) = \frac{\sqrt{3}}{2}f_x(x, y) + \frac{1}{2}f_y(x, y) = \frac{\sqrt{3}}{2}2x + \frac{1}{2}2y = \sqrt{3}x + y$$

(b) Since $\|3i - 4j\| = 5$, the unit vector in the direction of $3i - 4j$ is $u = \frac{1}{5}(3i - 4j) = \frac{3}{5}i - \frac{4}{5}j$. Appealing to Theorem 13.5, we then have:
$$D_u f(x, y) = \frac{3}{5}f_x(x, y) - \frac{4}{5}f_y(x, y) = \frac{3}{5}(2xy + 7y) - \frac{4}{5}(x^2 + 7x)$$

In particular:
$$D_u f(2, -3) = \frac{3}{5}[2(2)(-3) + 7(-3)] - \frac{4}{5}[2^2 + 7(2)] = -\frac{171}{5}$$

Interpretation: If you move a bit away from the point $(x_0, y_0) = (2, -3)$, in the direction of $3i - 4j$, then the function value will drop by approximately $171/5$ times that bit.

Answer: $\frac{\sqrt{3}}{2}$

CHECK YOUR UNDERSTANDING 13.7

Find the directional derivative of $f(x, y) = x\sin xy$ at $\left(1, \frac{\pi}{2}\right)$ in the direction of the unit vector u that makes an angle of $\frac{\pi}{6}$ with the positive x-axis.

∇f is read: "delta f" or "del f." It is worth repeating that ∇f is a vector-valued function.

DEFINITION 13.4
GRADIENT

The **gradient** of a differentiable function $z = f(x, y)$, denoted by $\nabla f(x, y)$, or ∇f, is the vector function:
$$\nabla f(x, y) = f_x(x, y)i + f_y(x, y)j$$

Returning to Theorem 13.5 we see that:

The directional derivative of a differentiable function $f(x, y)$ in the direction of the unit vector $u = ai + bj$ is given by:
$$D_u f(x, y) = u \cdot \nabla f(x, y)$$

CHECK YOUR UNDERSTANDING 13.8

Use the above vector equation to find the directional derivative of $f(x, y) = e^{x+y^2}$ at $(0, 1)$ in the direction of the vector $-i + j$.

The gradient vector is not merely a convenient notational device for representing the directional derivative of a function. A case in point:

In particular:
The value of $z = f(x, y)$ at (x_0, y_0) increases most rapidly in the direction of $\nabla f(x_0, y_0)$, and the value of $z = f(x, y)$ at (x_0, y_0) decreases most rapidly in the direction opposite that of $\nabla f(x_0, y_0)$.

THEOREM 13.6 If $z = f(x, y)$ is differentiable at (x_0, y_0) with $\nabla f(x_0, y_0) \neq 0$, then $D_u f(x_0, y_0)$ assumes its maximal value of $\|\nabla f(x_0, y_0)\|$ when the **unit** vector u is in the direction of the gradient vector $\nabla f(x_0, y_0)$; and $D_u f(x_0, y_0)$ assumes its minimum value of $-\|\nabla f(x_0, y_0)\|$ when u is in the direction of $-\nabla f(x_0, y_0)$.

PROOF: Let θ be the angle between $\nabla f(x_0, y_0)$ and an arbitrary **unit** vector u in the x-y plane. Then:

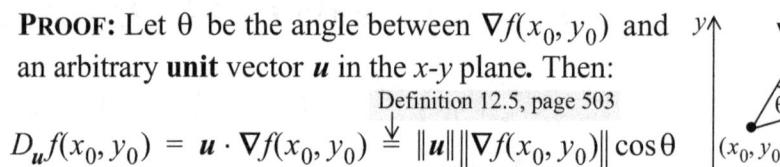

Definition 12.5, page 503
$$D_u f(x_0, y_0) = u \cdot \nabla f(x_0, y_0) = \|u\|\|\nabla f(x_0, y_0)\|\cos\theta$$
$$= \|\nabla f(x_0, y_0)\|\cos\theta$$

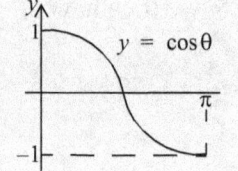

It follows that the maximum value of $D_u f(x_0, y_0)$ is $\|\nabla f(x_0, y_0)\|$, and that it occurs when $\cos\theta = 1$, or at $\theta = 0$ (see margin); which is to say: when u has the same direction as $\nabla f(x_0, y_0)$.

It also follows that the minimum value of $D_u f(x_0, y_0)$ is $-\|\nabla f(x_0, y_0)\|$, and that it occurs when $\theta = \pi$; which is to say, when u is in the direction opposite to $\nabla f(x_0, y_0)$.

EXAMPLE 13.7 Find the direction of greatest rate of increase and greatest rate of decrease for the function $f(x, y) = e^{xy}$ at the point $(1, 2)$.

SOLUTION: Turning to the gradient function:
$$\nabla f(x, y) = f_x(x, y)i + f_y(x, y)j = ye^{xy}i + xe^{xy}j$$
we conclude that the function values increase most rapidly in the direction of $\nabla f(1, 2) = f_x(1, 2)i + f_y(1, 2)j = 2e^2 i + e^2 j$ and decrease most rapidly in the direction of
$$-\nabla f(1, 2) = -(f_x(1, 2)i + f_y(1, 2)j) = -2e^2 i - e^2 j$$

CHECK YOUR UNDERSTANDING 13.9

Find the greatest value and smallest value of the directional derivative for the function $f(x, y) = x \sin y$ at the point $(2, 0)$.

FUNCTIONS OF THREE VARIABLES

The previous discussion extends to differentiable functions of three (or more) variables. In particular:

For $w = f(x, y, z)$ and unit vector $\boldsymbol{u} = a\boldsymbol{i} + b\boldsymbol{j} + c\boldsymbol{k}$:

$$D_{\boldsymbol{u}}f(x, y, z) = af_x(x, y, z) + bf_y(x, y, z) + cf_z(x, y, z) = \boldsymbol{u} \cdot \nabla f(x, y, z)$$

where: $\nabla f(x, y, z) = f_x(x, y, z)\boldsymbol{i} + f_y(x, y, z)\boldsymbol{j} + f_z(x, y, z)\boldsymbol{k}$

EXAMPLE 13.8 (a) Determine the directional derivative of $f(x, y, z) = xe^{yz}$ at $(1, 0, 1)$ in the direction $2\boldsymbol{i} + \boldsymbol{j} - \boldsymbol{k}$.

(b) Find the greatest value and smallest value of the directional derivative for $f(x, y, z) = xe^{yz}$ at $(1, 0, 1)$.

SOLUTION: (a) From:

$$f_x(x, y, z) = e^{yz}, \ f_y(x, y, z) = xze^{yz}, \ f_z(x, y, z) = xye^{yz}$$

we have: $\nabla f(x, y, z) = e^{yz}\boldsymbol{i} + xze^{yz}\boldsymbol{j} + xye^{yz}\boldsymbol{k}$

In particular: $\nabla f(1, 0, 1) = \boldsymbol{i} + \boldsymbol{j} + 0\boldsymbol{k}$

Since the unit vector in the direction of $2\boldsymbol{i} + \boldsymbol{j} - \boldsymbol{k}$ is $\boldsymbol{u} = \dfrac{1}{\sqrt{6}}(2\boldsymbol{i} + \boldsymbol{j} - \boldsymbol{k})$:

$$D_{\boldsymbol{u}}f(1, 0, 1) = \frac{1}{\sqrt{6}}(2\boldsymbol{i} + \boldsymbol{j} - \boldsymbol{k}) \cdot \nabla f(1, 0, 1)$$

$$= \left(\frac{2}{\sqrt{6}}\boldsymbol{i} + \frac{1}{\sqrt{6}}\boldsymbol{j} - \frac{1}{\sqrt{5}}\boldsymbol{k}\right) \cdot (\boldsymbol{i} + \boldsymbol{j} + 0\boldsymbol{k}) = \frac{2}{\sqrt{6}} + \frac{1}{\sqrt{6}} = \frac{3}{\sqrt{6}}$$

(b) By Theorem 13.6, the greatest value is:
$$\|\nabla f(1, 0, 1)\| = \|\boldsymbol{i} + \boldsymbol{j} + 0\boldsymbol{k}\| = \sqrt{2}$$
and the smallest value is
$$-\|\nabla f(1, 0, 1)\| = -\sqrt{2}$$

CHECK YOUR UNDERSTANDING 13.10

Find the greatest value and smallest value of the directional derivative for the function $f(x, y, z) = \ln(x^2 + y^2 + z^2)$ at $(1, 1, 2)$.

TANGENT PLANE

Just as the line in Figure 13.2(a) represents the tangent line to the graph of the function $y = f(x)$ at (x_0, y_0), so then we can agree that the plane in Figure 13.2(b) represents the tangent plane to the surface $z = f(x, y)$ at (x_0, y_0, z_0). Agreeing is all well and good, but what we really need is a formal definition.

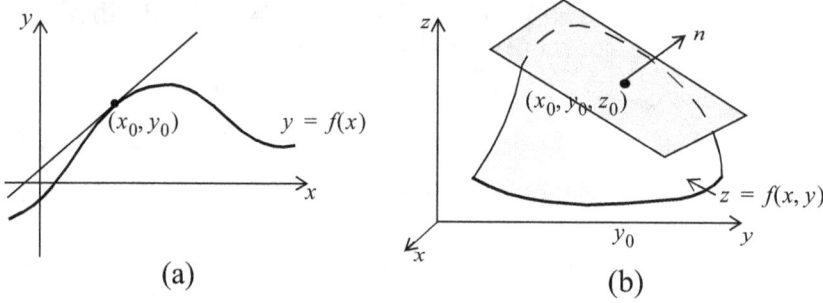

(a)   (b)

Figure 13.2

The task at hand boils down to that of finding a normal to the plane at the point $P_0 = (x_0, y_0, z_0)$ (reminiscent of finding the slope of the tangent line to a curve on page 71). Our intuition tells us that a normal n to the plane should be perpendicular to the tangent line T at P_0 on any curve obtained by intersecting the surface with a vertical plane through P_0 (see margin).

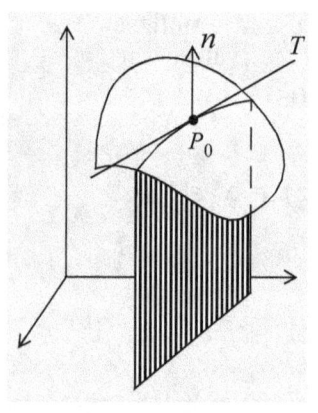

Lets begin by finding a tangent vector v, at the point P_0 on the curve C_{y_0} obtained by intersecting the surface $z = f(x, y)$ with the vertical plane $y = y_0$ (see adjacent figure). Turning to the line L in the xy-plane with parametric represesentation

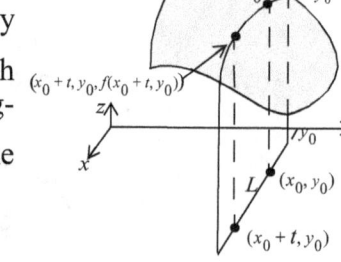

$$x = x_0 + t, y = y_0$$

we obtain the following parametric representation for C_{y_0}

$$x = x_0 + t, \quad y = y_0, \quad z = f(x_0 + t, y_0)$$

with position vector: $r(t) = (x_0 + t)i + y_0 j + f(x_0 + t, y_0)k$. Evaluating

$$r'(t) = (x_0 + t)'i + (y_0)'j + f'(x_0 + t, y_0)k = i + 0j + f'(x_0 + t, y_0)k$$

at $t = 0$, we obtain a vector tangent to the curve C_{y_0} at P_0:

$$v = i + 0j + f'(x_0, y_0)k$$

In a similar fashion one can show that the vector

$$u = 0i + j + f_y(x_0, y_0)k$$

is tangent at P_0 to the curve C_{x_0} obtained by intersecting the surface $z = f(x, y)$ with the plane $x = x_0$.

Taking the cross product of u and v we arrive at a vector perpendicular to both u and v:

$$n = u \times v = \det \begin{bmatrix} i & j & k \\ 0 & 1 & f_y(x_0, y_0) \\ 1 & 0 & f_x(x_0, y_0) \end{bmatrix} = f_x(x_0, y_0)i + f_y(x_0, y_0)j - k$$

In the exercises you are invited to show that the above vector is, in fact, perpendicular to the tangent line T at P_0 on any curve obtained by intersecting the surface with a vertical plane through P_0. Bringing us to:

DEFINITION 13.5
TANGENT PLANE

Let $z = f(x, y)$ be differentiable at (x_0, y_0). The **tangent plane** to the graph of f at $(x_0, y_0, f(x_0, y_0))$ is the plane passing through that point with normal vector $f_x(x_0, y_0)i + f_y(x_0, y_0)j - k$.

In particular, here is the scalar form equation of the plane (see page 516):
$$f_x(x_0, y_0)(x - x_0) + f_y(x_0, y_0)(y - y_0) - (z - z_0) = 0$$

EXAMPLE 13.9 Find an equation of the tangent plane to the graph of $f(x, y) = xye^{x+y}$ at the point $(1, 2, 2e^3)$.

SOLUTION: From $f(x, y) = xye^{x+y}$ we have:

$$f_x(x, y) \underset{\underset{y \text{ is held fixed}}{\uparrow}}{=} y(xe^{x+y} + e^{x+y}) \quad \text{and} \quad f_y(x, y) \underset{\underset{x \text{ is held fixed}}{\uparrow}}{=} x(ye^{x+y} + e^{x+y}) \text{ In}$$

particular:
$$f_x(1, 2) = 2(e^3 + e^3) = 4e^3 \quad \text{and} \quad f_y(1, 2) = 1(2e^3 + e^3) = 3e^3$$

Normal to the plane (Definition 13.5):
$$n = 4e^3 i + 3e^3 j - k$$
Consequently (see page 516):
$$\langle 4e^3, 3e^3, -1 \rangle \cdot \langle x - 1, y - 2, z - 2e^3 \rangle = 0 \quad \text{vector form}$$
$$4e^3(x - 1) + 3e^3(y - 2) - 1(z - 2e^3) = 0 \quad \text{scalar form}$$
$$4e^3 x + 3e^3 y - z = 8e^3 \quad \text{general form}$$

CHECK YOUR UNDERSTANDING 13.11

Find the general form equation of the tangent plane to the graph of $f(x, y) = 3y^2 - 2x^2 + x$ at the point $(2, -1, -3)$.

Answer: $7x + 6y + z = 5$

LEVEL CURVES AND LEVEL SURFACES

Though not labeled as such, level curves previously played a role in section 11.2. See, for example, Figures 11.1 and 11.2, page 436.

Consider the adjacent surface $z = f(x, y)$, along with the curve C_k obtained by cutting that surface with a horizontal plane of height k. **The projection of that curve onto the xy-plane**, is said to be the **k-level curve** of the function f; specifically:

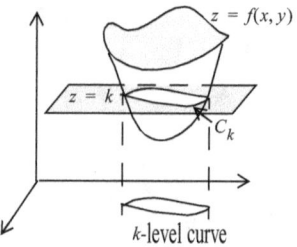

k-level curve

$$\{(x, y) | f(x, y) = k\}$$

Similarly, the **k-level surface** of $w = f(x, y, z)$ consists of all points (x, y, z) for which $f(x, y, z) = k$; specifically:

$$\{(x, y, z) | f(x, y, z) = k\}$$

Note that a level surface need not be level in the sense of being horizontal. In particular, the 4-level surface of

$$w = f(x, y, z) = x^2 + y^2 + z^2$$

is the sphere of radius 2 centered at the origin.

In CYU 13.12 you are invited to show that for $z = f(x, y)$:

$\nabla f(x_0, y_0)$ is normal to the k-level curve of f at (x_0, y_0).

Moving up a notch we have:

THEOREM 13.7   If $w = f(x, y, z)$ is a differentiable function, then $\nabla f(x_0, y_0, z_0)$ is perpendicular to the k-level surface of f at (x_0, y_0, z_0).

PROOF: Consider the k-level surface $S = \{(x, y, z) | f(x, y, z) = k\}$. We show that $\nabla f(x_0, y_0, z_0)$ is perpendicular to the tangent plane to S at $P_0 = (x_0, y_0, z_0)$ by showing that it is perpendicular to the tangent vector at P_0 (see Definition 12.9, page 527) of every curve C in S that passes through P_0:

Let C be such a curve, and let $r(t) = x(t)i + y(t)j + z(t)k$ be a parametrization of C with $r(t_0) = P_0$.

Differentiating both sides of $f(x(t), y(t), z(t)) = k$ with respect to t we have:

$$\frac{d}{dt} f(x(t), y(t), z(t)) = \frac{d}{dt}(k)$$

a constant

Chain Rule:
(Theorem 13.4(a), page 555)

$$\frac{\partial f}{\partial x}\frac{dx}{dt} + \frac{\partial f}{\partial y}\frac{dy}{dt} + \frac{\partial f}{\partial z}\frac{dz}{dt} = 0$$

$$\left(\frac{\partial f}{\partial x}i + \frac{\partial f}{\partial y}j + \frac{\partial f}{\partial z}k\right) \cdot \left(\frac{dx}{dt}i + \frac{dy}{dt}j + \frac{dz}{dt}k\right) = 0$$

In particular, $\nabla f(x_0, y_0, z_0) \cdot r'(t_0) = 0$ which establishes the fact that $\nabla f(x_0, y_0, z_0)$ is indeed perpendicular to the k-level surface of f at (x_0, y_0, z_0).

CHECK YOUR UNDERSTANDING 13.12

Answer: See page A-27.

Show that if $z = f(x, y)$ is differentiable, then $\nabla f(x_0, y_0)$ is perpendicular to the k-level curve of f at (x_0, y_0).

In the following example we show how Theorem 13.7 enables us to find the tangent plane to a 3-dimensional surface that is not the graph of a function $z = f(x, y)$. We also illustrate, in (b), how it can be used to determine the tangent plane to a surface that is of the form $z = f(x, y)$.

EXAMPLE 13.10   (a) Find an equation of the tangent plane to the sphere $x^2 + y^2 + (z-1)^2 = 1$ at the point $\left(\frac{1}{2}, \frac{1}{2}, 1 + \frac{1}{\sqrt{2}}\right)$.

See Example 13.9   (b) Find the normal to the tangent plane to the graph of $f(x, y) = xye^{x+y}$ at the point $(1, 2, 2e^3)$.

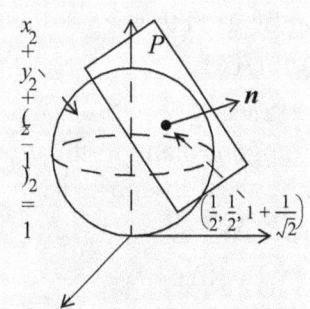

One can use Definition 13.5 to find a normal to the plane at the given point:
From the sphere's equation
$$x^2 + y^2 + (z-1)^2 = 1$$
we have:
$$z - 1 = \pm\sqrt{1 - x^2 - y^2}$$
In particular, the upper hemisphere is the graph of the **function**:
$$z = \sqrt{1 - x^2 - t^2} + 1$$
We leave it for you to verify that:
$$f_x\left(\frac{1}{2}, \frac{1}{2}\right)i + f_y\left(\frac{1}{2}, \frac{1}{2}\right)j - k$$
$$= \left(-\frac{\sqrt{2}}{2}\right)i + \left(-\frac{\sqrt{2}}{2}\right)j - k$$
which is parallel to:
$$\nabla F\left(\frac{1}{2}, \frac{1}{2}, 1 + \frac{1}{\sqrt{2}}\right)$$
$$= i + j + \sqrt{2}k$$

SOLUTION: (a) Since the sphere is not the graph of a function of two variables (see margin), we cannot proceed directly as in Example 13.9 to find the desired tangent plane. Taking a different approach, we consider the function of three variables:

$$F(x, y, z) = x^2 + y^2 + (z-1)^2$$

Note that the sphere in question is the 1-level surface of $F(x, y, z)$. That being the case (see Theorem 13.7):

$$\nabla F\left(\frac{1}{2}, \frac{1}{2}, 1 + \frac{1}{\sqrt{2}}\right) \text{ is normal to the tangent plane at } \left(\frac{1}{2}, \frac{1}{2}, 1 + \frac{1}{\sqrt{2}}\right)$$

Grinding away:

$$\nabla F(x, y, z) = F_x(x, y, z)i + F_y(x, y, z)j + F_z(x, y, z)k$$
$$= 2xi + 2yj + 2(z-1)k$$

we find that: $\nabla F\left(\frac{1}{2}, \frac{1}{2}, 1 + \frac{1}{\sqrt{2}}\right) = i + j + \sqrt{2}k$

Conclusion:

$$(i + j + \sqrt{2}k) \cdot \left[\left(x - \frac{1}{2}\right)i + \left(y - \frac{1}{2}\right)j + \left(x - \left(1 + \frac{1}{\sqrt{2}}\right)\right)k\right] = 0$$

is the tangent plane to the sphere $x^2 + y^2 + (z-1)^2 = 1$ at the point $\left(\frac{1}{2}, \frac{1}{2}, 1 + \frac{1}{\sqrt{2}}\right)$.

(b) We let the function $z = f(x, y) = xye^{x+y}$ direct us to the function: $F(x, y, z) = f(x, y) - z = xye^{x+y} - z$

Noting that the 0-level surface of F is $f(x, y) = xye^{x+y}$, we apply Theorem 13.7:

$$\nabla F(x, y, z) = F_x(x, y, z)\mathbf{i} + F_y(x, y, z)\mathbf{j} + F_z(x, y, z)\mathbf{k}$$

$$= y(xe^{x+y} + e^{x+y})\mathbf{i} + x(ye^{x+y} + e^{x+y})\mathbf{j} - 1\mathbf{k}$$

$$\nabla F(1, 2, 2e^3) = 2(e^3 + e^3)\mathbf{i} + (2e^3 + e^3)\mathbf{j} - \mathbf{k} = 4e^3\mathbf{i} + 3e^3\mathbf{j} - \mathbf{k}$$

We find, as we did in Example 13.9, that $\mathbf{n} = 4e^3\mathbf{i} + 3e^3\mathbf{j} - \mathbf{k}$ is the normal to the tangent plane at $(1, 2, 2e^3)$.

CHECK YOUR UNDERSTANDING 13.13

Answer: $x + 8y - z = 18$

Find the general equation of the tangent plane to the ellipsoid $x^2 + 4y^2 + z^2 = 18$ at the point $(1, 2, -1)$.

NUMERICAL APPROXIMATIONS USING TANGENT PLANES

Figure 13.3(a), previously appearing on page 82, displays how the tangent line to the graph of a function $y = f(x)$ can be used to approximate the change in function values $\Delta y = f(c + \Delta x) - f(c)$.

Specifically: $\Delta y \approx dy = f'(c)\Delta x$ (for Δx "small"_

Similarly, as is suggested in Figure 13.3(b), the tangent plane to the graph of a differentiable function $z = f(x, y)$ can be used to approximate changes in function values $\Delta z = f(x_0 + \Delta x, y_0 + \Delta y) - f(x_0, y_0)$.

Specifically: $\Delta z \approx dz = f_x(x_0, y_0)\Delta x + f_y(x_0, y_0)\Delta y$
(for $\Delta x, \Delta y$ small)

> Just as dy is called the **differential of** y, dz is called the **differential of** z. In addition, one often replaces Δx and Δy, with dx and dy, respectively; leading to the alternate form:
> $dz = f_x(x_0, y_0)dx + f_y(x_0, y_0)dy$

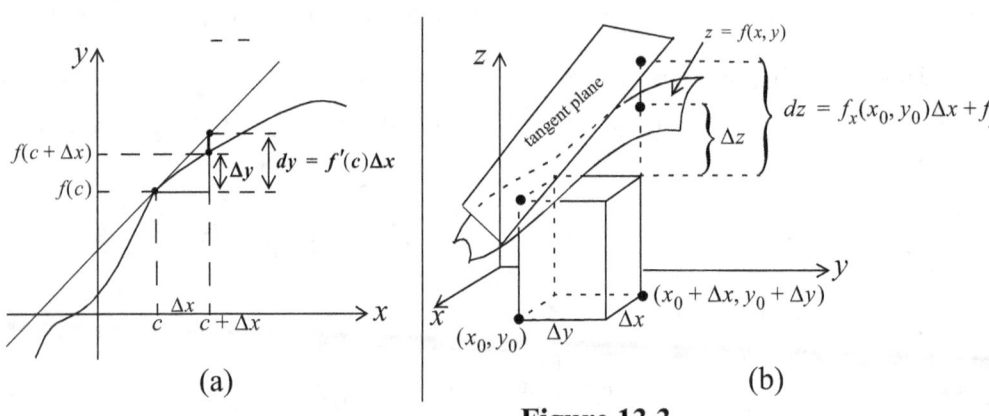

Figure 13.3

In support of $\Delta z \approx dz = f_x(x_0, y_0)\Delta x + f_y(x_0, y_0)\Delta y$:

Since f is differentiable (see Definition 13.2, page 553):

$$\Delta z = f(x_0 + \Delta x, y_0 + \Delta y) - f(x_0, y_0)$$

$$= f_x(x_0, y_0)\Delta x + f_y(x_0, y_0)\Delta y + \varepsilon_1\Delta x + \varepsilon_2\Delta y \quad (*)$$

Where $\varepsilon_1, \varepsilon_2 \to 0$ as $\Delta x, \Delta y \to 0$

The tangent plane to the graph of $z = f(x, y)$ passing through the point P is given by (see Definition 13.5):

$$f_x(x_0, y_0)\Delta x + f_y(x_0, y_0)\Delta y - (z - z_0) = 0$$

$$\text{Or: } z = z_0 + f_x(x_0, y_0)\Delta x + f_y(x_0, y_0)\Delta y$$

At that point, the tangent plane has height z_0, and when $x = x_0 + \Delta x$ and $y = y_0 + \Delta y$ it has height:

$$z = z_0 + f_x(x_0, y_0)\Delta x + f_y(x_0, y_0)\Delta y. \text{ So:}$$

$$z - z_0 = f(x_0 + \Delta x, y_0 + \Delta y) - f(x_0, y_0) = dz \text{ [see Figure 13.3(b)]}$$

Returning to (*) we then have:

$$\Delta z = dz + \varepsilon_1 \Delta x + \varepsilon_2 \Delta y$$

or that $dz \to \Delta z$ as Δx and Δy tend to 0.

Admittedly, nowadays there is really no need to approximate anything — just invoke a calculator. The main idea here, however, is to underline the fact that tangent planes can be used to approximate more complicated surfaces at points of interest, just as tangent lines can be used to approximate more complicated functions of a single variable.

EXAMPLE 13.11 Use a differential to approximate the change in the volume $V = \frac{1}{3}\pi r^2 h$ of a cone resulting from a change in radius from 10 cm to 10.01 cm, and a change in height from 20 cm to 19.97 cm.

SOLUTION: Turning to:

$$V(r, h) = \frac{1}{3}\pi r^2 h, \ (r, h) = (10, 20), \ \Delta r = 0.01 \text{ and } \Delta h = -0.03$$

we have:

$$\Delta V \approx dv = V_r(10, 20)(0.01) + V_h(10, 20)(-0.03)$$

From $V_r = \frac{\partial}{\partial r}\left(\frac{1}{3}\pi r^2 h\right) = \frac{\pi}{3}(2rh)$, $V_h = \frac{\partial}{\partial h}\left(\frac{1}{3}\pi r^2 h\right) = \frac{\pi}{3}r^2$:

$$V_r(10, 20) = \frac{\pi}{3}[2(10)(20)] = \frac{400\pi}{3}, \ V_h(10, 20) = \frac{\pi}{3}10^2 = \frac{100\pi}{3}$$

Thus: $\Delta V \approx dv = \frac{400\pi}{3}(0.01) + \frac{100\pi}{3}(-0.03) \approx 1.05 \text{ cm}^3$

CHECK YOUR UNDERSTANDING 13.14

Answer: -0.01

Let $z = f(x, y) = \sqrt{x^2 + y^2}$. Use a differential to approximate the change in z as (x, y) varies form $(3, 4)$ to $(3.01, 3.98)$.

	EXERCISES	

Exercises 1-2. Appeal directly to Definition 13.3 to verify that:

1. If $f(x, y) = x^2 + xy$ and if u is the unit vector making an angle $45°$ with the x-axis, then

$$D_u f(x, y) = \frac{3x + y}{\sqrt{2}}.$$

2. If $f(x, y) = y^2 + xy$ and if u is the unit vector making an angle $60°$ with the x-axis, then

$$D_u f(x, y) = \frac{\sqrt{3}x + (1 + 2\sqrt{3})y}{2}.$$

Exercises 3-25. Find the directional derivative of the given function at the indicated point and direction.

3. $f(x, y) = xy^2 + x^2$ at $(-1, 2)$ in the direction of $\theta = \frac{\pi}{3}$.

4. $f(x, y) = 2xy - y^2$ at $(-1, 3)$ in the direction of $\theta = \frac{2\pi}{3}$.

5. $f(x, y) = xe^y + ye^x$ at $(0, 0)$ in the direction of $\theta = \frac{\pi}{6}$.

6. $f(x, y) = ye^{-x}$ at $(0, 4)$ in the direction of $\theta = \frac{2\pi}{3}$.

7. $f(x, y) = xe^y - ye^x$ at $(0, 0)$ in the direction of $3i + 4j$.

8. $f(x, y) = 4x^3y^2$ at $(2, 1)$ in the direction of $4i - 3j$.

9. $f(x, y) = y^2 \ln x$ at $(1, 4)$ in the direction of $-3i + 3j$.

10. $f(x, y) = e^{4x - 7y}$ at $(0, \ln 2)$ in the direction of i.

11. $f(x, y) = (2x + 3y)^2$ at $(2, 2)$ in the direction of j.

12. $f(x, y) = e^{xy}e^{-y}$ at $(0, -1)$ in the direction of $-j$.

13. $f(x, y) = \frac{1}{x^2 + y^2}$ at $(2, 2)$ in the direction of j.

14. $f(x, y) = e^{xy}$ at $(1, 2)$ in the direction from $(1, 2)$ toward $(3, 0)$.

15. $f(x, y) = x^3 - x^2y + y^2$ at $(1, -1)$ in the direction from $(1, -1)$ toward $(4, 3)$.

16. $f(x, y) = \frac{\sin(xy)}{x + y}$ at $(1, 1)$ in the direction of $\frac{1}{\sqrt{2}}i - \frac{1}{\sqrt{2}}j$.

17. $f(x, y) = \sin^3(x+y)$ at $\left(\frac{\pi}{3}, -\frac{\pi}{6}\right)$ in the direction of $-i$.

18. $f(x, y) = \tan^{-1}\left(\frac{y}{x}\right)$ at $(1, 2)$ in the direction of $\theta = \frac{\pi}{4}$.

19. $f(x, y) = \tan^{-1}\left(\frac{y}{x}\right)$ at $(-2, 2)$ in the direction of $-i - j$.

20. $f(x, y) = xy$ at (a, b) in the direction from (a, b) toward $(0, 0)$.

21. $f(x, y, z) = xy + yz + zx$ at the point $(1, -1, 2)$ in the direction of the vector $3i + 6j - 2k$.

22. $f(x, y, z) = x \cos y \sin z$ at the point $\left(1, \pi, \frac{\pi}{4}\right)$ in the direction of the vector $2i - j + 4k$.

23. $f(x, y, z) = 3e^x \cos(yz)$ at the point $\left(1, 0, \frac{1}{2}\right)$ in the direction of the vector $i + 2j + 2k$.

24. $f(x, y, z) = z \tan^{-1}\left(\frac{y}{x}\right)$ at the point $(1, 1, 3)$ in the direction of the vector $i + j - k$.

25. $f(x, y, z) = (x + y^2 + z^3)^2$ at the point $(1, 2, 1)$ in the direction of the vector $i + j$.

Exercises 26-39. Find the greatest value and smallest value of the directional derivative for the given function at the given point.

26. $f(x, y) = x^2 + xy$ at $(1, -1)$

27. $f(x, y) = x^3 y^2 - xy$ at $\left(2, \frac{1}{2}\right)$

28. $f(x, y) = \frac{y^2}{x}$ at $(2, 4)$

29. $f(x, y) = xe^y + ye^x$ at $(0, 0)$

30. $f(x, y) = xe^x + ye$ at $(1, 1)$

31. $f(x, y) = \cos(3x - y)$ at $\left(\frac{\pi}{6}, \frac{\pi}{4}\right)$

32. $f(x, y) = x \ln y$ at $(5, 1)$

33. $f(x, y) = \frac{\sin(x+y)}{\cos(x-y)}$ at $\left(\frac{\pi}{2}, \frac{\pi}{4}\right)$

34. $f(x, y) = \sin(2x - y)$ at $\left(\frac{\pi}{4}, \frac{\pi}{4}\right)$

35. $f(x, y, z) = \frac{x}{y} - yz$ at $(4, 1, 1)$

36. $f(x, y, z) = \frac{y}{yx} - yz$ at $(1, 2, 3)$

37. $f(x, y, z) = \sqrt{x^2 + y^2 + z^2}$ at $(3, 6, -2)$

38. $f(x, y, z) = z \tan^{-1}\frac{y}{x}$ at $(1, 1, 3)$

39. $f(x, y, z) = \ln xy + \ln yz + \ln xz$ at $(1, 1, 1)$

Exercises 40-43. Find the general form equation of the tangent plane to the graph of the given function at the given point using: (a) Definition 13.5 (b) Theorem 13.7.

40. $z = 2x + 3xy^2$ at $(2, -1, 10)$

41. $z = xy - y^3 + x^2$ at $(1, 4, -59)$

42. $z = 2x^2 + y$ at $(1, 1, 3)$

43. $z = 4x^2 - y^2 + 2y$ at $(-1, 2, 4)$

Exercises 44-53. Find the general form equation of the tangent plane to the given surface at the given point.

44. $z = xe^{-y}$ at $(1, 0, 1)$

45. $z = y\cos(x - y)$ at $(2, 2, 2)$

46. $z = \ln(2 + x^2 + y^2)$ at $(1, 2, \ln 7)$

47. $z = e^{3y}\sin 3x$ at $\left(\frac{\pi}{6}, 0, 1\right)$

48. $x^2 + y^2 + z^2 = 49$ at $(2, -1, 1)$

49. $2x^2 + 3y^2 - z^2 = 10$ at $(2, 1, -1)$

50. $x^{2/3} + y^{2/3} + z^{2/3} = 9$ at $(1, 8, -8)$

51. $\frac{y}{x + z} = 2$ at $(1, 8, 3)$

52. $\cos(xyz) = 0$ at $\left(4, \pi, \frac{1}{8}\right)$

53. $\ln(1 + x + y + z) = 3$ at $(e^2 + 2, e^3 - e^2, -3)$

54. Use a differential to approximate, to two decimal places, the value of $\sqrt{(2.98)^2 + (4.03)^2}$.

55. Use a differential to approximate, to two decimal places, the value of $\sqrt{100^2 + 199^2 + 201^2}$.

56. Use differentials to estimate the change in $z = x^2y - 1$ from $(1, 2)$ to $(1.11, 1.92)$.

57. Use differentials to approximate, to two decimal places, the change in $z = e^x\ln xy$ from $(1, 2)$ to $(0.9, 2.1)$.

58. Use differentials to approximate, to two decimal places, the increase in area of a triangle if its base in increased from 2 to 2.05 centimeters and its altitude is increased from 5 to 5.1 centimeters.

59. Use differentials to approximate, to two decimal places, the increase in the area of a triangle if its base in increased from 2 to 2.05 cm, and its altitude is increased from 5 to 5.1 cm.

60. Use differentials to approximate, to two decimal places, the increase in the volume of a right circular cylinder if the height is increased from 2 to 2.1 cm, and the radius from0,5 to 0.51 cm.

61. Use differentials to approximate, to two decimal places, the change in $f(x, y, z) = 2xy^2z^3$ from $(1, -1, 2)$ to $(0.99, -1.02, 2.02)$.

62. Use differentials to approximate, to two decimal places, the change in
$$f(x, y, z) = x^2\cos\pi z - y^2\sin\pi z \text{ from } (2, 2, 2) \text{ to } (2.1, 1.9, 2.2).$$

63. The length, width, and height of a rectangular box are measured to be 3 cm, 4 cm, and 5 cm, respectively, each with a maximum error of 0.05 cm. Use differentials to approximate the maximum error in the calculated volume.

64. Use differentials to approximate the percentage error in $w = xy^2z^3$ if x, y, and z have errors of at most 1%, 2%2, and 3%, respectively.

65. Prove that if $\nabla f(x_0, y_0) = 0$, then all directional derivatives of f at (x_0, y_0) are 0.

66. Show that $\frac{xx_0}{a^2} + \frac{yy_0}{b^2} + \frac{zz_0}{c^2} = 1$ is the tangent plane to $\frac{x^2}{a^2} + \frac{y^2}{b^2} + \frac{z^2}{c^2} = 1$ at (x_0, y_0, z_0).

67. Show that $\frac{xx_0}{a^2} + \frac{yy_0}{b^2} - \frac{zz_0}{c^2} = 1$ is the tangent plane to $\frac{x^2}{a^2} + \frac{y^2}{b^2} - \frac{z^2}{c^2} = 1$ at (x_0, y_0, z_0).

68. The **normal line** at a point on a surface S is the line that is perpendicular to the tangent plane at that point. Two surfaces are said to be **orthogonal** at a point of intersection if their normal lines are perpendicular at that point. Prove that the surfaces $f(x, y, z) = 0$ and $g(x, y, z) = 0$ are orthogonal at a point of intersection (x_0, y_0, z_0) if and only if $f_x g_x + f_y g_y + f_z g_{xz} = 0$.

    (Assume that $\nabla f(x_0, y_0, z_0) \neq \mathbf{0}$ and that $\nabla g(x_0, y_0, z_0) \neq \mathbf{0}$.)

69. (Reminiscent of familiar derivative formulas) For $f(x, y, z)$ and $g(x, y, z)$ differentiable functions, and any number r, prove that:

    (a) $\nabla(rf) = r\nabla f$

    (b) $\nabla(f \pm g) = \nabla f \pm \nabla g$

    (c) $\nabla(fg) = f\nabla g + g\nabla f$

    (d) $\nabla\left(\dfrac{f}{g}\right) = \dfrac{g\nabla f - f\nabla g}{g^2}$

70. Let (x_0, y_0, z_0) be a point on a surface S with equation $z = f(x, y)$. Show that $f_x(x_0, y_0)\mathbf{i} + f_y(x_0, y_0)\mathbf{j} - \mathbf{k}$ is perpendicular to the tangent vector $\mathbf{r}'(t_0)$ to any smooth curve $\mathbf{r}(t) = x(t)\mathbf{i} + y(t)\mathbf{j} + z(t)\mathbf{k}$ lying on S that passes through (x_0, y_0, z_0); that is, for which $\mathbf{r}(t_0) = \langle x_0, y_0, z_0 \rangle$.

§3. Extreme Values

Definition 4.2, page 125, readily generalizes to accommodate functions of two (or more) variables:

DEFINITION 13.6
LOCAL EXTREMES

A function $z = f(x, y)$ has a **local maximum** at (x_0, y_0) in its domain, if $f(x_0, y_0) \geq f(x, y)$ for all (x, y) in its domain that are sufficiently close to (x_0, y_0).

A function f has a **local minimum** at (x_0, y_0) if $f(x_0, y_0) \leq f(x, y)$ for all (x, y) sufficiently close to (x_0, y_0).

> More precisely
> A local maximum occurs at (x_0, y_0) if there exist $\varepsilon > 0$ such that $f(x_0, y_0) \geq f(x, y)$ for every (x, y) with:
> $$\|(x, y) - (x_0, y_0)\| < \varepsilon$$

And just as the equation $f'(x) = 0$ is used to find the (local) maxima and minima points of a single-variable function [Figure 13.4(a)], so can the vector equation $\nabla f(x, y) = \mathbf{0}$ be used to locate the (local) extreme points of functions of two variables [Figure 13.4(b)]:.

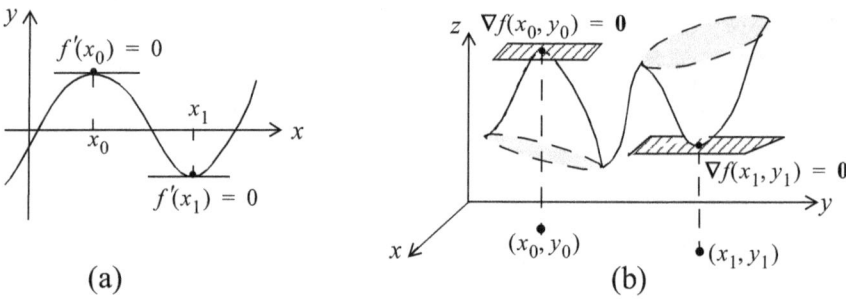

(a)   (b)

Figure 13.4

Why so? Because, If $z = f(x, y)$ has a local maximum (or minimum) at (x_0, y_0) then the intersection of the surface with the planes $x = x_0$ and $y = y_0$ have horizontal tangent lines at (x_0, y_0) (adjacent figure). It follows that $f_x(x_0, y_0) = 0$ and $f_y(x_0, y_0) = 0$.

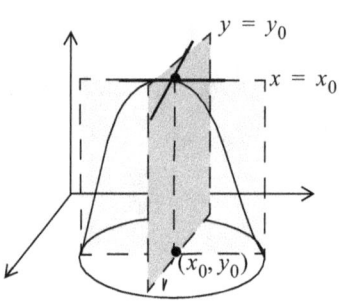

To summarize:

THEOREM 13.8

> A point (x_0, y_0) in an **interior point** of a set S if there exists $\varepsilon > 0$ such that all point (x, y) within ε units of (x_0, y_0) are contained in S.

If $z = f(x, y)$ has a local maximum or minimum at an interior point (x_0, y_0) of its domain, and if the partial derivatives exist at that point, then they both must be zero, and therefore:

$$\nabla f(x_0, y_0) = f_x(x_0, y_0)\mathbf{i} + f_y(x_0, y_0)\mathbf{j} = \mathbf{0}$$

Just as the single-variable function $y = f(x)$ is said to have a **critical point** at x_0 if $f'(x_0) = 0$ (or if $f'(x_0)$ does not exist), so then we say that (x_0, y_0) is a **critical point** of $z = f(x, y)$ if $f_x(x_0, y_0) = f_y(x_0, y_0) = 0$ (or if a partial derivative does not exist). And just as a local extremum need not occur at a critical point of $y = f(x)$ [see Figure 13.5(a)], so then a local extremum may not occur at a critical point of $z = f(x, y)$ [see Figure 13.5(b) and CYU 13.15]:

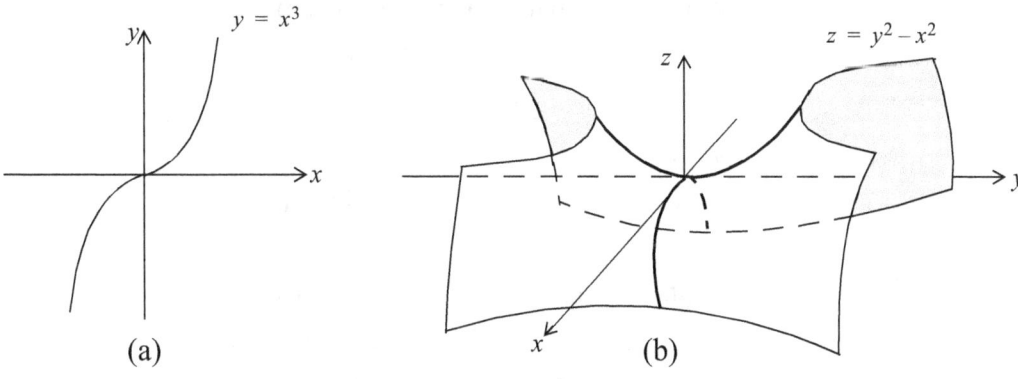

Figure 13.5

We note that (x_0, y_0) is said to be a **saddle point** of $z = f(x, y)$ if $f_x(x_0, y_0) = f_y(x_0, y_0) = 0$ and the function does not have an extremum at that point. In particular, $(0, 0)$ is a saddle point of the function in Figure 13.5(b).

CHECK YOUR UNDERSTANDING 13.15

Answer: See page A-27.

Verify that $(0, 0)$ is a saddle point of $f(x, y) = y^2 - x^2$.

The following result, offered without proof, can be used to identify the nature of critical points:

THEOREM 13.9

Compare with Theorem 4.8, page 137.

SECOND PARTIAL DERIVATIVE TEST

Let $f(x, y)$ have continuous second-order partial derivatives in an open region containing a critical point (x_0, y_0), and let

$$D = f_{xx}(x_0, y_0)f_{yy}(x_0, y_0) - [f_{xy}(x_0, y_0)]^2$$

Then:

(a) If $D > 0$ and $f_{xx}(x_0, y_0) > 0$: f has a local minimum at (x_0, y_0).

(b) If $D > 0$ and $f_{xx}(x_0, y_0) < 0$: f has a local maximum at (x_0, y_0).

(c) If $D < 0$: f has a saddle point at (x_0, y_0).

(d) If $D = 0$: Test is inconclusive.

Note: $D = f_{xx}(x, y)f_{yy}(x, y) - [f_{xy}(x, y)]^2$ is called the **discriminant of** f.

EXAMPLE 13.12 Identify the local extrema and saddle points of
$$f(x, y) = x^3 + y^3 + 3x^2 - 3y^2 - 8$$

SOLUTION: From:

$$f_x(x, y) = 3x^2 + 6x = 0 \qquad f_y(x, y) = 3y^2 - 6y = 0$$
$$3x(x + 2) = 0 \qquad\qquad 3y(y - 2) = 0$$
$$x = 0, x = -2 \qquad\qquad y = 0, y = 2$$

we conclude that f has four critical points:
$$(0, 0), \ (0, 2), \ (-2, 0), \ (-2, 2),$$
From:
$$f_{xx}(x, y) = 6x + 6 \qquad f_{yy} = 6y - 6 \qquad f_{xy} = 0$$

we have:

$$D = f_{xx}(x, y)f_{yy}(x, y) - [f_{xy}(x, y)]^2 = (6x + 6)(6y - 6)$$
$$= 36(x + 1)(y - 1)$$

Employing the Second Derivative Test:

$$D(0, 0) = -36 < 0 \Rightarrow f \text{ has a saddle pont at } (0, 0)$$

$$D(0, 2) = 36 > 0 \text{ and } f_{xx}(0, 2) = 6 > 0 \Rightarrow \text{local min. at } (0, 2)$$

$$D(-2, 0) = 36 > 0 \text{ and } f_{xx}(-2, 0) = -6 < 0 \Rightarrow \text{local max. at } -(2, 0)$$

$$D(-2, -2) = 108 > 0 \Rightarrow f \text{ has a saddle pont at } (-2, -2)$$

CHECK YOUR UNDERSTANDING 13.16

Identify the local extrema and saddle points of
$$f(x, y) = x^2 + 3y^2 + 3xy - 6x - 3y$$

Answer: Local minimum at $(9, -4)$.

How does this example differ from Exercise 22 of Section 4.3 (page 161)?

We modify the four step procedure of page 149:
Step 1: See the problem
Step 2: Express the quantity to be optimized in terms of any convenient number of variables.
BUT NOW:
Step 3: If necessary, use the given information to arrive at a function involving only two variables (as opposed to one).
Step 4: Find where the two partial derivatives are zero to locate the critical points of the function, and analyze the nature of those critical points.

EXAMPLE 13.13 Find the dimensions of a $4\,\text{ft}^3$ open rectangular box requiring the least amount of material.

SOLUTION: Let x, y, z denote the dimensions of the base and height of the box, respectively, and let A denote its surface area (amount of material used). Then:

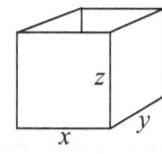

$$(*) \ V = xyz = 4 \text{ and } A = xy + 2yz + 2xz \ (**)$$

From (*): $z = \dfrac{4}{xy}$. Substituting in (**):

$$A = xy + \frac{8y}{xy} + \frac{8x}{xy} = xy + \frac{8}{x} + \frac{8}{y}$$

that which is to be minimized

Locating the critical points (see Theorem13.8):

$$\left.\begin{array}{l} A_x = y - \dfrac{8}{x^2} = 0 \\[3mm] A_y = x - \dfrac{8}{y^2} = 0 \end{array}\right\} \Rightarrow \left.\begin{array}{l} yx^2 = 8 \\[2mm] xy^2 = 8 \end{array}\right\} \Rightarrow yx^2 = xy^2 \Rightarrow yx^2 - xy^2 = 0$$

$$\Rightarrow xy(x-y) = 0$$

neither dimension x nor y can be zero: $\boldsymbol{x = y}$

So: $\left.\begin{array}{l} y^3 = 8 \\[2mm] x^3 = 8 \end{array}\right\} \Rightarrow x = y = 2$

$$\Rightarrow z = \frac{4}{xy} = 1$$

We now use Theorem 13.9 to make sure that a minimum occurs at the (only) critical point $(2, 2)$. Since $A_{xx} = \dfrac{16}{x^3}, S_{yy} = \dfrac{16}{y^3}, A_{xy} = 1$:

$$D = A_{xx}(2,2)A_{yy}(2,2) - [A_{xy}(2,2)]^2 = \left(\frac{16}{8}\right)\left(\frac{16}{8}\right) - 1^2 = 3 > 0$$

$$\text{and } A_{xx}(2,2) = 2 > 0$$

Conclusion: The box requiring the least material has a square base of length 2 ft, and a height of 1 ft.

CHECK YOUR UNDERSTANDING 13.17

Answer: $\dfrac{1}{3}, \dfrac{1}{3}, \dfrac{1}{3}$

Find the three positive numbers whose sum is 1 and such that the sum of their squares is as small a possible.

ABSOLUTE MAXIMUM AND MINIMUM VALUES

Consider Exercises 33-38 of Section 4.3 (page 146).

Let $z = f(x, y)$ be defined on a set S We say that f assumes its **absolute maximum** at (x_0, y_0) in S, if $f(x_0, y_0) \geq f(x, y)$ for every $(x, y) \in S$. Similarly, f is said to assume its **absolute minimum** at (x_0, y_0) if $f(x_0, y_0) \leq f(x, y)$ for every $(x, y) \in S$. We already know that if a differentiable function $z = f(x, y)$ has a local maximum or minimum at an interior point of S, then $\nabla f(x_0, y_0) = \mathbf{0}$ (Theorem 13.8). In order to locate the absolute extreme values of the function, however, we must also take into account the non-interior points of S. With this in mind, we direct your attention to the following particularly important result:

THEOREM 13.10 A real-valued continuous function f defined on a closed bounded set S assumes its maximum value and its minimum value at points in S.

While a proof of the above result lies outside the scope of this text, we can at least attempt to render it "understandable."

To begin with, we note that the symbol R^n denotes the Euclidean-n space; which is to say the set of all n-tuples. In particular:

R is the set of real numbers, R^2 is the set of two-tuples (the plane), and R^3 denotes three-dimensional space.

As for the **bounded** part:

A set S is bounded if there exists a number $M > 0$ such that every element of S falls within M units of the origin. For example:

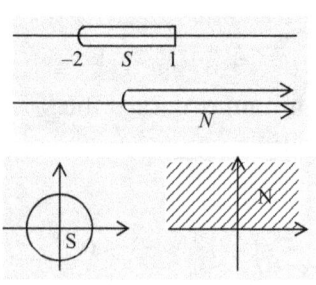

While the set $S = (-2, 1] = \{x \mid -2 < x \le 1\}$ is bounded in R (the set of real number), the set $N = (0, \infty)$ is not (see margin).

While the unit circle $S = \{(x, y) \mid x^2 + y^2 = 1\}$ is bounded in the plane, the set $N = \{(x, y) \mid y > 0\}$ is not (see margin).

As for the **closed** part:

A set S is closed if it contains all of its boundary points: those points which are "arbitrarily close" to both C and its complement; more precisely:

b is a **boundary point** of S if for any given $\varepsilon > 0$ there exists an element in S that is within ε units of b, and an element not in S that is also within ε units of b.

For example:

While the sets $[-2, 1]$ and $[3, \infty)$ are closed in R, the set $(-2, 1]$ is not (it does not contain the boundary point -2).

While the unit circle $\{(x, y) \mid x^2 + y^2 = 1\}$ is closed in the plane (it contains all of its boundary points), the (open) circle $\{(x, y) \mid (x^2 + y^2) < 1\}$ is not (it does not contain the boundary points on its rim).

EXAMPLE 13.14 Find the absolute maximum and minimum values of $f(x, y) = x^2 - 2y^3 - 3x + 2y$ on the adjacent region S.

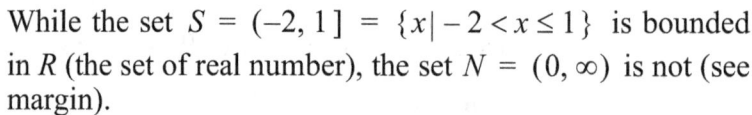

SOLUTION: We know that the continuous function f assumes both its maximum and minimum values on the bounded closed set S (Theorem 13.10). To find them, we will need to compare the values associated with the critical points in the interior of S, along with those on the boundary of S (consisting of the above line segments a, b, c and d).

Finding the critical points of $f(x, y) = x^2 - 2y^3 - 3x + 2y$ in the **interior** of S (see Theorem 13.8):

$$\nabla f(x, y) = f_x(x, y)\boldsymbol{i} + f_y(x, y)\boldsymbol{j}$$

$$= (2x - 3)\boldsymbol{i} + (-6y^2 + 2)\boldsymbol{j} = \boldsymbol{0} \Rightarrow x = \frac{3}{2}, y = \sqrt{\frac{1}{3}}$$

y is not negative in S

Conclusion: $\left(\frac{3}{2}, \sqrt{\frac{1}{3}}\right)$ is the only critical point of f in S's interior. Can it be a saddle point? Yes, and it is; for

$$D = f_{xx}(x, y)f_{yy}(x, y) - [f_{xy}(x, y)]^2 = 2(-12y) - 0^2$$

is negative at $\left(\frac{3}{2}, \sqrt{\frac{1}{3}}\right)$ (see Theorem 13.9).

It follows that the absolute maximum and minimum values of f must occur on the boundary of S.

Here are the critical points of $f(x, y) = x^2 - 2y^3 - 3x + 2y$ on the **boundary** of S:

On line a: $y = x$. And so we consider the single variable function:

$$g(x) = f(x, x) = x^2 - 2x^3 - 3x + 2x = -2x^3 + x^2 - x$$

Since the derivative $g'(x) = -6x^2 + 2x - 1$ is never zero, the only critical points of g on the line segment a occur at its endpoints. It follows that $(0, 0)$ and $(1, 1)$ are critical points of f on the boundary of S.

On line b: $y = 1$. Bringing us to:

$$g(x) = f(x, 1) = x^2 - 2 - 3x + 2 = x^2 - 3x$$

Since $g'(x) = 2x - 3$ is zero at $x = \frac{3}{2}$, the function f might assume an extreme value at $\left(\frac{3}{2}, 1\right)$, as well as at the b-end point $(2, 1)$ (note that $(1, 1)$ already made its presence felt on line segment a).

On line c: $y = -x + 3$. Bringing us to:

$$g(x) = f(x, -x + 3) = x^2 - 2(-x + 3)^3 - 3x + 2(-x + 3)$$

$$= 2x^3 - 17x^2 + 49x - 48$$

Since $g'(x) = 6x^2 - 34x + 49$ is never zero, we only pick up one new critical point for f; namely c's endpoint: $(3, 0)$.

On line d:

$g(x) = f(x, 0) = x^2 - 3x$ and $g'(x) = 2x - 3$. Yielding one additional critical point for f: $\left(\frac{3}{2}, 0\right)$.

Upon evaluating the function $f(x, y) = x^2 - 2y^3 - 3x + 2y$, at each of the above 6 boundary critical points, we found that:

$$f(0, 0) = f(3, 0) = 0 \qquad\qquad f(1, 1) = f(2, 1) = -2$$

$$f\left(\frac{3}{2}, 1\right) = f\left(\frac{3}{2}, 0\right) = -\frac{9}{4}$$

Conclusion: On the region S, f achieves its absolute maximum value of 0 at the points $(0, 0)$ and $(3, 0)$, and its absolute minimum value of $-\frac{9}{4}$ at $\left(\frac{3}{2}, 1\right)$ and $\left(\frac{3}{2}, 0\right)$.

CHECK YOUR UNDERSTANDING 13.18

Find the absolute maximum and minimum values of $f(x, y) = x^2 - 2y^3 - 3x + 2y$ on the adjacent region K.

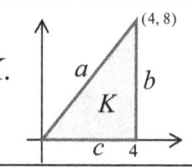

Answer: Max: $\dfrac{36 + 4\sqrt{3}}{9}$.

Min: -1004.

Joseph-Louis Lagrange aka Giuseppe Lodovicio (1736-1813).

LAGRANGE MULTIPLIERS

The following result can be used to determine the extreme values of a function when restricted, or constrained, to a subset of its domain:

THEOREM 13.11
Lagrange

Let $f(x, y)$ and $g(x, y)$ be differentiable functions. If f has an extreme value at a point $P_0 = (x_0, y_0)$ on the constraint curve $g(x, y) = k$, then either $\nabla g(x_0, y_0) = \mathbf{0}$ or there exists a constant λ such that:

$$\nabla f(x_0, y_0) = \lambda \nabla g(x_0, y_0)$$

The Greek letter λ (lambda), is said to be a **Lagrange multiplier**.

Similarly:

If $f(x, y, z)$ has an extreme value at a point $P_0 = (x_0, y_0, z_0)$ on the constraint surface $g(x, y, z) = k$, then either $\nabla g(x_0, y_0, z_0) = \mathbf{0}$ or there exists a constant λ such that:

$$\nabla f(x_0, y_0, z_0) = \lambda \nabla g(x_0, y_0, z_0)$$

PROOF: Assume that a maximum (or minimum) value of $f(x, y)$ occurs at (x_0, y_0). Let the vector function $r(t) = x(t)\mathbf{i} + y(t)\mathbf{j}$ trace out the curve $g(x, y) = k$. Employing CYU 13.12, page 567, we have:

Note: If $\nabla g(x, y) = 0$, then (*) will surely hold.

$$\nabla g(x_0, y_0) \cdot r'(t_0) = 0 \quad (*)$$

Now consider the function $h(t) = f[x(t), y(t)]$, and let t_0 be such that $h(t_0) = (x_0, y_0)$. Since an extremum occurs at t_0:

Theorem 13.4(a), page 555

$$h'(t_0) = f'[x(t), y(t)] \overset{\downarrow}{=} f_x(x_0, y_0)x'(t_0) + f_y(x_0, y_0)y'(t_0)$$

$$= \nabla f(x_0, y_0) \cdot \mathbf{r}'(t_0) = 0$$

Since $\nabla g(x_0, y_0)$ and $\nabla f(x_0, y_0)$ are both orthogonal to $\mathbf{r}'(t_0)$, those vectors must be parallel; which is to say:

$$\nabla f(x_0, y_0) = \lambda \nabla g(x_0, y_0) \text{ for some number } \lambda.$$

EXAMPLE 13.15 (a) Find the maximum and minimum values of the function $f(x, y) = x^2 - 3y^2$ on the ellipse $\dfrac{x^2}{2} + (y-1)^2 = 1$.

(b) Find the maximum value of the function $f(x, y, z) = xyz$ on the sphere $x^2 + y^2 + z^2 = 4$.

SOLUTION: (a) We apply Theorem 13.11 with:

$$f(x, y) = x^2 - 3y^2 \text{ and } g(x, y) = \frac{x^2}{2} + (y-1)^2 \quad.$$

(the constraint curve is $g(x, y) = 1$)

The vector equation $\nabla f(x, y) = \lambda \nabla g(x, y)$:

$$2x\mathbf{i} - 6y\mathbf{j} = \lambda[x\mathbf{i} + 2(y-1)\mathbf{j}] = \lambda x\mathbf{i} + (2\lambda y - 2\lambda)\mathbf{j}$$

reveals equations (1) and (2) below, while equation (3) stems from the constraint equation $\dfrac{x^2}{2} + (y-1)^2 = 1$:

As is the case here, the Lagrange method generally leads to a non-linear system of equations. Solving such a system can prove to be a challenging if not impossible task.

(1): $2x = \lambda x$

(2): $-6y = 2\lambda y - 2\lambda$

(3): $x^2 + 2y^2 - 4y = 0$

From (1): $x = 0$ or $\lambda = 2$.
Addressing both possibilities we have:

If $x = 0$, then, from (3):	If $\lambda = 2$, then, from (2):
$2y^2 - 4y = 0$	$-6y = 4y - 4 \Rightarrow y = \dfrac{2}{5}$
$2y(y-2) = 0 \Rightarrow y = 0 \text{ or } y = 2$	Substituting in (3):
	$x^2 + 2\left(\dfrac{2}{5}\right)^2 - 4\left(\dfrac{2}{5}\right) = 0 \Rightarrow x = \pm\dfrac{4\sqrt{2}}{5}$
Yielding the two critical points:	Yielding two additional critical points:
$(0, 0)$ and $(0, 2)$	$\left(\dfrac{4\sqrt{2}}{5}, \dfrac{2}{5}\right)$ and $\left(-\dfrac{4\sqrt{2}}{5}, \dfrac{2}{5}\right)$

Turning to $\nabla g(x, y) = \mathbf{0}$ with $g(x, y) = \dfrac{x^2}{2} + (y-1)^2$, we have:

$g_x(x, y) = 0$ and $g_y(x, y) = 0 \Rightarrow x = 0$ and $2y - 2 = 0$ $(y = 1)$

Note, however, that $(0, 1)$ is not a critical point as it does not lie of the ellipse $\dfrac{x^2}{2} + (y-1)^2 = 1$.

Evaluating $f(x, y) = x^2 - 3y^2$ at each of the four established critical points $(0, 0)$, $(0, 2)$, $\left(\dfrac{4\sqrt{2}}{5}, \dfrac{2}{5}\right)$, and $\left(-\dfrac{4\sqrt{2}}{5}, \dfrac{2}{5}\right)$, we have:

$$f(0, 0) = 0, f(0, 2) = -12, f\left(\frac{4\sqrt{2}}{5}, \frac{2}{5}\right) = \frac{4}{5}, \text{ and } f\left(-\frac{4\sqrt{2}}{5}, \frac{2}{5}\right) = \frac{4}{5}.$$

Conclusion (see margin):

We know that, when restricted to the ellipse, the function f must assume its maximum and minimum values at points on the ellipse (see Theorem 13.10)

The function $f(x, y) = x^2 - 3y^2$, when restricted to points on the ellipse $\dfrac{x^2}{2} + (y-1)^2 = 1$, assumes a maximum value of $\dfrac{4}{5}$ at $\left(\dfrac{4\sqrt{2}}{5}, \dfrac{2}{5}\right)$ and $\left(-\dfrac{4\sqrt{2}}{5}, \dfrac{2}{5}\right)$, and it assumes a minimum value of -12 at $(0, 2)$.

(b) We employ Theorem 13.11 with:
$$f(x, y, z) = xyz \text{ and } g(x, y, z) = x^2 + y^2 + z^2.$$
$$\text{(the constraint surface is } g(x, y, z) = 4)$$

The vector equation $\nabla f(x, y, z) = \lambda \nabla g(x, y, z)$:
$$yz\mathbf{i} + xz\mathbf{j} + xy\mathbf{k} = \lambda[2x\mathbf{i} + 2y\mathbf{j} + 2z\mathbf{k}]$$
reveals equations (1), (2) and (3) below, while (4) is the given constraint equation:

$$\text{(1): } yz = 2\lambda x \qquad \text{(2): } xz = 2\lambda y \qquad \text{(3): } xy = 2\lambda z$$
$$\text{(4): } x^2 + y^2 + z^2 = 4$$

Multiplying (1) by x, (2) by y, and (3) by z leads us to:
$$xyz = 2\lambda x^2 = 2\lambda y^2 = 2\lambda z^2 \text{ or: } \lambda x^2 = \lambda y^2 = \lambda z^2 \text{ (*)}$$

Note that the condition $\nabla g(x, y, z) = \mathbf{0}$:
$$\nabla(x^2 + y^2 + z^2) = \mathbf{0}$$
leads to the point $(0, 0, 0)$, which does not lie on the sphere $x^2 + y^2 + z^2 = 4$.

We can exclude the possibility that $\lambda = 0$, for $\lambda = 0$ would imply that y or z must be 0 [see (1)], and that would force $f(x, y, z) = xyz$ to be zero, which can not be the maximum value of f, as f assumes positive values on the sphere $x^2 + y^2 + z^2 = 4$. That being the case, (*) tells us that $x^2 = y^2 = z^2$. From (4) we then have:
$$x^2 + y^2 + z^2 = 3x^2 = 4 \Rightarrow x, y, z = \pm\frac{2}{\sqrt{3}}$$

As the maximum value of $f(x, y, z) = xyz$ is positive, there are only four critical points to consider: all three of the variables are positive or exactly two of them are negative.

Conclusion: On the sphere $x^2 + y^2 + z^2 = 4$, $f(x, y, z) = xyz$ assumes the maximum value of $\dfrac{8}{3\sqrt{3}}$ at the points:

$$\left(\frac{2}{\sqrt{3}}, \frac{2}{\sqrt{3}}, \frac{2}{\sqrt{3}}\right), \left(\frac{2}{\sqrt{3}}, -\frac{2}{\sqrt{3}}, -\frac{2}{\sqrt{3}}\right), \left(-\frac{2}{\sqrt{3}}, \frac{2}{\sqrt{3}}, -\frac{2}{\sqrt{3}}\right), \left(-\frac{2}{\sqrt{3}}, -\frac{2}{\sqrt{3}}, \frac{2}{\sqrt{3}}\right)$$

CHECK YOUR UNDERSTANDING 13.19

Find the maximum and minimum values of $f(x, y) = x^2 + 4y^2$ on the circle $x^2 + y^2 = 1$.

EXTREMA SUBJECT TO TWO CONSTRAINTS

It is sometimes necessary to determine the extreme values of a function $f(x, y, z)$ on the curve C that results from the intersection of two surfaces $g(x, y, z) = k_1$ and $h(x, y, z) = k_2$. In such a setting you can invoke a generalization of Theorem 13.11; specifically:

Both λ (lambda) and μ (mu) are said to be **Lagrange multipliers**.

$$\nabla f(x, y, z) = \lambda \nabla g(x, y, z) + \mu \nabla h(x, y, z)$$

(Extreme points of f on C satisfy the above vector equation.)

Consider the following Example.

EXAMPLE 13.16 Find the maximum and minimum values of $f(x, y, z) = x + 2y + 3z$ on the ellipse C stemming from the intersection of the plane $x - y + z = 1$ with the cylinder $x^2 + y^2 = 1$.

SOLUTION: For $f(x, y, z) = x + 2y + 3z$, $g(x, y, z) = x^2 + y^2$ and $h(x, y, z) = x - y + z$, the vector equation:

$$\nabla f(x, y, z) = \lambda \nabla g(x, y, z) + \mu \nabla h(x, y, z)$$

becomes: $\boldsymbol{i} + 2\boldsymbol{j} + 3\boldsymbol{k} = \lambda(2x\boldsymbol{i} + 2y\boldsymbol{j} + 0\boldsymbol{k}) + \mu(\boldsymbol{i} - \boldsymbol{j} + \boldsymbol{k})$.

Solving a system of equations, not all of which are linear, can be a tricky, if not impossible task.

Leading us to the following system of equations:

(1): $1 = 2\lambda x + \mu$ (2): $2 = 2\lambda y - \mu$ (3): $3 = \mu$

(4): $x^2 + y^2 = 1$ (5): $x - y + z = 1$

Since $\mu = 3$ [see (3)]: $x = -\dfrac{1}{\lambda}$ [see (1)], and $y = \dfrac{5}{2\lambda}$ [see (2)].

Consequently [see (4)]: $\dfrac{1}{\lambda^2} + \dfrac{25}{4\lambda^2} = 1 \Rightarrow 4\lambda^2 = 29 \Rightarrow \lambda = \pm\dfrac{\sqrt{29}}{2}$.

Returning to $x = -\dfrac{1}{\lambda}$, $y = \dfrac{5}{2\lambda}$, and $z = 1 - x + y$ [see (5)]:

$\lambda = \dfrac{\sqrt{29}}{2}: x = -\dfrac{2}{\sqrt{29}}, y = \dfrac{5}{\sqrt{29}}, z = 1 + \dfrac{2}{\sqrt{29}} + \dfrac{5}{\sqrt{29}} = 1 + \dfrac{7}{\sqrt{29}}$

$\lambda = -\dfrac{\sqrt{29}}{2}: x = \dfrac{2}{\sqrt{29}}, y = -\dfrac{5}{\sqrt{29}}, z = 1 - \dfrac{2}{\sqrt{29}} - \dfrac{5}{\sqrt{29}} = 1 - \dfrac{7}{\sqrt{29}}$

And so there are two critical points for f on the ellipse C; namely:

$$\left(-\frac{2}{\sqrt{29}}, \frac{5}{\sqrt{29}}, 1 + \frac{7}{\sqrt{29}}\right) \text{ and } \left(\frac{2}{\sqrt{29}}, -\frac{5}{\sqrt{29}}, 1 - \frac{7}{\sqrt{29}}\right)$$

A direct calculation shows that:

$$f\left(-\frac{2}{\sqrt{29}}, \frac{5}{\sqrt{29}}, 1 + \frac{7}{\sqrt{29}}\right) = \underset{\underset{\text{maximum value}}{\uparrow}}{3 + \sqrt{29}}, \text{ and } f\left(\frac{2}{\sqrt{29}}, -\frac{5}{\sqrt{29}}, 1 - \frac{7}{\sqrt{29}}\right) = \underset{\underset{\text{minimum value}}{\uparrow}}{3 - \sqrt{29}}$$

CHECK YOUR UNDERSTANDING 13.20

Answer: $\left(\frac{5}{2}, \frac{5}{4}, \frac{5}{8}\right)$

Find the critical points of $f(x, y, z) = xy + yz$ subject to the constraints $x + 2y = 5$ and $x - 4z = 0$.

	EXERCISES	

Exercises 1-12. Identify the local extrema and saddle points of the given function.

1. $f(x, y) = y^2 - xy + 2x + y + 1$

2. $f(x, y) = x^2 + 5y^2 - x + y$

3. $f(x, y) = xy + 3$

4. $f(x, y) = x^3 - 3xy - y^3$

5. $f(x, y) = xy - x^2 - y^2 - 2x - 2y$

6. $f(x, y) = xy - x^3 - y^2$

7. $f(x, y) = x^4 + y^4 - 4xy + 5$

8. $f(x, y) = 2x^2 - 4xy + y^4$

9. $f(x, y) = x^3 - 12xy + 8y^3$

10. $f(x, y) = 2x^3 - 2y^3 - 4xy + 1$

11. $f(x, y) = 2x^2 + 3xy + 4y^2 - 5x + 2y$

12. $f(x, y) = 3x^3 - 2xy + y^2 - 8y$

13. $f(x, y) = e^x \sin y$

14. $f(x, y) = x \sin y$

15. $f(x, y) = e^{-(x^2 + y^2 + 2x)}$

16. $f(x, y) = e^{-(x^2 + y^2)}$

17. $f(x, y) = \sin x + \sin y, \; 0 < x < \pi, \; 0 < y < \pi$

18. $f(x, y) = \sin x + \sin y + \sin(x + y), \; 0 < x < \dfrac{\pi}{2}, \; 0 < y < \dfrac{\pi}{2}$

Exercises 19-30. Find the absolute minimum and maximum values of the given function on the specified set S.

19. $f(x, y) = 3xy - 6x + 7; \; S: \; 0 \le x \le 3, \; 0 \le y \le 5$

20. $f(x, y) = x^2 + y^2 + x^2 y + 4; \; S: \; -1 \le x \le 1, \; -1 \le y \le 1$

21. $f(x, y) = x^3 - 3xy + 3y^2; \; S: \; \dfrac{1}{4} \le x \le 1, \; 0 \le y \le 2$

22. $f(x, y) = x^2 + 4y^2 - 2x^2 y + 4; \; S: \; -1 \le x \le 1, \; -1 \le y \le 1$

23. $f(x, y) = x^3 - xy + y^2 - x; \; S: \; x \ge 0, \; y \ge 0, \; x + y \le 2$

24. $f(x, y) = x^2 + 3y^2 - 4x - 6y \; ; \; S \; x \ge 0, \; y \ge 0, \; y \le \dfrac{3}{4} x + 3$

25. $f(x, y) = xe^y - x^2 - e^y; \; S: \; 0 \le x \le 2, \; 0 \le y \le 1$

26. $f(x, y) = -x^2 - y^2 + 1x + 2y + 1; \; S: \; x \ge 0, \; y \ge 0, \; x + y \le 2$

27. $f(x, y) = -2x^2 - y^2 + 4x + 3y + 5; \; S: \; y \le 2, \; y \ge -x, \; y \ge x$

28. $f(x, y) = x^3 - y^3 - 3x + 12y$; S: $0 \le x \le 2, -3x \le y \le 0$

29. $f(x, y) = 2x^2 - y^2 + 6y$; S: $x^2 + y^2 \le 16$

30. $f(x, y) = 2x^3 + y^4$; S: $x^2 + y^2 \le 1$

Exercises 31-42. Use Lagrange multipliers to find the minimum and maximum values of f, subject to the given constraint C.

31. $f(x, y) = x + y$; C: $x^2 + y^2 = 1$

32. $f(x, y) = xy$; C: $4x^2 + y^2 = 8$

33. $f(x, y) = xy + 14$; C: $x^2 + y^2 = 18$

34. $f(x, y) = x^2 + 2y^2 - 2x + 3$; C: $x^2 + y^2 = 10$

35. $f(x, y) = x^2 + 2y^2$; C: $x^2 + y^2 = 1$

36. $f(x, y) = x^2 + y^2$; C: $x^2 + y^2 - 2x - 4y = 0$

37. $f(x, y) = 2x^2 + 3y^2 - 4x - 5$; C: $x^2 + y^2 = 16$

38. $f(x, y) = 2x^2 - y^2$; C: $x^2 + 2y^2 - 4y = 0$

39. $f(x, y, z) = x + y + 2z$; C: $x^2 + y^2 + z^2 = 3$

40. $f(x, y, z) = x + y + z$; C: $x^2 + y^2 + z^2 = 1$

41. $f(x, y, z) = xyz$; C: $x + y + z = 1$ with $x, y, z \ge 0$

42. $f(x, y) = 8x - 4y + 2z$; C: $x^2 + y^2 + z^2 = 21$

Exercises 43-46. Find the absolute minimum and maximum values of the given function on the specified set S.

43. $f(x, y) = xy$; S: $4x^2 + y^2 \le 8$ (see exercise 32).

44. $f(x, y) = x^2 + 2y^2 - 2x + 3$; S: $x^2 + y^2 \le 10$ (see Exercise 34).

45. $f(x, y) = x^2 + 2y^2$; S: $x^2 + y^2 \le 1$ (see Exercise 35).

46. $f(x, y) = 2x^2 + 3y^2 - 4x - 5$; S: $x^2 + y^2 \le 16$ (see Exercise 37).

Exercises 47-51. Find the maximum and minimum values of f, subjected to the two given constraints.

47. $f(x, y, z) = 3x - y - 3z$; $x + y - z = 0$ and $x^2 + 2z^2 = 1$

48. $f(x, y, z) = x + 2y$; $x + y + z = 1$ and $y^2 + z^2 = 4$

49. $f(x, y, z) = xy + yz$; $xy = 1$ and $y^2 + z^2 = 1$

50. $f(x, y, z) = 2x + 2y + z^2 + 20$; $x^2 + y^2 + z^2 = 11$ and $x + y + z = 3$

51. $f(x, y, z) = 4y - 2z$; $2x - y - z = 2$ and $x^2 + y^2 = 1$

Exercises 52-57. Solve using (a) Theorems 13.8 and 13.9 and (b) Lagrange's Theorem.

52. Find three positive numbers x, y, and z such that $x + y + z = 12$ and x^2yz is maximum.

53. Find the points on the surface $z^2 = xy + 1$ that are closest to the origin.

54. Find the point in the plane $3x + 2y + z = 14$ that is nearest the origin.

55. Find dimensions of the most economical closed rectangular crate 96 cubic feet in volume if the base and lid costs 30 cents per square foot and the sides cost 10 cents per square foot.

56. Find the dimensions of a rectangular crate with maximum volume if the sum of the length of its 12 edges is 120 feet.

57. Find the shortest distance from $(1, 3, 4)$ to $2x - y + z = 1$.

58. Find the dimensions of a closed box of largest volume with 64 in^2 surface area.

59. Find the area of the largest rectangle that can be inscribed in the ellipse $\dfrac{x^2}{a^2} + \dfrac{y^2}{b^2} = 1$.

60. A triangular area is to enclose 100 square feet. What are the dimensions of the triangle requiring the least amount of fencing?

61. Assume that the combined cost of producing x units of one product and y units of another is given by $C(x, y) = 2x^2 + xy + y^2 + 500$. How many units of each product should be produced to minimize cost, given that a total of 200 units are to be manufactured?

62. The temperature at a point on the surface of the sphere $x^2 + y^2 + z^2 = 4$ is given by $T(x, y, z) = x^2yz$. Find the point of maximum temperature.

63. The sum of the three dimensions of a rectangular box is not to exceed 45 inches, with the length of one of its sides not to exceed half of the length of one of its other sides. Determine the dimensions of such a box of maximum volume.

64. Find three positive numbers x, y, and z such that $x + y + z = k$ and $x^a y^b z^c$ is maximum.

65. An open rectangular box has a fixed surface area S. Find the dimensions for maximum volume.

66. Find dimensions of the most economical open rectangular crate 96 cubic feet in volume if the base costs 30 cents per square foot and the sides cost 10 cents per square foot.

67. An open symmetrical irrigation channel is to have a perimeter of length l (see adjacent figure). Find the values of x and α to enable maximum flow.

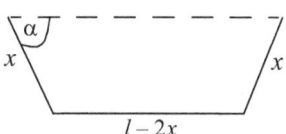

	CHAPTER SUMMARY
PARTIAL DERIVATIVES	If $z = f(x, y)$, then the **partial derivative of f with respect to x** at (x, y) is $$\frac{\partial z}{\partial x} = \lim_{h \to 0} \frac{f(x + h, y) - f(x, y)}{h} \qquad \frac{\partial z}{\partial y} = \lim_{h \to 0} \frac{f(x, y + h) - f(x, y)}{h}$$
THEOREM	If $z = f(x, y)$ and its partial derivatives f_x, f_y, f_{xy}, f_{yx} are defined and continuous in an open region containing (x_0, y_0), then $f_{xy}(x_0, y_0) = f_{yx}(x_0, y_0)$.
DIFFERENTIABLE	A function $z = f(x, y)$ is **differentiable** at (x_0, y_0) if $\Delta z = f(x_0 + \Delta x, y_0 + \Delta y) - f(x_0, y_0)$ can be expressed in the form: $$\Delta z = f_x(x_0, y_0)\Delta x + f_y(x_0, y_0)\Delta y + \varepsilon_1 \Delta x + \varepsilon_2 \Delta y$$ where $\varepsilon_1, \varepsilon_2 \to 0$ as $\Delta x, \Delta y \to 0$. A function f is said to be **differentiable** if it is differentiable at each point in its domain.
THEOREMS	If the partial derivatives f_x and f_y exist and are **continuous** in an open region about (x_0, y_0), then f is differentiable at (x_0, y_0). If $z = f(x, y)$ is differentiable at (x_0, y_0), then f is continuous at (x_0, y_0).
CHAIN RULES	(a) If $z = f(x, y)$ is differentiable, and if $x = g(t)$ and $y = h(t)$ are differentiable functions, then the single variable function $z = f[g(t), h(t)]$ is differentiable, and: $$\frac{dz}{dt} = \frac{\partial z}{\partial x}\frac{dx}{dt} + \frac{\partial z}{\partial y}\frac{dy}{dt}$$ (b) If $z = f(x, y)$ is differentiable, and if $x = g(s, t)$ and $y = h(s, t)$ are differentiable functions, then the two variable function $z = f[g(s, t), h(s, t)]$ is differentiable, and: $$\frac{\partial z}{\partial s} = \frac{\partial z}{\partial x}\frac{\partial x}{\partial s} + \frac{\partial z}{\partial y}\frac{\partial y}{\partial s} \qquad \frac{\partial z}{\partial t} = \frac{\partial z}{\partial x}\frac{\partial x}{\partial t} + \frac{\partial z}{\partial y}\frac{\partial y}{\partial t}$$
DIRECTIONAL DERIVATIVE	The **directional derivative** of f in the direction of the **unit** vector $\mathbf{u} = (\mathbf{a}, \mathbf{b}) = a\mathbf{i} + b\mathbf{j}$ is given by: $$D_{\mathbf{u}}f(x, y) = \lim_{h \to 0} \frac{f(x + ha, y + hb) - f(x, y)}{h}$$

GRADIENT	The **gradient** of a differentiable function $z = f(x, y)$, denoted by $\nabla f(x, y)$, or ∇f, is the vector function: $$\nabla f(x, y) = f_x(x, y)\mathbf{i} + f_y(x, y)\mathbf{j}$$ The directional derivative of a differentiable function $f(x, y)$ in the direction of the unit vector $\mathbf{u} = a\mathbf{i} + b\mathbf{j}$ is given by: $$D_{\mathbf{u}}f(x, y) = \nabla f(x, y) \cdot \mathbf{u}$$ $D_{\mathbf{u}}f(x_0, y_0)$ assumes its maximum value when the unit vector $\mathbf{u}$ is in the direction of the gradient vector $\nabla f(x_0, y_0)$, and its minimum value when $\mathbf{u}$ is in the direction of $-\nabla f(x_0, y_0)$.
TANGENT PLANE	The **tangent plane** to the graph of f at $(x_0, y_0, f(x_0, y_0))$ is the plane passing through that point with normal vector: $$f_x(x_0, y_0)\mathbf{i} + f_y(x_0, y_0)\mathbf{j} - \mathbf{k}$$
LEVEL SURFACE	For given $w = f(x, y, z)$ and constant k, the **k-level surface** of f consists of all points (x, y, z) for which $f(x, y, z) = k$. $\nabla f(x_0, y_0, z_0)$ is perpendicular to the tangent plane to the level surface of f at $P_0 = (x_0, y_0, z_0)$
LOCAL EXTREMES **SECOND DERIVATIVE TEST**	If $z = f(x, y)$ has a local maximum or minimum at an interior point (x_0, y_0) of its domain, then: $$\nabla f(x_0, y_0) = f_x(x_0, y_0)\mathbf{i} + f_y(x_0, y_0)\mathbf{j} = \mathbf{0}$$ Let $D = f_{xx}(x_0, y_0)f_{yy}(x_0, y_0) - [f_{xy}(x_0, y_0)]^2$, then: (a) If $D > 0$ and $f_{xx}(x_0, y_0) > 0$: f has a local minimum at (x_0, y_0). (b) If $D > 0$ and $f_{xx}(x_0, y_0) < 0$: f has a local minimum at (x_0, y_0). (c) If $D < 0$: f has a saddle point at (x_0, y_0). (d) If $D = 0$: Test is inconclusive.
LAGRANGE MULTIPLIERS	Maximum or minimum values of $f(x, y, z)$, subjected to the constraint surface $g(x, y, z) = k$, occur among the points (x, y, z) satisfying the vector equation: $$\nabla f(x, y, z) = \lambda \nabla g(x, y, z) \text{ or } \nabla g = 0$$

CHAPTER 14
Vector Calculus

§1. LINE (PATH) INTEGRALS

Our earlier development of the definite integral $\int_a^b f(x)\,dx$ is summarized in Figure 14.1(a) (see Definition 5.3, page 178). Figures 14.1(b) and (c) generalize the concept to accommodate functions defined on a curve C in the plane, and in three-space.

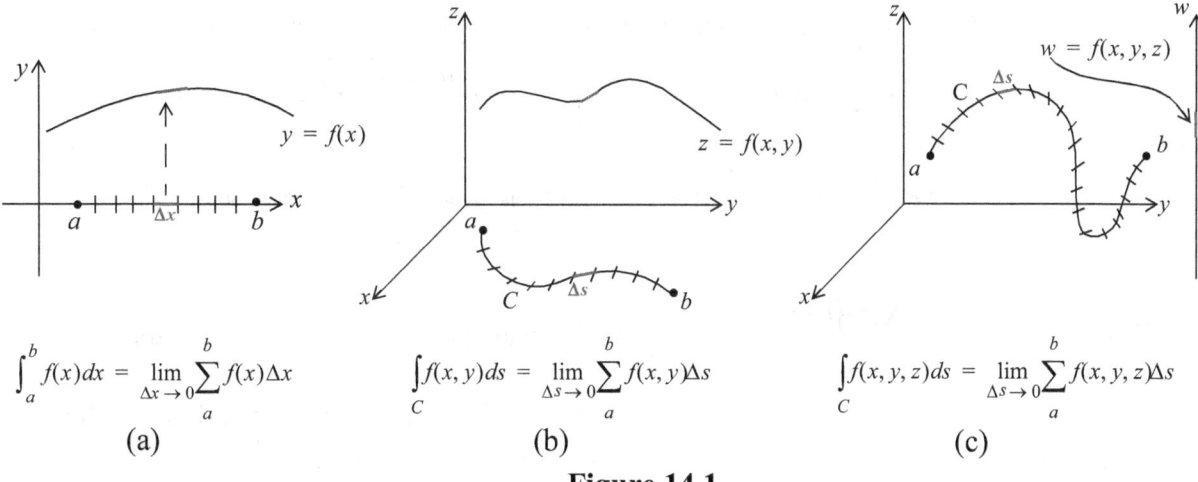

$$\int_a^b f(x)\,dx = \lim_{\Delta x \to 0} \sum_a^b f(x)\Delta x$$

(a)

$$\int_C f(x,y)\,ds = \lim_{\Delta s \to 0} \sum_a^b f(x,y)\Delta s$$

(b)

$$\int_C f(x,y,z)\,ds = \lim_{\Delta s \to 0} \sum_a^b f(x,y,z)\Delta s$$

(c)

Figure 14.1

As it is with the familiar integral in Figure 14.1(a), those in Figures 14.1 (b) and (c) also represent limits of Riemann sums. But while the Riemann sum in (a) involves pieces Δx of the interval $[a, b]$, those in (b) and (c) involve pieces Δs of the curves C [in the xy-plane in (b) and the xyz-space in (c)].

Formalizing:

We remind you that a smooth curve is a curve traced out by $r(t)$ for which $r'(t)$ is continuous and never 0.

DEFINITION 14.1
Line Integral
of a Scalar Function

Let the smooth curve C lie in the domain of a function $z = f(x, y)$ or $w = f(x, y, z)$. The **line** (or path) **integral** of f along C is given by:

$$\int_C f(x,y)\,ds = \lim_{\Delta s \to 0} \sum_a^b f(x,y)\Delta s \text{ or } \int_C f(x,y,z)\,ds = \lim_{\Delta s \to 0} \sum_a^b f(x,y,z)\Delta s$$

(providing the limit exists)

If $f(x, y) = 1$ [or $f(x, y, z) = 1$] and if the smooth curve C is traced out exactly once under the parametrization

$$x = x(t),\, y = y(t),\, \text{for } a \leq t \leq b$$

(or: $x = x(t),\ y = y(t)\,,\ z = z(t)\,,\ \text{for } a \leq t \leq b$)

then the line integral $\int_C \mathbf{1} \cdot ds$ denotes the length of the curve C and is given by (see page 534):

$$\int_C ds = \int_a^b \sqrt{\left(\frac{dx}{dt}\right)^2 + \left(\frac{dy}{dt}\right)^2}\, dt \quad \left[\text{or } \int_a^b \sqrt{\left(\frac{dx}{dt}\right)^2 + \left(\frac{dy}{dt}\right)^2 + \left(\frac{dz}{dt}\right)^2}\, dt \right]$$

Throwing a function f into the mix we have:

THEOREM 14.1 Let $x = x(t)$, $y = y(t)$, for $a \le t \le b$ be a parametrization of a smooth curve C. For $z = f(x, y)$ continuous on C:

$$\int_C f(x, y)\,ds = \int_a^b f[x(t), y(t)] \sqrt{\left(\frac{dx}{dt}\right)^2 + \left(\frac{dy}{dt}\right)^2}\, dt$$

For $w = f(x, y, z)$, with $x = x(t)$, $y = y(t)$, $z = z(t)$, for $a \le t \le b$:

$$\int_C f(x, y, z)\,ds = \int_a^b f[x(t), y(t), z(t)] \sqrt{\left(\frac{dx}{dt}\right)^2 + \left(\frac{dy}{dt}\right)^2 + \left(\frac{dz}{dt}\right)^2}\, dt$$

We note that $\int_C f(x, y)\,ds \left[\text{or } \int_a^b (x, y, z)\,ds \right]$ is independent of the parametrization of C.

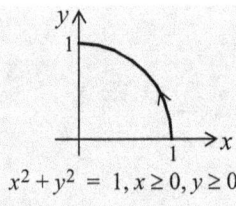

$x^2 + y^2 = 1, x \ge 0, y \ge 0$

EXAMPLE 14.1 (a) Evaluate $\int_C (x + xy)\,ds$ where C is the portion of the unit circle $x^2 + y^2 = 1$ lying in the first quadrant, directed from $(1, 0)$ to $(0, 1)$.

(b) Evaluate $\int_C (x + 2y^2 + 3z)\,ds$ where C is the line segment from $(0, 0, 0)$ to $(1, 2, 2)$.

SOLUTION: (a) Turning to the parametrization $x = \cos t$, $y = \sin t$, for $0 \le t \le \frac{\pi}{2}$:

$$\int_C (x + xy)\,ds = \int_0^{\frac{\pi}{2}} (\cos t + \cos t \sin t) \sqrt{\left(\frac{dx}{dt}\right)^2 + \left(\frac{dy}{dt}\right)^2}\, dt$$

$$= \int_0^{\frac{\pi}{2}} (\cos t + \cos t \sin t) \sqrt{(-\sin t)^2 + (\cos t)^2}\, dt$$

$$= \int_0^{\frac{\pi}{2}} (\cos t + \cos t \sin t)\,dt = \left(\sin t + \frac{1}{2}\sin^2 t \right)\Bigg|_0^{\pi/2}$$

$$= 1 + \frac{1}{2} = \frac{3}{2}$$

(b) Turning to the parametrization

$$x(t) = t,\ y(t) = 2t,\ z(t) = 2t,\ 0 \le t \le 1$$

we have:

$$\int_C (x + 2y^2 + 3z)ds = \int_0^1 (t + 8t^2 + 6t)\sqrt{\left(\frac{dx}{dt}\right)^2 + \left(\frac{dy}{dt}\right)^2 + \left(\frac{dz}{dt}\right)^2}\,dt$$

$$= \int_0^1 (8t^2 + 7t)\sqrt{1^2 + 2^2 + 2^2}\,dt$$

$$= 3\int_0^1 (8t^2 + 7t)dt = 3\left(\frac{8}{3}t^3 + \frac{7}{2}t^2\right)\Big|_0^1 = \frac{37}{2}$$

While $\int_C f(x, y)ds$ is independent of C's parametrization, it need not be independent of path. A case in point:

y
1
C of Example 14.1
$y = -x + 1$
$\overline{C}$
1
x

EXAMPLE 14.2 Evaluate $\int_{\overline{C}} (x + xy)ds$ where $\overline{C}$ is the line segment from $(0, 1)$ to $(1, 0)$.

SOLUTION: Turning to the $\overline{C}$ parametrization $x = t$, $y = -t + 1$, for $0 \le t \le 1$ we have:

$$\int_{\overline{C}} (x + xy)ds = \int_0^1 [t + t(-t + 1)]\sqrt{\left(\frac{dx}{dt}\right)^2 + \left(\frac{dy}{dt}\right)^2}\,dt$$

In Example 14.1 we observed that:
$$\int_C (x + xy)ds = \frac{3}{2}$$

$$= \int_0^1 (2t - t^2)\sqrt{1 + (-1)^2}\,dt = \sqrt{2}\left(t^2 - \frac{t^3}{3}\right)\Big|_0^1 = \frac{2\sqrt{2}}{3}$$

CHECK YOUR UNDERSTANDING 14.1

Evaluate $\int_C xy^2 ds$ where C is the helix with parametrization:

$$x = \cos t, \ y = \sin t, \ z = t \text{ for } 0 \le t \le 2\pi$$

Answer: 0

The following parametrization for a line segment from r_0 to r_1 is used in the next example:

$$r(t) = r_0 + t(r_1 - r_0) = (1 - t)r_0 + tr_1, \ 0 \le t \le 1$$

y
2
C_3
-1
2
x
C_1
C_2
-1

EXAMPLE 14.3 Evaluate $\int_C xy\,ds$ where C is the three-piece curve depicted in the margin.

SOLUTION: Using the parametrization:

$$C_1: \ r_1 = (1 - t)\langle -1, 0 \rangle + t\langle 0, -1 \rangle = \langle -1 + t, -t \rangle, 0 \le t \le 1$$

$$C_2: \ r_2 = (1 - t)\langle 0, -1 \rangle + t\langle 2, 0 \rangle = \langle 2t, -1 + t \rangle, 0 \le t \le 1$$

$$C_3: \ r_3 = (1 - t)\langle 2, 0 \rangle + t\langle 2, 2 \rangle = \langle 2, 2t \rangle, 0 \le t \le 1$$

we have:

$$\int_C xy\,ds = \int_{C_1} xy\,ds + \int_{C_2} xy\,ds + \int_{C_3} xy\,ds$$

$$= \int_0^1 (-1+t)(-t)\sqrt{1^2+(-1)^2}\,dt + \int_0^1 2t(-1+t)\sqrt{2^2+1^2}\,dt + \int_0^1 2(2t)\sqrt{2^2}\,dt$$

$$= \sqrt{2}\int_0^1 (t-t^2)\,dt + 2\sqrt{5}\int_0^1 (-t+t^2)\,dt + 8\int_0^1 t\,dt$$

$$= \sqrt{2}\left(\frac{t^2}{2}-\frac{t^3}{3}\right)\Big|_0^1 + 2\sqrt{5}\left(-\frac{t^2}{2}+\frac{t^3}{3}\right)\Big|_0^1 + 8\left(\frac{t^2}{2}\right)\Big|_0^1$$

$$= \sqrt{2}\left(\frac{1}{6}\right) + 2\sqrt{5}\left(-\frac{1}{6}\right) + 4 = \frac{1}{6}(\sqrt{2}-2\sqrt{5}+24)$$

Answer:
$\frac{1}{6}(5\sqrt{2}-2\sqrt{5}+24+6\sqrt{13})$

CHECK YOUR UNDERSTANDING 14.2

Evaluate $\int_C xy\,ds$ where C is the closed curve obtained by adding the line segment from $(2,2)$ to the point $(-1,0)$ in the curve of Example 14.3.

MASS

Double integrals were employed to address the mass of thin flat objects or laminas (see Example 11.14, page 451). Line integrals enable us to do the same for "thin" curves. For example:

If the density at a point (x,y) of the wire C in the adjacent figure is given by $\delta(x,y)$, then its mass, M, is (naturally) defined to be:

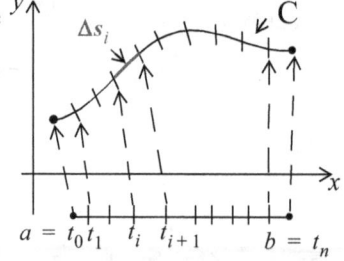

$$M = \lim_{\Delta s \to 0} \sum_a^b \delta(x,y)\Delta s = \int_C \delta(x,y)\,ds.$$

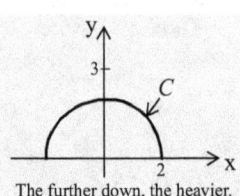

The further down, the heavier.

EXAMPLE 14.4 Find the mass of a wire C in the shape of the semicircle $x^2+y^2 = 4$, $y \geq 0$, with density $\delta(x,y) = k(3-y)$.

SOLUTION: Turning to the C parametrization $x = 2\cos t$, $y = 2\sin t$, for $0 \leq t \leq \pi$ we have:

$$M = \int_C \delta(x,y)ds = k\int_C (3-y)ds$$

$$= k\int_0^\pi (3-2\sin t)\sqrt{(-2\sin t)^2 + (2\cos t)^2}\,dt$$

$$= 2k\int_0^\pi (3-2\sin t)dt = 2k(3t+2\cos t)\Big|_0^\pi = 2k(3\pi-4)$$

CHECK YOUR UNDERSTANDING 14.3

Find the mass of the "slinky" $C(t) = (\cos t, \sin t, t)$ for $0 \le t \le 6\pi$, with density function $\delta(x,y,z) = 1+x+z$

LINE INTEGRALS OF VECTOR-VALUED FUNCTIONS

A **vector-valued function** (in two- or three-space) is a function that assigns a vector (as opposed to a scalar) to elements in its domain; as is the case with the function:

$$F(x,y) = xi + yj \text{ in } R^2, \text{ and } G(x,y,z) = 0i + 0j + zk \text{ in } R^3$$

If you position the vector $F(x,y)$ directly at the point (x,y) and $G(x,y,z)$ at (x,y,z), then you arrive at a visual representation of what are called the **vector fields** associated with the given vector functions F and G (see Figure 14.2).

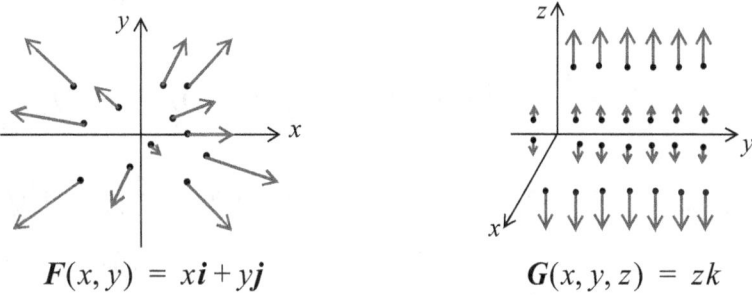

$$F(x,y) = xi + yj \qquad\qquad G(x,y,z) = zk$$

Figure 14.2

Our next goal is to generalize the concept of work defined on page 211:

$$W = \lim_{\Delta x \to 0} \sum_a^b f(x_i)\Delta x_i = \int_a^b f(x)dx$$

to arrive at the definition of the work done by a force $F(x,y) = g(x,y)i + h(x,y)j$ in moving an object from point A to point B, along a smooth curve $r(t) = x(t)i + y(t)j$, $a \le t \le b$. Once again, we **DIVIDE AND CONQUER:**

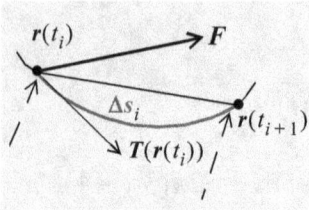

Note that $F \cdot T$ is the scalar component of F in the direction of the curve's unit tangent vector T (see Definition 12.5, page 503).

In three space:
For $F(x, y, z)$ and
$r(t) = x(t)i + y(t)j + z(t)k$:
$$a \le t \le b$$
$$W = \int_a^b F(r(t)) \cdot r'(t)dt$$

For Δs small, the particle's movement along Δs can be approximated by a movement in the direction of the **unit tangent vector** $T(r(t_i))$. That being the case, the work done by F in moving the particle from $r(t_i)$ to $r(t_{i+1})$ can be approximated by:

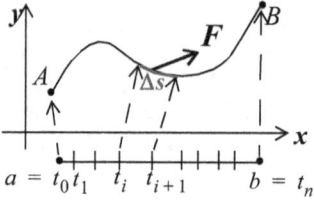

$$F(r(t_i)) \cdot [\Delta s_i T(r(t_i))] = [F(r(t_i)) \cdot T(r(t_i))]\Delta s_i$$

(see Exercises 49-52, page 512)

Leading us to the definition: $W = \int_C F \cdot T ds$

As for the rest of the story:

THEOREM 14.2   The work W done by a continuous force field $F(x, y)$ along a smooth curve C with parametrization $r(t) = x(t)i + y(t)j, a \le t \le b$ is given by:

$$W = \int_C F \cdot T ds = \int_a^b F(r(t)) \cdot r'(t)dt$$

(providing the integral exists)

We also write: $W = \int_C F \cdot dr = \int_a^b F(r(t)) \cdot r'(t)dt$, where, symbolically: $dr = Tds$

PROOF:

$$W = \int_C [F(x, y) \cdot T(x, y)]ds$$

Theorem 14.1:
$$= \int_a^b F(r(t)) \cdot T(r(t)) \sqrt{\left(\frac{dx}{dt}\right)^2 + \left(\frac{dy}{dt}\right)^2} dt$$

$$= \int_a^b F(r(t)) \cdot \frac{r'(t)}{\|r'(t)\|} \|r'(t)\| dt = \int_a^b F(r(t)) \cdot r'(t)dt$$

unit tangent vector in direction of $T(r(t))$

EXAMPLE 14.5   Find the work done by the planar force field $F(x, y) = xi + (x + y)j$ in moving a particle along the path $r(t) = t^2 i + t^3 j, 0 \le t \le 1$.

SOLUTION:

Note:
For $r(t) = x(t)i + y(t)j$, the symbol $F(r(t))$ represents $F(x(t), y(t))$.

$$W = \int_a^b F(r(t)) \cdot r'(t)dt = \int_0^1 F(t^2, t^3) \cdot (2ti + 3t^2 j)dt$$

$$= \int_0^1 (t^2 i + (t^2 + t^3)j) \cdot (2ti + 3t^2 j)dt$$

$$= \int_0^1 (2t^3 + 3t^4 + 3t^5)dt$$

$$= \left(\frac{t^4}{2} + \frac{3t^5}{5} + \frac{t^6}{2}\right)\Bigg|_0^1 = \frac{8}{5}$$

CHECK YOUR UNDERSTANDING 14.4

Find the work done by the force field $F(x, y, z) = xi + (x + y)j + zk$ in moving a particle along the path $r(t) = t^2i + t^3j + tk$, $0 \le t \le 1$.

Different applications and interpretations of line integrals are rampant throughout mathematics and the sciences.
    In particular:
If F is a force field, then the line integral denotes the work done by F along C.

If F denotes the velocity field of a fluid, then the line integral represents the flow of the fluid along the curve C.

Replacing the force field F in Theorem 14.2 with a more general vector field brings us to:

DEFINITION 14.2
LINE INTEGRAL
OF A VECTOR
FUNCTION

Let $r(t)$, $a \le t \le b$ be a parametrization of the smooth curve C, and let F be a continuous vector function defined on C. The **line integral** (or **path integral**) of F over C, is given by:

$$\int_C F \cdot T ds = \int_C F \cdot dr = \int_a^b F(r(t)) \cdot r'(t) dt$$

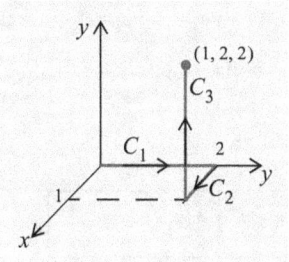

EXAMPLE 14.6 Evaluate

$$\int_C (yi + 2xj + 3zk) \cdot dr$$

for the curve C depicted in the margin.

SOLUTION: Turning to the line-segment parametrization

$$r(t) = (1 - t)r_0 + tr_1, \ 0 \le t \le 1$$

we have:

$$C_1: \ r_1 = (1 - t)\langle 0, 0, 0 \rangle + t\langle 0, 2, 0 \rangle = \langle 0, 2t, 0 \rangle, 0 \le t \le 1$$

$$C_2: \ r_2 = (1 - t)\langle 0, 2, 0 \rangle + t\langle 1, 2, 0 \rangle = \langle t, 2, 0 \rangle, 0 \le t \le 1$$

$$C_3: \ r_3 = (1 - t)\langle 1, 2, 0 \rangle + t\langle 1, 2, 2 \rangle = \langle 1, 2, 2t \rangle, 0 \le t \le 1$$

Hence:

$$\int_C F \cdot dr = \int_{C_1} F \cdot dr_1 + \int_{C_2} F \cdot dr_2 + \int_{C_3} F \cdot dr_3$$

On C_1. For $0 \le t \le 1$: $x = 0, y = 2t$, and $z = 0$. Thus:

$$\int_{C_1} F \cdot dr_1 = \int_0^1 (yi + 2xj + 3zk) \cdot dr_1$$

$$= \int_0^1 (2ti + 0j + 0k) \cdot \langle 0, 2t, 0 \rangle' dt$$

$$= \int_0^1 \langle 2t, 0, 0 \rangle \cdot \langle 0, 2, 0 \rangle dt = \int_0^1 0 dt = 0$$

On C_2. For $0 \le t \le 1$: $x = t, y = 2$, and $z = 0$. Thus:

$$\int_{C_2} F \cdot dr_2 = \int_0^1 (yi + 2xj + 3zk) \cdot dr_2$$

$$= \int_0^1 (2i + tj + 0k) \cdot \langle t, 2, 0 \rangle'$$

$$= \int_0^1 \langle 2, t, 0 \rangle \cdot \langle 1, 0, 0 \rangle = \int_0^1 2\,dt = 2t \Big|_0^1 = 2$$

On C_3. For $0 \le t \le 1$: $x = 1$, $y = 2$, and $z = 2t$. Thus:

$$\int_{C_3} F \cdot dr_2 = \int_0^1 (yi + 2xj + 3zk) \cdot dr_3$$

$$= \int_0^1 (2i + 1j + 6tk) \cdot \langle 1, 2, 2t \rangle'$$

$$= \int_0^1 \langle 2, 1, 6t \rangle \cdot \langle 0, 0, 2 \rangle = \int_0^1 12t\,dt = 6t^2 \Big|_0^1 = 6$$

Hence: $\displaystyle\int_C F \cdot dr = \int_{C_1} F \cdot dr_1 + \int_{C_2} F \cdot dr_2 + \int_{C_3} F \cdot dr_3 = 0 + 2 + 6 = 8$

CHECK YOUR UNDERSTANDING 14.5

Evaluate $\displaystyle\int_C F \cdot dr$ for $F(x, y, z) = yi + 2xj + 3zk$ and C the line segment from the origin to the point $(1, 2, 2)$ (compare with above example).

Answer: 9

If C is a curve with parametrization $r(t)$, $a \le t \le b$, then $-C$ will denote the curve with parametrization $\bar{r}(t) = r(a + b - t)$, $a \le t \le b$. Note that $\bar{r}(t)$ traces out C, but in the opposite direction of $r(t)$:

$$\bar{r}(t) \text{ starts at } \bar{r}(a) = r(a + b - a) = r(b)$$

$$\text{and ends at } \bar{r}(b) = r(a + b - b) = r(a).$$

Reminiscent of:
$$\int_a^b f(x)\,dx = -\int_b^a f(x)\,dx$$

THEOREM 14.3 If $r(t)$, $a \le t \le b$ is a parametrization for the smooth curve C and if F is a continuous vector function defined on C, then:

$$\int_C F \cdot dr = -\int_{-C} F \cdot dr$$

From the chain rule:

$\frac{d}{dt}[\bar{r}'(t)]$

$= \frac{d}{dt}[r(a+b-t)]$

$= [r'(a+b-t)](a+b-t)'$

$= r'(a+b-t)(-1)$

PROOF:

$$\int_{-C} F \cdot dr = \int_a^b F(\bar{r}(t)) \cdot \bar{r}'(t)dt$$

$$\underset{\text{see margin} \nearrow}{=} -\int_a^b F(r(a+b-t)) \cdot r'(a+b-t)dt$$

$$= \int_b^a F(r(a+b-t)) \cdot r'(a+b-t)dt$$

$u = a+b-t$
$du = -dt \text{ or } dt = -du \rightarrow$
$t = b \Rightarrow u = a$
$t = a \Rightarrow u = b$

$$= -\int_a^b F(r(u)) \cdot r'(u)du = -\int_C F \cdot dr$$

CHECK YOUR UNDERSTANDING 14.6

Referring to the function F and curve C of Example 14.6, show, directly, that:

$$\int_C F \cdot dr = -\int_{-C} F \cdot dr$$

Answer: See page A-32.

ALTERNATE NOTATION FOR LINE INTEGRALS

Let $F(x, y) = P(x, y)i + Q(x, y)j$ be a vector field defined on a curve C with parametrization $r(t) = x(t)i + y(t)j$, $a \le t \le b$. Then:

$$\int_C F \cdot dr = \int_a^b \left[P(x, y)\frac{dx}{dt} + Q(x, y)\frac{dy}{dt} \right]dt$$

$$= \int_a^b P(x, y)\frac{dx}{dt}dt + \int_a^b Q(x, y)\frac{dy}{dt}dt$$

$$= \int_a^b P(x, y)dx + \int_a^b Q(x, y)dy$$

Imposing a more compact form:

$$\int_C F \cdot dr = \int_a^b P\,dx + Q\,dy$$

A mnemonic device:

$\int F \cdot dr$

$= \int (Pi + Qj) \cdot (dxi + dyj)$

$= \int P\,dx + Q\,dy$

EXAMPLE 14.7 Determine $\int_C xy\,dx + y\,dy$, where C has parametrization $r(t) = t^2 i + 5t j$, $1 \le t \le 2$.

SOLUTION: For $x(t) = t^2$ and $y(t) = 5t$ we have $\frac{dx}{dt} = 2t$ and $\frac{dy}{dt} = 5$. Consequently:

Earlier notation:

$$\int_C F \cdot dr = \int_0^1 [(5t^3 i + 5tj) \cdot r'(t)] dt$$

$$= \int_0^1 [(5t^3 i + 5tj) \cdot (2ti + 5j)] dt$$

$$= \int_0^1 (10t^4 + 25t) dt = \frac{29}{2}$$

Same→

notation for

$$\int_C xy\,dx + y\,dy = \int_0^1 \left[(t^2 \cdot 5t)\frac{dx}{dt} + 5t\frac{dy}{dt} \right] dt$$

$$= \int_0^1 [5t^3(2t) + 5t(5)] dt = \int_0^1 (10t^4 + 25t) dt$$

$$= \left(2t^5 + \frac{25t^2}{2} \right) \Bigg|_0^1 = \frac{29}{2}$$

CHECK YOUR UNDERSTANDING 14.7

Answer: $\frac{535}{2}$

Determine $\displaystyle\int_C xy\,dx + y\,dy + yz\,dz$, where C has parametrization $r(t) = t^2 i + 5tj - 2t^2 k$, $1 \le t \le 2$.

As previously noted, line integral of a vector-valued function F defined on a curve C with parametrization $r(t)$ for $a \le t \le b$ can be represented in several forms:

$$\int_C F \cdot T\,ds \quad \text{or} \quad \int_C F \cdot dr \quad \text{or} \quad \int_a^b F \cdot \frac{dr}{dt} dt \quad \text{or} \quad \int_a^b F(r(t)) \cdot r'(t)\,dt$$

In addition, we have:

For $F(x, y) = P(x, y)i + Q(x, y)j$ and C in the plane:	For $F(x, y, z) = P(x, y, z)i + Q(x, y, z)j + R(x, y, z)k$ and C in three-space:
$\displaystyle\int_a^b \left[P(x, y)\frac{dx}{dt} + Q(x, y)\frac{dy}{dt} \right] dt$ or: $\displaystyle\int_a^b P\,dx + Q\,dy$	$\displaystyle\int_a^b \left[P(x, y, z)\frac{dx}{dt} + Q(x, y, z)\frac{dy}{dt} + R(x, y, z)\frac{dz}{dt} \right] dt$ or: $\displaystyle\int_a^b P\,dx + Q\,dy + R\,dz$

	EXERCISES	

Exercises 1-11. Evaluate $\int_C f(x, y)\,ds$ for the given function f and given path C.

1. $f(x, y) = x^2 + y$, and C is the line segment from $(0, 0)$ to $(1, 1)$.

2. $f(x, y) = x + y^2$ and C is the line segment from $(1, 1)$ to $(4, 5)$.

3. $f(x, y) = 2 + x^2 y$, and C is the upper half of the unit circle $x^2 + y^2 = 1$, traversed in a counterclockwise direction.

4. $f(x, y) = 2 + x^2 y$, and C is the lower half of the unit circle $x^2 + y^2 = 1$, traversed in a counterclockwise direction.

5. $f(x, y) = xy^4$, and C is the right half of the unit circle $x^2 + y^2 = 4$, traversed in a counterclockwise direction.

6. $f(x, y) = 1 + x^2 y$, and C is the right half of the ellipse $\dfrac{x^2}{2} + y^2 = 1$, traversed in a counterclockwise direction.

7. $f(x, y) = 3x$, and C is the curve $y = x^2$ from $(0, 0)$ to $(3, 9)$.

8. $f(x, y) = xy$, and C is the curve $x = y^2$ from $(0, 0)$ to $(3, 9)$.

9. $f(x, y) = xy$, and C consists of the line segment from $(0, 0)$ to $(1, 1)$ followed by the line segment from $(1, 1)$ to $(1, 2)$.

10. $f(x, y) = \dfrac{x}{1 + y^2}$, and C is given parametrically by $x = 1 + 2t, y = t$ for $0 \le t \le 1$.

11. $f(x, y) = 4x^3$, and C consists of the line segment from $(-2, -1)$ to $(0, -1)$ followed by the graph of $y = x^3 - 1$ for $0 \le x \le 1$, followed by the line segment from $(1, 0)$ to $(1, 2)$.

Exercises 12-16. Evaluate $\int_C f(x, y, z)\,ds$ for the given function f and given path C.

12. $f(x, y, z) = z + y^2$, and C is the line segment from $(0, 0, 0)$ to $(1, 4, 2)$.

13. $f(x, y, z) = x - 3y^2 + z$, and C is the line segment from $(0, 0, 0)$ to $(1, 1, 1)$.

14. $f(x, y, z) = x - 3y + z$, and C is consists of the line segment from $(0, 0, 0)$ to $(1, 1, 1)$ followed by the line segment from $(1, 1, 1)$ to $(1, 1, 3)$.

15. $f(x, y, z) = xyz$, and C is the helix given by $x(t) = \cos t$, $y(t) = \sin t$, $z(t) = 3t$ for $0 \le t \le 4\pi$.

16. $f(x, y, z) = -\sqrt{x^2 + z^2}$, and C is given parametrically by $x(t) = 0$, $y(t) = a\cos t$, $z(t) = a\sin t$ for $0 \le t \le 2\pi$.

17. Find the mass of a wire in the shape of a semicircle $x(t) = 2\cos t$, $y(t) = 2\sin t$ for $0 \le t \le \pi$, with density function $\delta(x, y) = y + 2$.

18. Find the mass of a wire in the shape of the curve $x(t) = t$, $y(t) = 2t$, $z(t) = \dfrac{2t^{3/2}}{3}$ for $0 \le t \le 1$, with density function $\delta = 3\sqrt{5+t}$.

19. Find the mass of a wire in the shape of the helix $x(t) = 3\cos t$, $y(t) = 3\sin t$, $z(t) = 4t$ for $0 \le t \le \pi$, with density the square of the distance from the origin.

Exercises 20-31. Evaluate $\displaystyle\int_C F \cdot dr$ for the given function F and curve C with given parametrization $r(t)$.

20. $F(x, y) = (x + 2y)i + (2x + y)j$, $r(t) = ti + t^2j$ for $0 \le t \le 1$.

21. $F(x, y) = xyi + y^2j$, $r(t) = \cos ti + \sin tj$ for $0 \le t \le \dfrac{\pi}{3}$.

22. $F(x, y) = xyi + \sin yj$, $r(t) = e^t i + e^{-t^2}j$ for $1 \le t \le 2$.

23. $F(x, y) = (e^x + y)i + xj$, $r(t) = ti + t^2j$ for $0 \le t \le 1$.

24. $F(x, y) = e^{x-1}i + xyj$, $r(t) = t^2i + t^3j$ for $0 \le t \le 1$.

25. $F(x, y) = x^2i - yj$, $r(t) = ti + e^tj$ for $0 \le t \le 2$.

26. $F(x, y, z) = xi + yj - zk$, $r(t) = ti + 3t^2j + 2t^3k$ for $0 \le t \le 1$.

27. $F(x, y, z) = 8x^2yzi + 5zj - 4xyk$, $r(t) = ti + t^2j + t^3k$ for $0 \le t \le 1$.

28. $F(x, y, z) = (x + y)i + (y - z)j - z^2k$, $r(t) = t^2i + t^3j + t^2k$ for $0 \le t \le 1$.

29. $F(x, y, z) = (2xz + y^2)i + 2xyj + (x^2 + 3z^2)k$, $r(t) = t^2i + (t + 1)j + (2t - 1)k$ for $0 \le t \le 1$.

30. $F(x, y, z) = e^y i + xe^y j - (z + 1)e^z k$, $r(t) = ti + t^2j + t^3k$ for $0 \le t \le 1$.

31. $F(x, y, z) = \sin xi + \cos yj + xzk$, $r(t) = t^3i + t^2j + tk$ for $0 \le t \le 1$.

32. Find the work done by the force field $F(x, y) = yi + xj$ in moving a particle along the path $r(t) = ti + t^2j$, $1 \le t \le 2$.

33. Find the work done by the force field $F(x, y) = xyi + x^3j$ in moving a particle along the path $r(t) = t^{1/2}i + t^{1/4}j$, $1 \le t \le 16$.

34. Find the work done by the force field $F(x, y) = x\sin y\, i + y j$ in moving a particle along the parabola $y = x^2$ from $(-1, 1)$ to $(2, 4)$.

35. Find the work done by the force field $F(x, y) = x i + (y + 2) j$ in moving a particle along the path $r(t) = (t - \sin t) i + (1 - \cos t) j,\ 0 \le t \le 2\pi$.

36. Find the work done by the force field $F(x, y, z) = (y + z) i + (x + z) j + (x + y) k$ in moving a particle along the line segment from $(1, 0, 0)$ to $(2, 1, 5)$.

37. Find the work done by the force field $F(x, y, z) = (y - x^2) i + (z - y^2) j + (x - z^2) k$ in moving a particle along the path $r(t) = t i + t^2 j + t^3 k,\ 0 \le t \le 1$.

38. Find the work done by the force field $F(x, y, z) = z i + x j + y k$ in moving a particle along the path $r(t) = (\sin t) i + (\cos t) j + t k,\ 0 \le t \le 2\pi$.

39. Find the work done by the force field $F(x, y, z) = x^2 i + y^2 j + z^2 k$ in moving a particle along the path $r(t) = (1 + t^2) i + t^2 j + [2 + \sin(\pi t)] k,\ 0 \le t \le 1$.

40. Find the work done by the force field $F(x, y, z) = (x + y) i + (x - y) j + 4z k$ in moving a particle along the path $r(t) = \sin(3t) i + \cos(3t) j + 3t k,\ 0 \le t \le 1$.

§2. CONSERVATIVE FIELDS AND PATH-INDEPENDENCE

Here is one of the main characters of this section.

DEFINITION 14.3
PATH-INDEPENDENT VECTOR FIELDS

A vector field F is said to be **path-independent** in a region S if for any two points p_0 and p_1 in S, and any two smooth curves C_1 and C_2 in S from p_0 to p_1:

$$\int_{C_1} F \cdot dr = \int_{C_2} F \cdot dr$$

As it turn out, to challenge a vector field for path-independence one need only consider closed paths (see margin):

A curve C with parametrization $r(t)$, $a \le t \le b$ is said to be **closed** if $r(a) = r(b)$:

A closed curve:

The symbol $\oint_C F \cdot dr$ is used to indicate that the path C is closed.

THEOREM 14.4 A vector field F is path-independent in a region S if and only if $\oint_C F \cdot dr = 0$ for every (smooth) closed curve C in S.

PROOF: Assume that F is path-independent. Consider the closed path C in Figure 14.3(a) (with indicated orientation). Choose two points p_0 and p_1 on C. Break C into two pieces C_1, C_2, oriented so that they both start at P_0 and end at P_1, as is indicated in Figure 14.3(b).

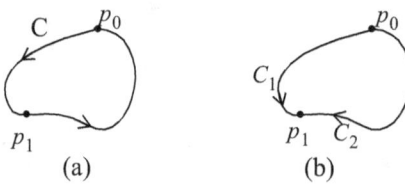

(a)   (b)

Figure 14.3

We then have:

$$\oint_C F \cdot dr = \int_{C_1} F \cdot dr + \int_{-C_2} F \cdot dr = \int_{C_1} F \cdot dr - \int_{C_2} F \cdot dr = 0$$

Theorem 14.3, page 599   by path independence

As for the converse (your turn):

CHECK YOUR UNDERSTANDING 14.8

Show that if $\oint_C F \cdot dr = 0$ for every closed curve C in S, then F is path-independent in S.

Answer: See page A-33.

In general, vector fields tend **not** to be path-independent; even "nice" ones, like the vector field $F(x, y) = xi + (x+y)j$ of Example 14.5, page 628. Theorem 14.5 below, often called the ***Fundamental Theorem for Line Integrals***, identifies an important class of path-independent fields:

We remind you that:
$$\nabla f(x,y) = f_x(x,y)\mathbf{i} + f_y(x,y)\mathbf{j}$$
and:
$$\nabla f(x,y,z) = f_x\mathbf{i} + f_y\mathbf{j} + f_z\mathbf{k}$$

DEFINITION 14.4
CONSERVATIVE FIELD

A vector field $\mathbf{F}$ is said to be **conservative** on a set S if there exists a scalar-valued function f, called a **potential function** for $\mathbf{F}$, such that for every $p \in S$:
$$\mathbf{F}(p) = \nabla f(p)$$

Conservative fields are path independent. Indeed:

Compare:
If $\mathbf{F} = \nabla f$, then:
$$\int_C \mathbf{F} \cdot d\mathbf{r} = f(p_1) - f(p_0)$$
with the Fundamental Theorem of Calculus (page 180):
If $f(x) = g'(x)$, then:
$$\int_a^b f(x)dx = g(b) - g(a)$$

THEOREM 14.5 If $\mathbf{F} = \nabla f$ and if f has continuous first order partial derivatives in an open region S, then:
For any two points p_0, p_1 in S, and any piecewise-smooth curve C from p_0 to p_1:
$$\int_C \mathbf{F} \cdot d\mathbf{r} = f(p_1) - f(p_0)$$

A proof for piecewise-smooth curves can be obtained by considering its pieces separately.

PROOF: We establish the proof for a smooth curve C in the plane. Let $r(t)$, $a \le t \le b$, be a parametrization of C from p_0 to p_1, then:

$$\int_C \mathbf{F} \cdot d\mathbf{r} = \underset{\substack{\uparrow \\ \text{Definition 14.2, page 597}}}{\int_a^b \mathbf{F}[r(t)] \cdot r'(t)dt} = \int_a^b \nabla f(r(t)) \cdot r'(t)dt$$

$$\nabla f(r(t)) \cdot r'(t) = f_x\frac{dx}{dt} + f_y\frac{dy}{dt} \leftarrow$$

From Theorem 13.4(a) page 555:
$$\frac{d}{dt}[f(x(t),y(t)] = f_x\frac{dx}{dt} + f_y\frac{dy}{dt} \leftarrow$$

$$\text{margin:} = \int_a^b \frac{d}{dt}[f(r(t))]dt$$

$$= [f(r(t))]\Big|_a^b$$

$$= f(r(b)) - f(r(a))$$

$$= f(p_1) - f(p_0)$$

The above theorem tells us that line integrals of conservative vector fields are path independent. In most cases the converse also holds; specifically:

A region S is connected if any two points p_0, p_1 in S can be joined by a (smooth) curve contained in S.

THEOREM 14.6 Let $\mathbf{F}$ be a continuous vector field on an open connected region S (see margin). If $\mathbf{F}$ is independent of path in S, then it is conservative (i.e: $\mathbf{F} = \nabla f$ for some f).

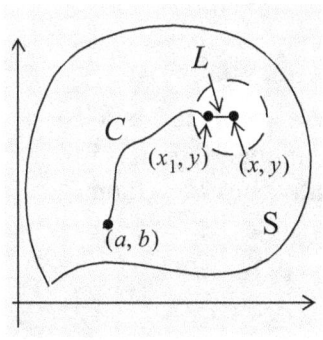

Proof: We offer a proof for a two-dimensional region S.

Fix a point (a,b) in S and let (x,y) be any other point in S. Since S is an open region, we can choose a circle, centered at (x,y), that is contained in S, along with a point (x_1,y) in the circle that lies directly to the left of (x,y) (see margin). We then define

$$f(x,y) = \int_C \mathbf{F} \cdot d\mathbf{r} + \int_L \mathbf{F} \cdot d\mathbf{r}$$

where C is a smooth path from (a,b) to (x_1,y), and L is the horizontal line segment from (x_1,y) to (x,y). Note that f is well-defined, as $\mathbf{F}$ is independent of path.

We now show that $F = Pi + Qj = \nabla f$:

While we do not know the value of $\int_C F \cdot dr$, we do know that it

is solely a function of its endpoints (a, b) and (x_1, y). As such,

$\int_C F \cdot dr$ is independent of x, and therefore: $\dfrac{\partial}{\partial x}\displaystyle\int_C F \cdot dr = 0$.

Turning to $\displaystyle\int_L F \cdot dr$, with parametrization $r(t) = ti + yj$,

$x_1 \le t \le x$, we have:

$$\frac{\partial}{\partial x}\int_L F \cdot dr = \frac{\partial}{\partial x}\int_{x_1}^x [P(t,y)i + Q(t,y)j] \cdot (i)dt = \frac{\partial}{\partial x}\int_{x_1}^x P(t,y)dt$$

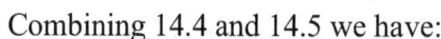

Theorem 5.7, page 178: $= P(x, y)$

Note that, since y is held fixed, $\dfrac{\partial}{\partial x}\displaystyle\int_{x_1}^x P(t,y)dt = \dfrac{d}{dx}\displaystyle\int_{x_1}^x P(t,y)dt$

At this point we know that:

$$\frac{\partial}{\partial x}f(x,y) = \frac{\partial}{\partial x}\left[\int_C F \cdot dr + \int_L F \cdot dr\right] = P(x,y)$$

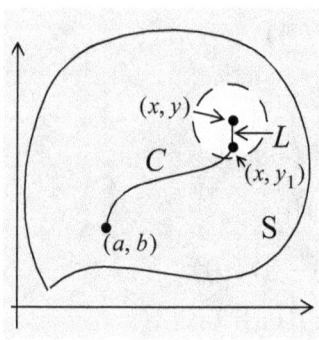

A similar argument with the vertical line L depicted in the margin can be used to show that:

$$\frac{\partial}{\partial y}f(x,y) = Q(x,y)$$

And so we have: $F = Pi + Qj = \nabla f$.

Combining 14.4 and 14.5 we have:

A continuous vector field F on an open connected region S is path-independent **if and only if** it is conservative [i.e: $F = \nabla f$].

We also have:

THEOREM 14.7   (a) If $F = P(x,y)i + Q(x,y)j$ is a conservative vector field on an open region S for which the second partial derivatives of P and Q are continuous, then:

$$\frac{\partial P}{\partial y} = \frac{\partial Q}{\partial x}$$

(b) If $F = P(x,y,z)i + Q(x,y,z)j + R(x,y,z)k$ is a conservative vector field on an open region S for which the second partial derivatives of P, Q, and R are continuous, then:

$$\frac{\partial P}{\partial y} = \frac{\partial Q}{\partial x}, \quad \frac{\partial P}{\partial z} = \frac{\partial R}{\partial x}, \quad \frac{\partial Q}{\partial z} = \frac{\partial R}{\partial y}$$

PROOF: (a) If F is conservative with potential function f, then:
$$F = Pi + Qj = \nabla f(x, y) = f_x i + f_y j$$

$$\Rightarrow P = f_x \quad \text{and} \quad Q = f_y$$

$$\Rightarrow \frac{\partial P}{\partial y} = f_{xy} \overline{\uparrow} f_{yx} = \frac{\partial Q}{\partial x}$$

Theorem 13.1, page 552

As for (b):

CHECK YOUR UNDERSTANDING 14.9

Answer: See page A-33.

Establish part (b) of Theorem 14.7.

As you might anticipate: S is an open region in three-space if for any point (x_0, y_0, z_0) in S there exists $r > 0$ such that the sphere of radius r centered at (x_0, y_0, z_0) is contained in S.

EXAMPLE 14.8 (a) Show that $F(x, y) = xyi + yxj$ is not conservative on any open planar region S.

(b) Show that $F(x, y, z) = yi + 2xj + 3zk$ is not conservative on any open spacial region S.

SOLUTION: (a) For $F = Pi + Qj = xyi + yxj$:

$$\frac{\partial P}{\partial y} = x \quad \text{and} \quad \frac{\partial Q}{\partial x} = y$$

Since the open region S must contain a point (x, y) with $x \neq y$, and since $\frac{\partial P}{\partial y} \neq \frac{\partial Q}{\partial x}$ for any such point, F is not conservative on S.

We previously observed that this vector field F is not independent of path (see Example 14.6 and CYU 14.5, pages 597 and 598.

(b) For $F = Pi + Qj + Rk = yi + 2xj + 3zk$:

$$\frac{\partial P}{\partial y} = 1 \quad \text{and} \quad \frac{\partial Q}{\partial x} = 2 .$$

It follows, from Theorem 14.7, that F is not conservative on S.

CHECK YOUR UNDERSTANDING 14.10

Answer: See page A-33.

Show that the vector field $F(x, y, z) = xi + zj + yzk$ is not conservative on any open spacial region S.

The converse of Theorem 14.7 need not hold:

EXAMPLE 14.9 Show that the vector field

$$F = \frac{y}{x^2 + y^2} i + \frac{-x}{x^2 + y^2} j$$

satisfies the condition $\frac{\partial P}{\partial y} = \frac{\partial Q}{\partial x}$ but fails to be conservative on the open region $S = \{(x, y) | (x, y) \neq (0, 0)\}$.

SOLUTION: Note that F satisfies the given condition:

$$\left[\frac{y}{x^2+y^2}\right]_y = \frac{(x^2+y^2)-(y)(2y)}{(x^2+y^2)^2} = \frac{x^2-y^2}{(x^2+y^2)^2} \longleftarrow$$

and: $\left[\frac{-x}{x^2+y^2}\right]_x = \frac{(x^2+y^2)(-1)-(-x)(2x)}{(x^2+y^2)^2} = \frac{x^2-y^2}{(x^2+y^2)^2} \longleftarrow$

We now show that F is not conservative on S by demonstrating that it is not path independent in S (see Theorem 14.4):

Let C and $\bar{C}$ be the curves from $(1, 0)$ to $(-1, 0)$ with parametrizations:

$$r(t) = (\cos t)i + (\sin t)j, \, 0 \le t \le \pi$$

and $\bar{r}(t) = (\cos t)i - (\sin t)j, \, 0 \le t \le \pi,$ respectively

Then:

$$\int_C F \cdot dr = \int_0^\pi F(r(t)) \cdot r'(t)dt$$

note that $x^2 + y^2 = \cos^2 t + \sin^2 t = 1$

$$= \int_0^\pi [(\sin t)i + (-\cos t)j] \cdot [(-\sin t)i + (\cos t)j]dt$$

$$= \int_0^\pi (-\sin^2 t - \cos^2 t)dt = -\int_0^\pi 1\,dt = -\pi$$

On the other hand:

$$\int_{\bar{C}} F \cdot dr = \int_0^\pi F(\bar{r}(t)) \cdot \bar{r}'(t)dt$$

$$= \int_0^\pi [(-\sin t)i + (-\cos t)j] \cdot [(-\sin t)i + (-\cos t)j]dt$$

$$= \int_0^\pi (\sin^2 t + \cos^2 t)dt = \int_0^\pi 1\,dt = \pi$$

We maintain that the culprit in the above example is not the vector field F, but is the region S, which contains a "hole": it is missing the point $(0, 0)$. Indeed (proof omitted):

Roughly speaking, S is **simply connected** if it has "no holes" in it. To put it another way: the interior of every closed curve in S that does not intersect itself anywhere between its endpoints is entirely contained in S. Visually speaking:

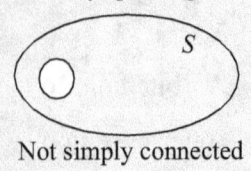

Not simply connected

THEOREM 14.8 Let $F = P(x, y)i + Q(x, y)j$ be a continuously differentiable vector field on a **simply connected open region** S (see margin). If $\frac{\partial P}{\partial y} = \frac{\partial Q}{\partial x}$ in S, then there exists a function f with continuous partial derivatives such that $F = \nabla f$ (i.e. F is conservative).

Merging previous results we come to:

THEOREM 14.9   Let F be a continuous vector field on a simply connected open region S.

The following properties are **equivalent**:

(i) $\oint F \cdot dr = 0$ for any smooth closed curve C in S.

(ii) F is a conservative vector field on S, i.e. $F = \nabla f$.

(iii) $F = \nabla f$ is path-independent on S with

$$\int_C F \cdot dr = f(p_1) - f(p_0)$$

for any smooth curve C from p_0 to p_1.

In the event that we are dealing with a vector field with continuous second partial derivatives, we can throw in a fifth equivalent condition:

(iv) For $F = P(x, y)i + Q(x, y)j$:

$$\frac{\partial P}{\partial y} = \frac{\partial Q}{\partial x}$$

For $F = P(x, y, z)i + Q(x, y, z)j + R(x, y, z)k$:

$$\frac{\partial P}{\partial y} = \frac{\partial Q}{\partial x}, \quad \frac{\partial P}{\partial z} = \frac{\partial R}{\partial x}, \quad \frac{\partial Q}{\partial z} = \frac{\partial R}{\partial y}$$

We now turn our attention to the task of finding potential functions for conservative vector fields:

EXAMPLE 14.10   Show that

$$F(x, y) = (x - y^2)i + (-2xy + \sin y)j$$

is a conservative vector field, and find a potential function for F.

SOLUTION: Theorem 14.9(iv) assures us that F is conservative:

$$(x - y^2)_y = -2y \quad \text{and} \quad (-2xy + \sin y)_x = -2y$$

To find f such that $F = \nabla f$ we consider:

$$(x - y^2)i + (-2xy + \sin y)j = \nabla f(x, y) = f_x(x, y)i + f_y(x, y)j$$

Which brings us to:

(1): $f_x(x, y) = x - y^2$ and   (2): $f_y(x, y) = -2xy + \sin y$

Treating y as a **constant** in (1), we integrate $x - y^2$ with respect to x, realizing that the resulting constant of integration will be a function of y:

$$(3): f(x, y) = \int (x - y^2)dx = \frac{x^2}{2} - xy^2 + g(y)$$

Our next goal is to find $g(y)$. To do so, we take the partial derivative of (3) with respect to y to arrive at:

$$f_y(x, y) = -2xy + [g(y)]_y = -2xy + g'(y)$$
$$\underbrace{}_{g(y) \text{ contains no } x}$$

Bringing (2) into the picture, we have:
$$-2xy + \sin y = -2xy + g'(y) \Rightarrow g'(y) = \sin y$$

So: $g(y) = \int \sin y \, dy = -\cos y + C$

The appearance of the constant C tells us that the potential function f in (3) is not unique.

Letting $C = 0$ we have an answer:

$$F(x, y) = (x - y^2)i + (-2xy + \sin y)j = \nabla f = \nabla\left(\frac{x^2}{2} - xy^2 - \cos y\right)$$

EXAMPLE 14.11

Show that
$$F(x, y, z) = (2xz)i + (-2yz)j + (x^2 - y^2)k$$
is a conservative vector field, and find a potential function for F.

For $P(x,y,z)i + Q(x,y,z)j + R(x,y,z)k$:
$P_y = Q_x, P_z = R_x, Q_z = R_y$

SOLUTION: Verifying (iv) of Theorem 14.9 (see margin):
$$P_y = (2xz)_y = 0 \quad \text{and} \quad Q_x = (-2yz)_x = 0$$

$$P_z = (2xz)_z = 2x \quad \text{and} \quad R_x = (x^2 - y^2)_x = 2x$$

$$Q_z = (-2yz)_z = -2y \quad \text{and} \quad R_y = (x^2 - y^2)_y = -2y$$

At this point we know that $F = \nabla f$ for some f. To find such a potential function we turn to:
$$(2xz)i + (-2yz)j + (x^2 - y^2)k = f_x(x, y, z)i + f_y(x, y, z)j + f_z(x, y, z)k$$
Bringing us to:
$$(1): f_x(x, y, z) = 2xz, \quad (2): f_y(x, y, z) = -2yz, \quad (3): f_z(x, y, z) = x^2 - y^2$$

Treating y and z as **constants** in (1), we integrate $2xz$ with respect to x, realizing that the resulting constant of integration will be a function $g(y, z)$:

$$(4): f(x, y, z) = \int 2xz \, dx = x^2 z + g(y, z)$$

Our next goal is to find $g(y, z)$. To do so, we take the partial derivative of (4) with respect to y to arrive at:
$$f_y(x, y, z) = [g(y, z)]_y$$

From (2): $-2yz = [g(y, z)]_y$

So: $g(y, z) = \int (-2yz) \, dy = -y^2 z + h(z)$

Returning to (4), we have:
$$(5): f(x, y, z) = x^2 z - y^2 z + h(z)$$
Taking the partial derivative with respect to z:
$$f_z(x, y, z) = x^2 - y^2 + h'(z)$$

From (3): $x^2 - y^2 = x^2 - y^2 + h'(z)$

$$h'(z) = 0 \Rightarrow h(z) = C$$

Letting $C = 0$ and returning to (5) we have an answer:

$$F(x, y, z) = \nabla(x^2z - y^2z)$$

Check:

$$\nabla(x^2z - y^2z) = (x^2z - y^2z)_x \mathbf{i} + (x^2z - y^2z)_y \mathbf{j} + (x^2z - y^2z)_z \mathbf{k}$$

$$= (2xz)\mathbf{i} - (2yz)\mathbf{j} + (x^2 - y^2)\mathbf{k}$$

CHECK YOUR UNDERSTANDING 14.11

(a) Show that $F(x, y) = 2xy\mathbf{i} + x^2\mathbf{j}$ is a conservative vector field, and find a potential function for F.

(b) Show that $F(x, y) = (2xyz)\mathbf{i} + (x^2z)\mathbf{j} + (x^2y)\mathbf{k}$ is a conservative vector field, and find a potential function for F.

Answer: See page A-33.

EXAMPLE 14.12 Evaluate $\int_C F \cdot dr$ where:

$$F(x, y, z) = (2xz)\mathbf{i} + (-2yz)\mathbf{j} + (x^2 - y^2)\mathbf{k}$$

and where the curve C has parametrization:

$$r(t) = (t\cos t)\mathbf{i} + t\mathbf{j} + (\sin^2 t)\mathbf{k}, \ 0 \le t \le \frac{\pi}{4}.$$

SOLUTION: Approach (a): We showed, in Example 14.11, that F is a conservative field with potential function $f(x, y, z) = x^2z - y^2z$.

Since $r(0) = (0, 0, 0)$ and $r\left(\frac{\pi}{4}\right) = \left(\frac{\pi}{4}\cos\frac{\pi}{4}, \frac{\pi}{4}, \sin^2\frac{\pi}{4}\right)$ the curve C

starts at the point $(0, 0, 0)$ and ends at the point $\left(\frac{\pi}{4\sqrt{2}}, \frac{\pi}{4}, \frac{1}{2}\right)$.

Employing Theorem 14.9(iii) we conclude that:

$$\int_C F \cdot dr = f\left(\frac{\pi}{4\sqrt{2}}, \frac{\pi}{4}, \frac{1}{2}\right) - f(0, 0, 0)$$

$$= \underbrace{\left[\left(\frac{\pi}{4\sqrt{2}}\right)^2\left(\frac{1}{2}\right) - \left(\frac{\pi}{4}\right)^2\left(\frac{1}{2}\right)\right]}_{x^2z - y^2z} - 0 = -\frac{\pi^2}{64}$$

Approach (b): The above approach hinged on the lengthy solution of Example 14.11. Here is a more independent solution:

Rather than performing the tedious integration $\int_C F \cdot dr$ along the path C, we chose to take the path $\overline{C} = C_1 \cup C_2 \cup C_3$ from

$(0, 0, 0)$ to $\left(\frac{\pi}{4\sqrt{2}}, \frac{\pi}{4}, \frac{1}{2}\right)$ (see margin).

Turning to the line-segment parametrization

$$r(t) = (1-t)r_0 + tr_1, \ 0 \le t \le 1$$

we have:

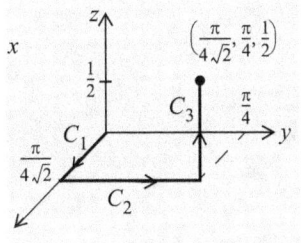

$$C_1: \quad r_1 = (1-t)(0,0,0) + t\left(\frac{\pi}{4\sqrt{2}}, 0, 0\right) = \left(\frac{\pi}{4\sqrt{2}}t, 0, 0\right), 0 \le t \le 1$$

$$C_2: \quad r_2 = (1-t)\left(\frac{\pi}{4\sqrt{2}}, 0, 0\right) + t\left(\frac{\pi}{4\sqrt{2}}, \frac{\pi}{4}, 0\right) = \left(\frac{\pi}{4\sqrt{2}}, \frac{\pi}{4}t, 0\right), 0 \le t \le 1$$

$$C_3: \quad r = (1-t)\left(\frac{\pi}{4\sqrt{2}}, \frac{\pi}{4}, 0\right) + t\left(\frac{\pi}{4\sqrt{2}}, \frac{\pi}{4}, \frac{1}{2}\right) = \left(\frac{\pi}{4\sqrt{2}}, \frac{\pi}{4}, \frac{t}{2}\right), 0 \le t \le 1$$

For $F(x,y,z) = (2xz)i + (-2yz)j + (x^2 - y^2)k$ we then have:

$$\int_{\overline{C}} F \cdot dr = \int_{C_1} F \cdot dr_1 + \int_{C_2} F \cdot dr_2 + \int_{C3} F \cdot dr_3$$

$$= \int_0^1 \left\{ \left[\left(\frac{\pi}{4\sqrt{2}}t\right)^2 k\right] \cdot \left(\frac{\pi}{4\sqrt{2}}\right)i \right\} dt + \int_0^1 \left\{ \left[\left(\frac{\pi}{4\sqrt{2}}\right)^2 - \left(\frac{\pi}{4}t\right)^2\right]k \cdot \left(\frac{\pi}{4}\right)j \right\} dt$$

$$+ \int_0^1 \left\{ \left[2\left(\frac{\pi}{4\sqrt{2}}\right)\left(\frac{t}{2}\right)i - 2\left(\frac{\pi}{4}\right)\left(\frac{t}{2}\right)j + \left(\left(\frac{\pi}{4\sqrt{2}}\right)^2 - \left(\frac{\pi}{4}\right)^2\right)k\right] \cdot \left(\frac{1}{2}\right)k \right\} dt$$

$$= \int_0^1 0\, dt + \int_0^1 0\, dt + \int_0^1 \frac{1}{2}\left(\frac{\pi^2}{32} - \frac{\pi^2}{16}\right) dt = \left(-\frac{\pi^2}{64}t\right)\Big|_0^1 = -\frac{\pi^2}{64}$$

CHECK YOUR UNDERSTANDING 14.12

Use both of the approaches of the previous example to evaluate $\int_C F \cdot dr$ for $F(x,y) = 2xyi + x^2 j$ and where the curve C has parametrization: $r(t) = (t\cos^2 t)i + t^2 j$ for $0 \le t \le \frac{\pi}{4}$.

Answer: $\dfrac{\pi^4}{1024}$

CONSERVATIVE FORCE FIELDS

By now, the term **conservative force field** should be self-explanatory. That being the case:

EXAMPLE 14.13 Find the work done by a particle in moving from the origin to the point $(3,6)$, while subjected to the force field $F(x,y) = e^y i + xe^y j$.

Note that F is conservative:
$(e^y)_y = (xe^y)_x = e^y$

SOLUTION: We find a potential function f for F:

$$e^y i + xe^y j = f_x(x,y)i + f_y(x,y)j$$

$$f_x(x, y) = e^y \qquad\qquad f_y(x, y) = xe^y$$

integrating with respect to x, holding y fixed: integrating with respect to y, holding x fixed:

$$f(x, y) = xe^y + g(y) \qquad\qquad f(x, y) = xe^y + h(x)$$

It follows that $f(x, y) = xe^y$ is a potential function for $\boldsymbol{F}$. So:

$$\int\limits_C \boldsymbol{F} \cdot \boldsymbol{dr} = f(3, 6) - f(0, 0) = 3e^6$$

CHECK YOUR UNDERSTANDING 14.13

Show that the work done in moving a particle from the $(0, 0, 0)$ to (x_0, y_0, z_0) when subjected solely to the force of gravity $-mg$, is equal to $-mgz_0$, independently of the chosen path taken from $(0, 0, 0)$ to (x_0, y_0, z_0).

Answer: See page A-35.

	EXERCISES	

Exercises 1-12. Determine if the vector field F is conservative on the region S. If it is, find a function f such that $F = \nabla f$.

1. $F(x, y) = (x - y)i + (x - 2)j$, $S = R^2$.

2. $F(x, y) = 2xy^2 i + (1 + 3x^2 y^2)j$, $S = R^2$.

3. $F(x, y) = 2xy^3 i + (1 + 3x^2 y^2)j$, $S = R^2$.

4. $F(x, y) = (3 + 2xy)i + (x^2 - 3y^2)j$, $S = R^2$.

5. $F(x, y) = (y \cos x + 2xe^y)i + (\sin x + xe^y + 4)j$, $S = R^2$.

6. $F(x, y) = (y \cos x + 2xe^y)i + (\sin x + x^2 e^y + 4)j$, $S = R^2$.

7. $F(x, y, z) = yz^2 i + (2xyz - y)j + (xz^2 - z)k$, $S = R^3$.

8. $F(x, y, z) = yz^2 i + (xz^2 - z)j + (2xyz - y)k$, $S = R^3$.

9. $F(x, y, z) = (e^x \cos y + yz)i + (xz - e^x \sin y)j + (xy + z)k$, $S = R^3$.

10. $F(x, y, z) = (xz - e^x \sin y)i + (e^x \cos y + yz)j + (xy + z)k$, $S = R^3$.

11. $F(x, y, z) = y^2 i + (2xy + e^{3z})j + 3ye^{3z}k$, $S = R^3$.

12. $F(x, y, z) = y^2 i + (2xy + e^{3z})j + ye^{3z}k$, $S = R^3$.

Exercises 13-16. Verify that the vector field F is conservative. Evaluate $\int_C F \cdot dr$ using both Approach (a) and Approach (b) of Example 14.12.

13. $F(x, y) = yi + xj$ and C with parametrization $r(t) = \sin t\, i + 2t j$ for $\frac{\pi}{2} \leq t \leq \pi$.

14. $F(x, y) = (xy)i + (x - y)j$ and C with parametrization $r(t) = t^2 i + 2t^3 j$ for $0 \leq t \leq 1$.

15. $F(x, y, z) = (yz + 1)i + (xz + 2)j + (xy + 3)k$ where C is any smooth curve from the origin to the origin to the point $(1, 2, 3)$.

16. $F(x, y, z) = e^y i + xe^y j + (z + 1)e^z k$ and $r(t) = ti + t^2 j + t^3 k$ for $0 \leq t \leq 1$.

Exercises 17-25. Verify that the vector field F is conservative and then evaluate $\int_C F \cdot dr$ where C has parametrization r.

17. $F(x, y) = (3 + 2xy)i + (x^2 - 3y^2)j$, $r(t) = e^t i + \dfrac{1}{t^2 + 2}j$ for $0 \leq t \leq 1$.

18. $F(x, y) = (3 + 2xy)i + (x^2 - 3y^2)j$, $r(t) = \sin t i + 2t^3 j$ for $0 \leq t \leq 1$.

19. $F(x, y) = xy^2 i + x^2 y j$, $r(t) = \left(t + \sin\dfrac{\pi t}{2}\right)i + \left(t + \cos\dfrac{\pi t}{2}\right)j$ for $0 \leq t \leq 1$.

20. $F(x, y) = (x + 2xy)i + (x^2 - 3y^2)j$, $r(t) = \sin 2t i + t^3 j$ for $0 \leq t \leq 1$.

21. $F(x, y) = (3 + 2xy)i + (x^2 - 3y^2)j$, $r(t) = e^t \sin t i + e^t \cos t j$ for $0 \leq t \leq \pi$.

22. $F(x, y, z) = yz^2 i + (xz^2 - z)j + (2xyz - y)k$, $r(t) = t^2 i + t^3 j + \dfrac{1}{t^2 + 1}k$ for $0 \leq t \leq 1$.

23. $F(x, y, z) = yz^2 i + (xz^2 - z)j + (2xyz - y)k$, $r(t) = \dfrac{1}{t^2 + 2}i + t^2 j + k$ for $0 \leq t \leq 1$.

24. $F(x, y, z) = (yz + 1)i + (xz + 2)j + (xy + 3)k$, $r(t) = t^2 i + t^3 j + \dfrac{1}{t^2 + 1}k$ for $0 \leq t \leq 1$.

25. $F(x, y, z) = (yz + 1)i + (xz + 2)j + (xy + 3)k$, $r(t) = e^t i + t^3 j + e^t t^2 k$ for $0 \leq t \leq 1$.

Exercises 26-29. Determine a function $K(x, y)$ for which the field F is conservative in $S = R^2$.
(no unique solutions)

26. $F(x, y) = (x + 2y)i + K(x, y)j$

27. $F(x, y) = (2xy^2)i + K(x, y)j$

28. $\quad F(x, y) = K(x, y)i + xy^3 j$

29. $F(x, y) = K(x, y)i + (x^2 e^y)j$

Exercises 30-32. Determine functions $M(x, y, z)$ and $N(x, y, z)$ for which the field F is conservative in $S = R^3$.

30. $F(x, y, z) = 2xy^3 i + M(x, y, z)j + N(x, y, z)k$

31. $F(x, y, z) = M(x, y, z)i + 2xy^3 j + N(x, y, z)k$

32. $F(x, y, z) = M(x, y, z)i + N(x, y, z)j + 2xy^3 k$

33. Verify that $F(x, y) = e^y i + xe^y j$ is a conservative force field. Determine the work done if a particle subjected to F moves around the circle $r(t) = \cos t i + \sin t j$ for $0 \leq t \leq 2\pi$.

34. Verify that $F(x, y) = e^y i + xe^y j$ is a conservative force field. Determine the work done if a particle subjected to F moves over the semicircle $r(t) = \cos t i + \sin t j$ for $0 \leq t \leq \pi$.

35. Verify that $F(x, y, z) = (x^2 + y)i + (y^2 + x)j + (ze^z)k$ is a conservative force field. Determine the work done if a particle subjected to F moves over the helix
$$r(t) = \cos ti + \sin tj + \frac{t}{2\pi}k \text{ for } 0 \le t \le 2\pi.$$

36. Verify that $F(x, y, z) = e^{yz}i + (xze^{yz} + z\cos y)j + (xye^{yz} + \sin y)k$ is a conservative force field. Determine the work done if a particle subjected to F moves over the curve $r(t) = ti + t^2j + t^3k$ for $-1 \le t \le 1$.

37. Prove that any two potential functions, $f(x, y)$ and $h(x, y)$, for a conservative vector field $F(x, y) = P(x, y)i + Q(x, y)j$ can only differ by a constant.

§3. GREEN'S THEOREM

George Green (1793-1841). a self-taught British mathematician.

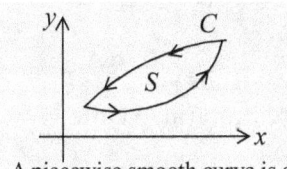

A piecewise smooth curve is a curve that is composed of finitely many smooth curves.

Note that Green's Theorem relates a double integral over a region D to a line integral on the boundary C of that region. It is analogous to the Fundamental Theorem of Calculus that relates an integral over $[a, b]$ to the values of f at its endpoints (boundary of the interval).

$$\int_a^b g'(x)dx = g(b) - g(a)$$

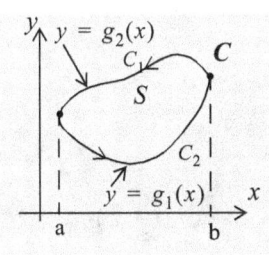

Green's Theorem relates a line integral along a closed curve to a double integral over the region enclosed by that curve.

Here is one form of that important result (proof omitted):

THEOREM 14.10

GREEN'S THEOREM
FORM $F \cdot T$

Let S be a simply connected region in R^2 bounded by a closed counterclockwise curve C that is parameterized by a piecewise smooth one-to-one vector-valued function $r(t)$, $a \le t \le b$ (see margin). If

$$F = P(x, y)i + Q(x, y)j$$

has continuous first-order partial derivatives on some open region containing S, then:

$$\oint_C F \cdot T\, ds = \oint_C P\, dx + Q\, dy = \iint_S \left(\frac{\partial Q}{\partial x} - \frac{\partial P}{\partial y} \right) dA$$

PROOF: The claim will be established once we show that:

$$(1): \int_C P\, dx = -\iint_D \frac{\partial P}{\partial y} dA \quad \text{and} \quad (2): \int_C Q\, dy = \iint_D \frac{\partial Q}{\partial x} dA$$

We content ourselves by establishing (1) when the curve C is bounded below and above by continuous functions $g_1(x)$ and $g_2(x)$, respectively — giving rise to the decomposition of C into two pieces C_1 and C_2 (see margin):

$$-\iint_D \frac{\partial P}{\partial y} dA = -\int_a^b \int_{g_1(x)}^{g_2(x)} \frac{\partial P}{\partial y} dy\, dx = -\int_a^b [P(x, y)] \Big|_{g_1(x)}^{g_2(x)} dx$$

$$= -\int_a^b [P(x, g_2(x)) - P(x, g_1(x))] dx$$

$$= -\int_a^b P(x, g_2(x))dx + \int_a^b P(x, g_1(x))dx$$

$$= -\int_{-C_1} P(x, y)dx + \int_{C_2} P(x, y)dx$$

$$= \int_{C_1} P(x, y)dx + \int_{C_2} P(x, y)dx = \int_C P\, dx$$

A similar argument can be used to establish the validity of (2) when the curve C is bounded on the left and on the right by continuous functions $f_1(y)$ and $f_2(y)$, respectively.

EXAMPLE 14.14 Verify Green's Theorem for:

$$\text{(a) } F(x, y) = y\boldsymbol{i} - x\boldsymbol{j}$$

$$\text{(b) } \oint_C 2y\,dx - 3x^2\,dy$$

where: $C = \{(x, y) \mid x^2 + y^2 = 1\}$.

SOLUTION: The parametrization $r(t) = (\cos t)\boldsymbol{i} + (\sin t)\boldsymbol{j}$, $0 \le t \le 2\pi$ traverses the unit circle C in the counterclockwise direction (also called a **positive orientation**). We then have:

(a) $\displaystyle \oint_C F \cdot T\,ds = \oint_C F \cdot d\boldsymbol{r} = \int_0^{2\pi} [(\sin t\,\boldsymbol{i} - \cos t\,\boldsymbol{j}) \cdot (-\sin t\,\boldsymbol{i} + \cos t\,\boldsymbol{j})]\,dt$

$$= \int_0^{2\pi} [-\sin^2(t) - \cos^2(t)]\,dt = \int_0^{2\pi} -1\,dt = -2\pi$$

Green's Theorem asserts that $\displaystyle \iint_S \left(\frac{\partial Q}{\partial x} - \frac{\partial P}{\partial y}\right) dA$ must also equal -2π,

where Q and P stem from $F(x, y) = P(x, y)\boldsymbol{i} + Q(x, y)\boldsymbol{j} = y\boldsymbol{i} - x\boldsymbol{j}$ and where $S = \{(x, y) \mid x^2 + y^2 \le 1\}$. Let's check it out:

$$\iint_S \left(\frac{\partial Q}{\partial x} - \frac{\partial P}{\partial y}\right) = \iint_S \left(\frac{\partial}{\partial x}(-x) - \frac{\partial}{\partial y}(y)\right) dA = -2\iint_S dA \underset{\substack{\uparrow \\ \text{a circle of radius 1 has area } \pi}}{=} -2\pi$$

(b)

$\displaystyle \oint_C 2y\,dx - 3x^2\,dy$

$\displaystyle = \int_0^{2\pi} [(2\sin t)(\cos t)' - (3\cos^2 t)(\sin t)']\,dt$

$\displaystyle = \int_0^{2\pi} [-2\sin^2 t - 3\cos^3 t]\,dt$

$\displaystyle = \int_0^{2\pi} \left[-2\left(\frac{1 - \cos 2t}{2}\right) - 3(1 - \sin^2 t)\cos t\right]dt$

$\displaystyle = \left[-t + \frac{\sin 2t}{2} - 3\left(\sin t - \frac{1}{3}(\sin^3 t)\right)\right]\Big|_0^{2\pi}$

$\displaystyle = -2\pi$

$\displaystyle \iint_S \left(\frac{\partial Q}{\partial x} - \frac{\partial P}{\partial y}\right) dA = \iint_S (-6x - 2)\,dA$

Theorem 11.6 page 457: $\displaystyle = \int_0^{2\pi}\int_0^1 (-6r\cos\theta - 2)r\,dr\,d\theta$

$\displaystyle = \int_0^{2\pi}\int_0^1 (-6r^2\cos\theta - 2r)\,dr\,d\theta$

$\displaystyle = \int_0^{2\pi} (-2r^3\cos\theta - r^2)\Big|_0^1\,d\theta$

$\displaystyle = \int_0^{2\pi} (-2\cos\theta - 1)\,d\theta$

$\displaystyle = (-2\sin\theta - \theta)\Big|_0^{2\pi} = -2\pi$

CHECK YOUR UNDERSTANDING 14.14

Answer: See page A-34.

For $F(x, y) = x^2\boldsymbol{i} + xy\boldsymbol{j}$, evaluate $\displaystyle \oint_C F \cdot T\,ds$ both directly and by using Green's Theorem.

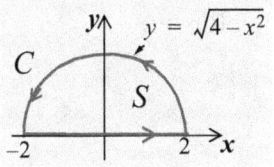

EXAMPLE 14.15 Find the work done by the force:

$$F(x, y) = 2x\mathbf{i} + (x^3 + 3xy^2)\mathbf{j}$$

in moving a particle once along the directed curve C in the margin.

SOLUTION: The work is given by:

$$W = \oint_C F \cdot T ds = \iint_S \left(\frac{\partial Q}{\partial x} - \frac{\partial P}{\partial y}\right) dA = \iint_S (3x^2 + 3y^2 - 0) dA$$

Changing to polar coordinates (page 457) we have:

$$W = \iint_S (3x^2 + 3y^2) dA = \int_0^\pi \int_0^2 3r^2 r\, dr\, d\theta$$

$$= \int_0^\pi \left(\frac{3r^4}{4}\right)\Big|_0^2 d\theta = \int_0^\pi 12\, d\theta = 12\pi$$

Though the particle moved back to its starting point, the work done is not zero. Not an issue, since the force field is not conservative.

CHECK YOUR UNDERSTANDING 14.15

Find the work done by the force field $F(x, y) = y\mathbf{i} - 2xy\mathbf{j}$ in moving a particle once along the indicated curve C.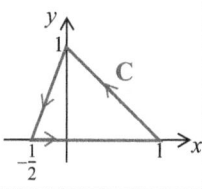

Answer: $-\dfrac{5}{4}$

Green's theorem provides a formula for the area of planar regions:

THEOREM 14.11 If the conditions of Theorem 14.10 are met for the region S bounded by the closed curve C, then:

$$\text{Area of } S: \quad A = \frac{1}{2}\oint_C -y\, dx + x\, dy$$

PROOF: Letting $P = -\dfrac{y}{2}$ and $Q = \dfrac{x}{2}$ in Green's Theorem we have:

$$\oint_C -\frac{y}{2} dx + \frac{x}{2} dy = \iint_S \left[\left(\frac{x}{2}\right)_x - \left(-\frac{y}{2}\right)_y\right] dA = \iint_S \left(\frac{1}{2} - \left(-\frac{1}{2}\right)\right) dA$$

$$= \iint_S dA = A$$

Or: $\dfrac{1}{2}\oint_C -y\, dx + x\, dy = A$

EXAMPLE 14.16 Find the area enclosed by the ellipse:

$$\frac{x^2}{a^2} + \frac{y^2}{b^2} = 1$$

SOLUTION: Choosing the parametrization:

$$r(t) = (a\cos t)i + (b\sin t)j, \; 0 \le t \le 2\pi$$

for the ellipse, we have:

$$A = \frac{1}{2}\oint_C -y\,dx + x\,dy = \frac{1}{2}\int_0^{2\pi} [-b\sin t(a\cos t)' + a\cos t(b\sin t)']dt$$

$$= \frac{1}{2}\int_0^{2\pi} (ba\sin^2 t + ab\cos^2 t)dt = \frac{1}{2}\int_0^{2\pi} abdt = \pi ab$$

CHECK YOUR UNDERSTANDING 14.16

Answer: $\frac{3}{8}\pi$

Use Theorem 14.11 to find the area enclosed by the closed curve with parametrization $r(t) = (\cos^3 t)i + (\sin^3 t)j, \; 0 \le t \le 2\pi$.

As you know, the work W done by a force field F in moving a particle along a smooth curve C is given by $W = \int_C F \cdot T\,ds$ (Theorem 14.2, page 596. Replacing the unit tangent vector T with the unit normal vector n we arrive at another important concept:

DEFINITION 14.5
Flux Across a
Closed Curve

Let $r(t)$ trace out a smooth closed curve C in a counterclockwise direction exactly once, with C contained in the domain of a differentiable vector field F. The **flux of F across C** is given by:

$$\text{Flux}(F_C) = \oint_C F \cdot n\,ds$$

where n is the outward pointing unit vector to the curve.

Just as $F \cdot T$ denotes the scalar component of F tangent to the curve, so then does $F \cdot n$ represent the scalar component of F normal to the curve. In particular, if $F = P(x,y)i + Q(x,y)j$ is the velocity field of a fluid flowing in a planar region containing the closed curve C, then the integral $\oint_C F(x,y) \cdot T(x,y)ds$ represents the fluid's flow along the curve, while $\oint_C F(x,y) \cdot n(x,y)ds$ yields the net outward flow of the liquid across that curve.

Our next goal is to arrive at a method for evaluating the flux-integral. As a first step, we observe that (see margin):

If the planar curve is traversed in a counterclockwise direction, then its outward unit normal n at a point on the curve is given by the cross product
$$n = T \times k$$
where T is the unit tangent vector at the point, and k is the unit coordinate vector along the z-axis.

$$n = T \times k = \frac{r'(t)}{\|r'(t)\|} \times k$$

$$= \frac{1}{\|r'(t)\|}[(x'(t)i + y'(t)j + 0k) \times (0i + 0j + 1k)]$$

$$= \frac{1}{\|r'(t)\|}\left(\det\begin{bmatrix} i & j & k \\ x'(t) & y'(t) & 0 \\ 0 & 0 & 1 \end{bmatrix}\right) = \frac{1}{\|r'(t)\|}[y'(t)i - x'(t)j]$$

Proceeding as in Theorem 14.2, page 596, we then have:

$$\text{Flux}(F_C) = \oint_C F(x, y) \cdot n(x, y)ds$$

Theorem 14.1: $$= \int_a^b F(r(t)) \cdot n(r(t)) \sqrt{\left(\frac{dx}{dt}\right)^2 + \left(\frac{dy}{dt}\right)^2}\, dt$$

$$= \int_a^b F(r(t)) \cdot \frac{y'(t)i - x'(t)j}{\|r'(t)\|}\|r'(t)\|\, dt$$

$$= \int_a^b F(r(t)) \cdot [y'(t)i - x'(t)j]\, dt$$

Summarizing:

THEOREM 14.12 If $r(t) = x(t)i + y(t)j, a \le t \le b$ traces out a smooth closed curve C in a counterclockwise direction exactly once, with C in the domain of a differentiable vector field $F = Pi + Qj$ then:

$$\oint_C F \cdot n ds = \int_a^b (Pi + Qj) \cdot \left(\frac{dy}{dt}i - \frac{dx}{dt}j\right) dt = \oint_C P dy - Q dx$$

EXAMPLE 14.17 Find the flux of $F(x, y) = (x + y)i + yj$ across the circle $x^2 + y^2 = 1$.

SOLUTION: The parametrization $r(t) = (\cos t)i + (\sin t)j, 0 \le t \le 2\pi$ traces the unit circle C exactly once in the counterclockwise direction. Appealing to Theorem 14.12, with $P = x + y = \cos t + \sin t$ and $Q = y = \sin t$ we have:

$$\text{Flux}(F_C) = \int_0^{2\pi} [(\cos t + \sin t)(\sin t)' - (\sin t)(\cos t)']dt$$

$$= \int_0^{2\pi} (\cos^2 t + \cos t \sin t + \sin^2 t)dt = \int_0^{2\pi} (1 + \cos t \sin t)dt$$

$$= \left(t + \frac{1}{2}\sin^2 t\right)\Big|_0^{2\pi} = 2\pi$$

CHECK YOUR UNDERSTANDING 14.17

Find the flux of $F(x, y) = (y^2 - x^2)i + (x^2 + y^2)j$ across the triangle bounded by $y = x$, $x = 3$, and $y = 0$.

Tweaking Theorem 14.12 a bit we arrive at another form of Green's Theorem:

THEOREM 14.13

GREEN'S THEOREM FORM $F \cdot n$

Let C be a piecewise-smooth counterclockwise oriented simple closed curve in the plane. If the region S consisting of C and its interior is simply connected, and if $F = P(x, y)i + Q(x, y)j$ has continuous first-order partial derivatives on some open region containing D, then:

$$\oint_C F \cdot n \, ds = \oint_C P \, dy - Q \, dx = \iint_S \left(\frac{\partial P}{\partial x} + \frac{\partial Q}{\partial y} \right) dA$$

PROOF: Applying Theorem 14:10:

$$\oint_C P \, dx + Q \, dy = \iint_S \left(\frac{\partial Q}{\partial x} - \frac{\partial P}{\partial y} \right) dA$$

to the vector function $F_2 = -Qi + Pj$ we have:

$$\oint_C -Q \, dx + P \, dy = \iint_S \left[\frac{\partial P}{\partial x} - \left(\frac{\partial Q}{\partial y} \right) \right] dA = \iint_S \left(\frac{\partial P}{\partial x} + \frac{\partial Q}{\partial y} \right) dA$$

EXAMPLE 14.18 Use Green's Theorem to find the flux of $F(x, y) = (x + y)i + yj$ across the circle $x^2 + y^2 = 1$.

Compare with Example 14.17,

SOLUTION: Turning to Theorem 14.13 with

$$F(x, y) = P(x, y)i + Q(x, y)j = (x + y)i + yj$$

we have: $\displaystyle \oint_C F \cdot n \, ds = \iint_D \left(\frac{\partial P}{\partial x} + \frac{\partial Q}{\partial y} \right) dA = \iint_D (1 + 1) \, dA = 2\pi$

CHECK YOUR UNDERSTANDING 14.18

Use Green's Theorem to address CYU 14.17:

	EXERCISES	

Exercises 1-11. Verify both forms of Green's Theorem (14.10).

1. $F(x, y) = (x - y)i + xj$, $D = \{(x, y) | x^2 + y^2 \le 1\}$.

2. $F(x, y) = (xy)i + x^2 j$, D is the rectangular region with vertices $(0, 0), (2, 0), (2, 3)$, and $(0, 3)$.

3. $F(x, y) = (xy)i + (x^2 - y^2)j$, D is the rectangular region with vertices $(0, 0), (0, 1), (1, 0)$, and $(1, 1)$.

4. $F(x, y) = (-xy^2)i + xy^2 j$, $D = \{(x, y) | x^2 + y^2 \le a^2\}$

5. $F(x, y) = x^2 i - xy^2 j$, D is the triangular region with vertices $(0, 0), (0, 1)$, and $(1, 1)$.

6. $F(x, y) = xi + yj$, D is the region bounded by the line segments from $(0, 0)$ to $(0, 1)$ and from $(0, 0)$ to $(1, 0)$, along with the parabola $y = 1 - x^2$ from $(1, 0)$ to $(0, 1)$.

7. $F(x, y) = -yi + xj$, D is the region bounded by the line segments $(0, 0)$ to $(0, 1)$ and from $(0, 0)$ to $(1, 0)$, along with the quarter-circle $y = \sqrt{1 - x^2}$ in the first quadrant.

8. $\oint_C y^2 dx + x^2 dy$ and C is the triangle with vertices $(0, 0), (1, 0)$, and $(0, 1)$.

9. $\oint_C xy dx + x^2 y^3 dy$ and C is the triangle with vertices $(0, 0), (1, 0)$, and $(1, 2)$.

10. $\oint_C (x^2 - y^2) dx + x dy$, and C is the circle of radius 3 centered at the origin.

11. $\oint_C y^3 dx - x^3 dy$ and C is the circle of radius 2 centered at the origin.

Exercises 12-17. Use Green's Theorem to evaluate the line integral along the given positively oriented curve C.

12. $\oint_C y^4 dx + 2xy^3 dy$ and C is the circle $x^2 + y^2 = 9$.

13. $\oint_C y^4 dx + 2xy^3 dy$ and C is the ellipse $\dfrac{x^2}{2} + y^2 = 1$.

14. $\oint_C \cos y\,dx + x^2 \sin y\,dy$ and C is the rectangle with vertices at $(0, 0)$, $(2, 0)$, $(2, 3)$, and $(0, 3)$.

15. $\oint_C y\,dx - x\,dy$ and C consists of the line segments joining $(0, 0)$ to $(0, 1)$, joining $(0, 0)$ to $(1, 0)$, and the parabola $y = 1 - x^2$ from $(1, 0)$ to $(0, 1)$.

16. $\oint_C -2y^3\,dx + 2x^3\,dy$ and C is the circle of radius 3 centered at the origin.

17. $\oint_C y\cos x\,dx + x^2\,dy$ and C is the boundary of the unit square in the first quadrant with a vertex at the origin.

Exercises 18-22. Find the work done by the force $\boldsymbol{F}$ in moving a particle once along the given positively oriented curve C.

18. $\boldsymbol{F}(x, y) = 2xy^3\boldsymbol{i} + 4x^2y^2\boldsymbol{j}$ and C is the region in the first quadrant enclosed by the x-axis, the line $x = 1$ and the curve $y = x^3$.

19. $\boldsymbol{F}(x, y) = (x^2 + xy)\boldsymbol{i} + xy^2\boldsymbol{j}$ and C is the triangle with vertices $(0, 0)$, $(1, 0)$, and $(0, 1)$.

20. $\boldsymbol{F}(x, y) = (4x - 2y)\boldsymbol{i} + (2x - 4y)\boldsymbol{j}$ and C is the circle $(x - 2)^2 + (y - 2)^2 = 4$.

21. $\boldsymbol{F}(x, y) = 7x\boldsymbol{i} + (x^3 + 3xy^2)\boldsymbol{j}$ and C is bounded by the line segment from $(-2, 0)$ to $(2, 0)$ and the semicircle $y = \sqrt{4 - x^2}$.

22. $\boldsymbol{F}(x, y) = -2y^3\boldsymbol{i} + 2x^3\boldsymbol{j}$ and C is the circle of radius 3 centered at the origin.

Exercises 23-27. Use Theorem 14.11 to find the area of the region D.

23. $D = \{(x, y)|x^2 + y^2 \le r^2\}$.

24. $D = \{(x, y)|x^2 + 2y^2 \le 2\}$

25. D is the region between the curves $y = x^2$ and $y = x^3$.

26. D is the curve $\boldsymbol{r}(t) = t^2\boldsymbol{i} + \left(\dfrac{t^3}{3} - t\right)\boldsymbol{j}$, $-\sqrt{3} \le t \le \sqrt{3}$.

27. D is the region lying between the x-axis and one arch of the cycloid with parametric equations $x = a(t - \sin t)$ and $y = a(1 - \cos t)$.

Exercises 28-31. Verify both forms of Green's Theorem (Theorem 14.13). .

28. $F(x, y) = 2xy^3\mathbf{i} + 4x^2y^2\mathbf{j}$ and C is the region in the first quadrant enclosed by the x-axis, the line $x = 1$ and the curve $y = x^3$.

29. $F(x, y) = (x - y)\mathbf{i} + x\mathbf{j}$ and $C = \{(x, y) \mid x^2 + y^2 = 1\}$.

30. $F(x, y) = (x + y)\mathbf{i} - (x^2 + y^2)\mathbf{j}$ and C is the triangular region with vertices $(-1, 0)$, $(0, 1)$, and $(1, 0)$.

31. $F(x, y) = (4x - 2y)\mathbf{i} + (2x - 4y)\mathbf{j}$ and C is the circle $(x - 2)^2 + (y - 2)^2 = 4$.

§4. CURL AND DIV

In this section we introduce two functions on vector fields that play important roles in numerous applications. We begin with the curl-function that assigns vectors to elements in the field:

DEFINITION 14.6

curl(F)

Let $F(x, y) = P(x, y)i + Q(x, y)j$. The **curl of F**, denoted by curl(F) is the (spacial) vector field:

$$\text{curl}(F) = \left(\frac{\partial Q}{\partial x} - \frac{\partial P}{\partial y}\right)k \quad (*)$$

For $F(x, y, z) = P(x, y, z)i + Q(x, y, z)j + R(x, y, z)k$:

$$\text{curl}(F) = \left(\frac{\partial R}{\partial y} - \frac{\partial Q}{\partial z}\right)i + \left(\frac{\partial P}{\partial z} - \frac{\partial R}{\partial x}\right)j + \left(\frac{\partial Q}{\partial x} - \frac{\partial P}{\partial y}\right)k \quad (**)$$

(assuming the indicated partial derivatives exist)

To help us recall the expression in (**) we introduce the following "delta-operator:"

$$\nabla = \frac{\partial}{\partial x}i + \frac{\partial}{\partial y}j + \frac{\partial}{\partial z}k$$

Symbolically, we then have:

The ∇-symbol was first introduced in the definition of the gradient function (page 561):

$\nabla f(x, y) = f_x(x, y)i + f_y(x, y)j$

In that role, ∇f assigned vectors to points in the domain of a real-valued function f.

Here, the delta-operator assigns vectors to vectors in a vector field.

$$\text{curl}(F) = \nabla \times F \underset{\substack{\uparrow \\ \text{see page 506}}}{=} \det \begin{vmatrix} i & j & k \\ \frac{\partial}{\partial x} & \frac{\partial}{\partial y} & \frac{\partial}{\partial z} \\ P & Q & R \end{vmatrix}$$

see page 505: $$= \det \begin{vmatrix} \frac{\partial}{\partial y} & \frac{\partial}{\partial z} \\ Q & R \end{vmatrix} i - \det \begin{vmatrix} \frac{\partial}{\partial x} & \frac{\partial}{\partial z} \\ P & R \end{vmatrix} j + \det \begin{vmatrix} \frac{\partial}{\partial x} & \frac{\partial}{\partial y} \\ P & Q \end{vmatrix} k$$

$$= \left(\frac{\partial R}{\partial y} - \frac{\partial Q}{\partial z}\right)i + \left(\frac{\partial P}{\partial z} - \frac{\partial R}{\partial x}\right)j + \left(\frac{\partial Q}{\partial x} - \frac{\partial P}{\partial y}\right)k$$

Note that the delta operator can also be used to generate (*) in Definition 14.6. Simply move $F(x, y) = P(x, y)i + Q(x, y)j$ up a notch:

$$F(x, y, z) = P(x, y)i + Q(x, y)j + 0k$$

Then: $$\text{curl}(F) = \nabla \times F = \det \begin{vmatrix} i & j & k \\ \frac{\partial}{\partial x} & \frac{\partial}{\partial y} & \frac{\partial}{\partial z} \\ P & Q & 0 \end{vmatrix} \underset{\substack{\uparrow \\ \text{since } \frac{\partial Q}{\partial z} = \frac{\partial P}{\partial z} = 0}}{=} \left(\frac{\partial Q}{\partial x} - \frac{\partial P}{\partial y}\right)k$$

EXAMPLE 14.19   Find curl(F) for:
$$F(x, y, z) = xyz\boldsymbol{i} + 2yz\boldsymbol{j} - 3\boldsymbol{k}$$

SOLUTION:

$$\text{curl}(F) = \nabla \times F = \det \begin{bmatrix} \boldsymbol{i} & \boldsymbol{j} & \boldsymbol{k} \\ \dfrac{\partial}{\partial x} & \dfrac{\partial}{\partial y} & \dfrac{\partial}{\partial z} \\ xyz & 2yz & -3 \end{bmatrix}$$

$$= \left(\frac{\partial}{\partial y}(-3) - \frac{\partial}{\partial z}(2yz) \right)\boldsymbol{i} - \left(\frac{\partial}{\partial x}(-3) - \frac{\partial}{\partial z}(xyz) \right)\boldsymbol{j} + \left(\frac{\partial}{\partial x}(2yz) - \frac{\partial}{\partial y}(xyz) \right)\boldsymbol{k}$$

$$= -2y\boldsymbol{i} + xy\boldsymbol{j} - xz\boldsymbol{k}$$

CHECK YOUR UNDERSTANDING 14.19

Answer:
$-xe^{xz}\boldsymbol{i} + (ze^{xz} - x\cos xy)\boldsymbol{k}$

Find curl(F) for:
$$F(x, y, z) = \sin(xy)\boldsymbol{i} + e^{xz}\boldsymbol{j} + \ln(z^2 + 1)\boldsymbol{k}$$

Recalling the gradient function of page 561:
$$\nabla f(x, y, z) = f_x(x, y, z)\boldsymbol{i} + f_y(x, y, z)\boldsymbol{j} + f_z(x, y, z)\boldsymbol{k}$$
we have:

THEOREM 14.14   If $w = f(x, y, z)$ has continuous second-partial derivatives, then:
$$\text{curl}[\nabla f(x, y, z)] = \boldsymbol{0}$$

PROOF:

$$\text{curl}[\nabla f(x, y, z)] = \nabla \times \nabla f(x, y, z) = \det \begin{bmatrix} \boldsymbol{i} & \boldsymbol{j} & \boldsymbol{k} \\ \dfrac{\partial}{\partial x} & \dfrac{\partial}{\partial y} & \dfrac{\partial}{\partial z} \\ \dfrac{\partial f}{\partial x} & \dfrac{\partial f}{\partial y} & \dfrac{\partial f}{\partial z} \end{bmatrix}$$

$$= (f_{zy} - f_{yz})\boldsymbol{i} - (f_{zx} - f_{xz})\boldsymbol{j} + (f_{yx} - f_{xy})\boldsymbol{k}$$

Theorem 13.1:
(page 552)
$$= 0\boldsymbol{i} - 0\boldsymbol{j} + 0\boldsymbol{k} = \boldsymbol{0}$$

In particular, the vector field:
$F(x, y, z) = xyz\boldsymbol{i} + 2yz\boldsymbol{j} - 3\boldsymbol{k}$
of Example 14.19 is not conservative.

COROLLARY   If F is a conservative vector field with continuous second partial derivatives, then:
$$\text{curl}(F) = \boldsymbol{0}$$

PROOF: To say that $F(x, y, z)$ is a conservative vector field is to say that there exists a scalar function $f(x, y, z)$ such that $F = \nabla f$. Consequently: $\text{curl}(F) = \text{curl}[\nabla f(x, y, z)] = \boldsymbol{0}$.

Let's modify the all important Theorem 14.9 of page 609:

THEOREM 14.15   Let F be a continuous vector field on a **simply connected** open region S.

The following four properties are equivalent:

(i) $\oint_C F \cdot dr = 0$ for any smooth closed curve C in S.

(ii) F is a conservative vector field on S, i.e. $F = \nabla f$.

(iii) $F = \nabla f$ is path-independent on S with

$$\int_C F \cdot dr = f(p_1) - f(p_0)$$

for any smooth curve C from p_0 to p_1.

(iv) $\text{curl}(F) = \mathbf{0}$

Note that, the old-(iv) of Theorem 14.9:

For $F = P(x,y)i + Q(x,y)j$: $\dfrac{\partial P}{\partial y} = \dfrac{\partial Q}{\partial x}$

For $F = P(x,y,z)i + Q(x,y,z)j + R(x,y,z)k$:

$$\frac{\partial P}{\partial y} = \frac{\partial Q}{\partial x}, \quad \frac{\partial P}{\partial z} = \frac{\partial R}{\partial x}, \quad \frac{\partial Q}{\partial z} = \frac{\partial R}{\partial y}$$

has now been replaced by one compact statement: $\text{curl}(F) = \mathbf{0}$.

How so? Your turn:

CHECK YOUR UNDERSTANDING 14.20

(a) Verify that Theorem 14.15(iv) is equivalent to Theorem 14.9(iv).

(b) Show that the converse of the Corollary to Theorem 14.14 need not hold.
(Suggestion: Consider Example14.9, page 607)

Answer: See page A-37.

DIVERGENCE

Unlike $\text{curl}(F)$ that assigns vectors to elements in the vector field F, the following function assigns scalars (real numbers) to the vectors in F:

DEFINITION 14.7

$\text{div}(F)$

Let $F(x,y) = P(x,y)i + Q(x,y)j$. The **divergence of** F, denoted by $\text{div}(F)$ is given by:

$$\text{div}(F) = \frac{\partial P}{\partial x} + \frac{\partial Q}{\partial y}$$

For $F(x,y,z) = P(x,y,z)i + Q(x,y,z)j + R(x,y,z)k$:

$$\text{div}(F) = \frac{\partial P}{\partial x} + \frac{\partial Q}{\partial y} + \frac{\partial R}{\partial z}$$

(assuming the indicated partial derivatives exist)

That is: $\text{div}(F) = \nabla \cdot F$

EXAMPLE 14.20 Find $\text{div}(F)$ for:

(a) $F(x, y) = yx^2 i + e^{xy} j$

(b) $F(x, y, z) = xz i + xy j - yz^2 k$

SOLUTION:

(a) $\text{div}[yx^2 i + e^{xy} j] = \nabla \cdot F$

$$= \left(\frac{\partial}{\partial x} i + \frac{\partial}{\partial y} j + \frac{\partial}{\partial z} k\right) \cdot [yx^2 i + e^{xy} j + 0k]$$

$$= (yx^2)_x + (e^{xy})_y = 2xy + xe^{xy}$$

(b) $\text{div}[xz i + xy j - yz^2 k] = \nabla \cdot F$

$$= \left(\frac{\partial}{\partial x} i + \frac{\partial}{\partial y} j + \frac{\partial}{\partial z} k\right) \cdot [xz i + xy j - yz^2 k]$$

$$= (xz)_x + (xy)_y - (yz^2)_z = z + x - 2yz$$

CHECK YOUR UNDERSTANDING 14.21

Answers: (a) 0 (b) No

(a) Find $\text{div}[\text{curl}(F)]$ for $F(x, y, z) = xz i + e^{xz} j - (\sin z) k$.

(b) Is the expression $\text{curl}[\text{div} F]$ meaningful? Justify your answer.

That $\text{div}[\text{curl}(F)]$ turned out to be 0 in part (a) of the above CYU is no fluke; for:

THEOREM 14.16 Let $F = Pi + Qj + Rk$. If P, Q, and R have continuous second partial derivatives, then:

$$\text{div}[\text{curl}(F)] = 0$$

PROOF:

$\text{div}[\text{curl}(F)] = \nabla \cdot (\nabla \times F)$

$$= \frac{\partial}{\partial x}\left(\frac{\partial R}{\partial y} - \frac{\partial Q}{\partial z}\right) + \frac{\partial}{\partial y}\left(\frac{\partial P}{\partial z} - \frac{\partial R}{\partial x}\right) + \frac{\partial}{\partial z}\left(\frac{\partial Q}{\partial x} - \frac{\partial P}{\partial y}\right)$$

$$= R_{yx} - Q_{zx} + P_{zy} - R_{xy} + Q_{xz} - P_{yz}$$

$$= (R_{yx} - R_{xy}) + (Q_{xz} - Q_{zx}) + (P_{zy} - P_{yz}) \underset{\uparrow}{=} 0$$

Theorem 13.1, page 584

CHECK YOUR UNDERSTANDING 14.22

Answer: See page A-37.

Establish the validity of Theorem 14.16 for:

$$F(x, y) = P(x, y)i + Q(x, y)j$$

VECTOR FORM OF GREEN'S THEOREM

In the previous section you encountered two versions of Green's Theorem involving a vector field $F = P(x, y)i + Q(x, y)j$; both of which can be expressed in vector form;

Theorem 14.10, page 617	Theorem 14.13, page 622

$$\oint_C F \cdot T\, ds = \oint_C P\, dx + Q\, dy = \iint_D \left(\frac{\partial Q}{\partial x} - \frac{\partial P}{\partial y}\right) dA \qquad \oint_C F \cdot n\, ds = \oint_C P\, dy - Q\, dx = \iint_D \left(\frac{\partial P}{\partial x} + \frac{\partial Q}{\partial y}\right) dA$$

Vector Form	Vector Form

$$\oint_C F \cdot T\, ds = \iint_D (\text{curl } F) \cdot k\, dA$$

$$\left[= \iint_D (\nabla \times F) \cdot k\, dA \right] \quad (*)$$

$$\oint_C F \cdot n\, ds = \iint_D (\text{div } F)\, dA$$

$$\left[= \iint_D (\nabla \cdot F)\, dA \right] \quad (**)$$

For:	For:

$$(\text{curl } F) \cdot k = \left(\frac{\partial Q}{\partial x} - \frac{\partial P}{\partial y}\right) k \cdot k = \left(\frac{\partial Q}{\partial x} - \frac{\partial P}{\partial y}\right) \qquad \text{div } F = \frac{\partial P}{\partial x} + \frac{\partial Q}{\partial y}$$

Why do we need two forms?
For the sake of applications and interpretations:

Curl and divergence play important roles in a variety of vector field applications. Both are best understood by thinking of the vector field as representing the flow of a liquid. Roughly speaking, divergence measures the tendency of the fluid to flow into or disperse from a point — it is a scalar quantity. The curl, on the other hand, measures the tendency of a fluid to swirl around a point — its magnitude measures how much the fluid is swirling, and its direction indicates the axis around which it tends to swirl.

A glance into the near future:

The **F** · **T**-Greens Theorem:

$$\oint_C \mathbf{F} \cdot \mathbf{T}\, ds = \iint_D (\text{curl } \mathbf{F}) \cdot \mathbf{k}\, dA$$

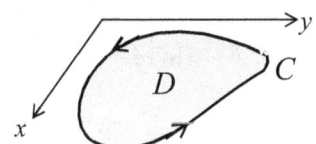

will, in Section 14.6, evolve into Stoke's Theorem, which relates the integral over a closed curve C in three-space to a double integral involving the surface S bounded by that curve:

STOKE'S THEOREM

$$\oint_C \mathbf{F} \cdot \mathbf{T}\, ds = \iint_S \text{curl}(\mathbf{F}) \cdot \mathbf{n}\, dS$$

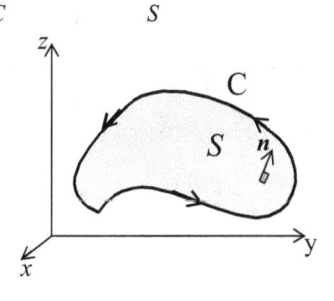

The **F** · **n**-Greens Theorem:

$$\oint_C \mathbf{F} \cdot \mathbf{n}\, ds = \iint_D (\text{div } \mathbf{F})\, dA$$

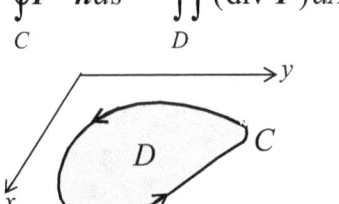

will, in Section 14.7, evolve into the Divergence Theorem, which relates the integral over a three-dimensional region E to a double integral involving the surface bounded by that curve:

DIVERGENCE THEOREM

$$\iint_S \mathbf{F} \cdot \mathbf{n}\, dS = \iiint_E (\text{div } \mathbf{F})\, dV$$

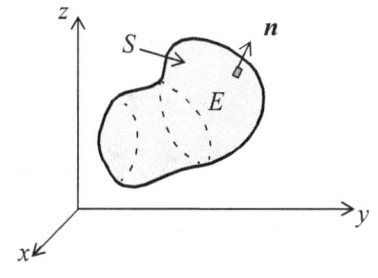

	EXERCISES	

Exercises 1-14. Find curl(F).

1. $F(x, y) = (x^2 - y)i + (xy - y^2)j$

2. $F(x, y) = (3x^2 - xy)i + (y - x^2)j$

3. $F(x, y) = \sin(xy)i + \cos(xy)j$

4. $F(x, y) = e^{xy}i + y\ln x j$

5. $F(x, y, z) = y^2 zi - x^3 j + xy k$

6. $F(x, y, z) = x^2 zi - y^3 j + xy^2 k$

7. $F(x, y, z) = xzi + xyzj - y^2 k$

8. $F(x, y, z) = \sin x i - \cos y j + xyz k$

9. $F(x, y, z) = \ln(x + z)i - e^{yz}j + xy k$

10. $F(x, y, z) = e^{xy} - \cos y j + \ln z^2 k$

11. $F(x, y, z) = \sin(yz)i + \sin(zx)j - \sin(xy)k$

12. $F(x, y, z) = e^x i + \ln y j - \sin z k$

13. $F(x, y, z) = \tan(yz)i + \cos(xy)j - \sin(xy)k$

14. $F(x, y, z) = ye^x i + yz^2 j - \sin z k$

Exercises 15-22. Find div(F).

15. $F(x, y) = x^2 yi + \sin x j$

16. $F(x, y) = xy^2 i + \cos xy j$

17. $F(x, y, z) = xyi + yzj - 2xz k$

18. $F(x, y, z) = yz^2 i + xzj + xzy^2 k$

19. $F(x, y, z) = e^{x-z}i + e^{z-y}j - e^{y-x}k$

20. $F(x, y, z) = e^{xz}i + e^{yz}j - e^{\sin z}k$

21. $F(x, y, z) = \cos(xy)i - \sin(yz)j + \cos y \sin x k$

22. $F(x, y, z) = \sqrt{xz}i + \ln(x)j - \sin z k$

Exercises 23-26. Show that F is conservative.

23. $F(x, y) = x^2 yi + \dfrac{x^3}{3}j$

24. $F(x, y) = (3 + 2xy)i + (x^2 - 3y^2)j$

25. $F(x, y, z) = y^2 z^3 i + 2xyz^3 j + 3xy^2 z^2 k$

26. $F(x, y, z) = (x^2 + y)i + (y^2 + x)j + (ze^z)k$

Exercises 27-34. Let f and g be scalar functions, and F and G be vector fields. Establish the given identity, assuming that all partial derivatives exist and are continuous.

27. div $(fF) = f(\text{div } F) + F \cdot \nabla f$

28. $\nabla(gf) = g\nabla f + f\nabla g$

29. $\text{div}(\nabla f \times \nabla g) = 0$

30. $\nabla \times (gF) = (\nabla g) \times F + g(\nabla \times F)$

31. $\text{div}(F \times G) = G \cdot \text{curl}(F) - F \cdot \text{curl}(G)$

32. $\nabla \times (g\nabla f + f\nabla g) = \mathbf{0}$

33. $\nabla \cdot [\nabla(fg)] = f[\nabla \cdot (\nabla g)] + g[\nabla \cdot (\nabla f)] + 2(\nabla f \cdot \nabla g)$

34. $\nabla \times (g\nabla f - f\nabla g) = 2\nabla g \times \nabla f$

Exercises 35-41. Let f be a scalar function, and F a vector field. Indicate if the given expression is meaningful. If not, state why not. If so, then indicate whether the output is a scalar field or a vector field.   (all from A-3)

35. ∇F

36. $\nabla(\text{div} f)$

37. $\text{curl}(\text{curl} F)$

38. $\text{div}(\text{div} F)$

39. $\text{div}(\nabla f)$

40. $\nabla f \times \text{div} F$

41. $\text{div}[\text{curl}(\nabla f)]$

§5. Surface Integrals

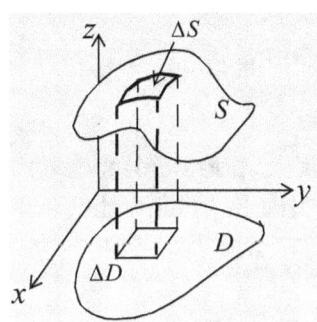

How are we to define the area $A(S)$ of a surface S that is defined by the equation $z = f(x, y)$ over a region D in the plane? We turn to a familiar development, and:

(1) Find an adequate approximation for the surface area of the depicted region ΔS.

(2) Sum all of those bits of surface areas.

(3) Take the limit of the evolving Riemann sums, as the areas of the ΔD's tend to zero.

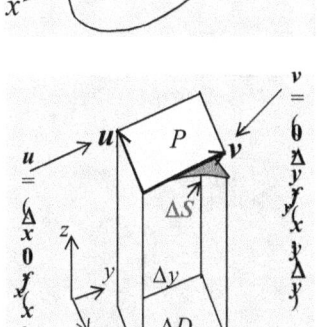

As for (1): Choose a corner (x, y) in the rectangular region ΔD with sides parallel to the x- and y-axis in the margin and consider the parallelogram P that is tangent to S at $(x, y, f(x, y))$ — a parallelogram with side-vectors $u = \langle \Delta x, 0, f_x(x, y) \Delta x \rangle$ and $v = \langle 0, \Delta y, f_y(x, y) \Delta y \rangle$.

Here is how we arrived at the above expression for u:
(a similar development leads to the expression for v)

u is the vector from: $(x, y, f(x, y))$ to $(x + \Delta x, y, f(x + \Delta x, y))$;

which is to say: $u = (x + \Delta x - x, y - y, f(x + \Delta x, y) - f(x, y))$.

Since $f(x + \Delta x, y) - f(x, y) \approx f_x(x, y) \Delta x$: $u = (\Delta x, 0, f_x(x, y) \Delta x)$

Noting that P has area $\| u \times v \|$ (see CYU 12.18, page 509), we have:

$$\| u \times v \| = \left\| \det \begin{bmatrix} i & j & k \\ \Delta x & 0 & f_x(x, y) \Delta x \\ 0 & \Delta y & f_y(x, y) \Delta y \end{bmatrix} \right\|$$

$$= \| -f_x(x, y) \Delta x \Delta y \, i - (f_y(x, y) \Delta x \Delta y) j + \Delta x \Delta y \, k \|$$

$$= \sqrt{[f_x(x, y)]^2 + [f_y(x, y)]^2 + 1} \, \Delta x \Delta y$$

As for (2):

$$A(S) \approx \sum_D \sqrt{1 + [f_x(x, y)]^2 + [f_y(x, y)]^2} \, \Delta x \Delta y$$

(3) Bringing us to:

DEFINITION 14.8
SURFACE AREA

Let S be the surface $z = f(x, y)$, where f is a differentiable function defined on a region D. The **surface area** of S, denoted by $A(S)$, is given by:

$$A(S) = \iint_D \sqrt{1 + [f_x(x, y)]^2 + [f_y(x, y)]^2} \, dA$$

EXAMPLE 14.21   (a) Find the area of that portion of the plane $3x + 2y + z = 4$ that lies above the disk $D = \{(x, y)|x^2 + y^2 \le 1\}$.

(b) Find the surface area S of the cone:

$$z = \sqrt{x^2 + y^2}, \; 0 \le z \le 1$$

(c) Find the surface area of the portion of the paraboloid $z = x^2 + y^2$ that lies below the plane $z = 4$.

SOLUTION: (a) Since $z = f(x, y) = 4 - 3x - 2y$, we have:

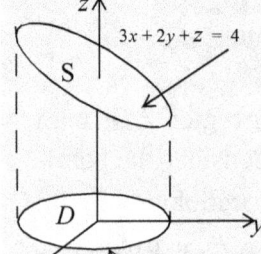

$$A(S) = \iint_D \sqrt{1 + [f_x]^2 + [f_y]^2}\, dA = \iint_D \sqrt{1 + (-3)^2 + (-2)^2}\, dA$$

$$= \sqrt{14} \iint_D dA = \sqrt{14}\,\pi \quad \leftarrow \text{Area of } D$$

(b) Since $z = f(x, y) = \sqrt{x^2 + y^2}$:

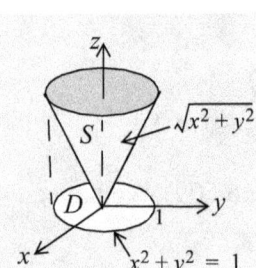

$$A(S) = \iint_D \sqrt{1 + [f_x]^2 + [f_y]^2}\, dA$$

$$= \iint_D \sqrt{1 + \left(\frac{x}{\sqrt{x^2 + y^2}}\right)^2 + \left(\frac{y}{\sqrt{x^2 + y^2}}\right)^2}\, dA$$

$$= \iint_D \sqrt{1 + \frac{x^2 + y^2}{x^2 + y^2}}\, dA = \sqrt{2} \iint_D dA = \sqrt{2}\,\pi$$

(c) Since $z = f(x, y) = x^2 + y^2$:

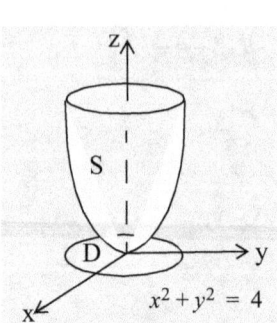

$$A(S) = \iint_D \sqrt{1 + [f_x]^2 + [f_y]^2}\, dA = \iint_D \sqrt{1 + (2x)^2 + (2y)^2}\, dA$$

$$= \iint_D \sqrt{1 + 4(x^2 + y^2)}\, dA$$

Converting to polar coordinates (see page 457), we have:

$$A(S) = \int_0^{2\pi} \int_0^2 \sqrt{1 + 4r^2}\, r\, dr\, d\theta = \int_0^{2\pi} d\theta \int_0^2 r\sqrt{1 + 4r^2}\, dr$$

$$\begin{aligned} u &= 1 + 4r^2 \\ du &= 8r\, dr \end{aligned} = \frac{1}{8}\int_0^{2\pi} d\theta \int_1^{17} u^{1/2}\, du = \frac{1}{8}\int_0^{2\pi} \left(\frac{2}{3} u^{3/2}\right)\Bigg|_1^{17} d\theta$$

$$= \frac{\pi}{6}\left(17^{\frac{3}{2}} - 1\right)$$

$$= \frac{\pi}{6}(17\sqrt{17} - 1)$$

CHECK YOUR UNDERSTANDING 14.23

Find the surface area of the portion of the cone $z = \sqrt{x^2 + y^2}$ that lies above the region $D = \{(x, y) \mid 1 \leq x^2 + y^2 \leq 4\}$.

Extending Definition 14.8 to allow for a real-valued function g defined on a surface S in three space, we come to:

DEFINITION 14.9

SURFACE INTEGRAL

Let S be the surface $z = f(x, y)$, where f is a differentiable function defined on a region D in the xy-plane, and let $g(x, y, z)$ be a continuous function on S. The **surface integral** of g over S, denoted by

$$\iint_S g(x, y, z)dS \text{ is given by:}$$

$$\iint_S g(x, y, z)dS = \iint_D g[x, y, f(x, y)]\sqrt{1 + [f_x(x, y)]^2 + [f_y(x, y)]^2}\,dA$$

Note that while f is defined on a region D in the plane, the function g is defined on a surface S in three-space. Because the points in S are of the form $[x, y, f(x, y)]$, the expression $g[x, y, f(x, y)]$, as opposed to $g(x, y, z)$, appears in the above integral.

EXAMPLE 14.22 Find the mass M of that portion S of the (thin) plane $x + y + z = 9$ that lies above the disk $D = \{(x, y) \mid x^2 + y^2 \leq 4\}$, if the density at each point of the plane equals its vertical distance from D.

SOLUTION: As $x + y + z = 9$, $z = 9 - x - y = f(x, y)$. For a point p in S: $\delta(p) = z = 9 - x - y$. Hence:

$$M = \iint_D \delta[x, y, f(x, y)]\sqrt{1 + [f_x(x, y)]^2 + [f_y(x, y)]^2}\,dA$$

$$= \iint_D (9 - x - y)\sqrt{1 + (-1)^2 + (-1)^2}\,dA$$

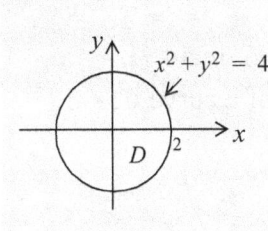

polar form: $\displaystyle = \sqrt{3}\int_0^{2\pi}\int_0^2 (9 - r\cos\theta - r\sin\theta)r\,dr\,d\theta$

$$= \sqrt{3}\int_0^{2\pi} \left[\frac{9}{2}r^2 - \frac{1}{3}r^3(\cos\theta + \sin\theta)\right]\Bigg|_0^2 d\theta$$

$$= \sqrt{3}\int_0^{2\pi} \left[18 - \frac{8}{3}(\cos\theta + \sin\theta)\right]d\theta$$

$$= \sqrt{3}\left[18\theta - \frac{8}{3}(\sin\theta - \cos\theta)\right]\Bigg|_0^{2\pi} = 36\sqrt{3}\pi$$

CHECK YOUR UNDERSTANDING 14.24

Answer: $\frac{\pi}{560}(13\sqrt{2}+8)$

Find the mass of the portion of the paraboloid $z = x^2 + y^2$ that lies below the plane $z = \frac{1}{4}$, if the density at each point of the surface is equal to the square of its distance from the origin D.

SURFACE INTEGRAL OF FLUX

The concept of flux across a closed curve C was defined on page 620. Removing the closed-restriction we have:

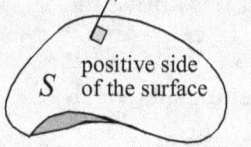

DEFINITION 14.10

FLUX ACROSS A CURVE

The flux of a two-dimensional vector field F across a curve C is given by:

$$\text{Flux}(F_C) = \int_C F \cdot n\, ds$$

where, at any point on C, n is the unit normal pointing $90°$ in a clockwise direction from the tangent vector.

Note that if C is a closed curve, oriented in a counterclockwise direction, then n is the outward pointing unit vector to the curve.

Typically, a surface S has two sides, and a normal vector to the surface can point in one of two directions. Once the direction is chosen, the surface is said to be oriented, and the side from which n sprouts is said to be its positive side. That said:

DEFINITION 14.11

FLUX ACROSS A SURFACE

The **flux** of a three-dimensional vector field F across an oriented surface S is given by:

$$\text{Flux}(F_S) = \iint_S F \cdot n\, dS$$

Note that the ds in Definition 14.10 "grew into" a dS in 14.11 — a reflection of the fact that while Δs represents a piece of a curve C, ΔS represent a piece of a surface S:

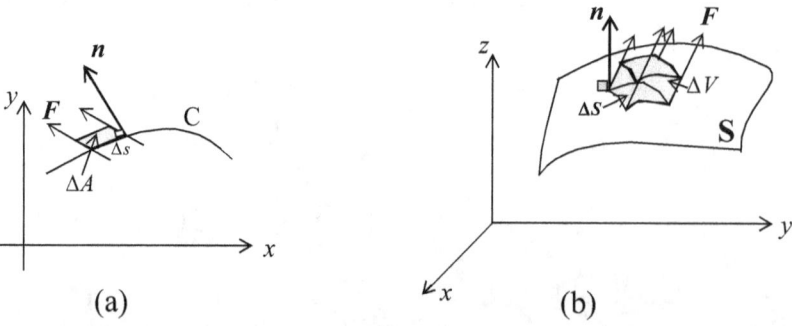

(a)   (b)

Figure 14.4

As for the rest of the story assume that, in both instances, F represents the velocity field of a fluid. If so, then the area ΔA in Figure 14.4(a) is an approximation of the amount of fluid crossing Δs in a unit of time: $\Delta A \approx (F \cdot n)\Delta s$; while the volume ΔV in Figure 14.4(b) is an approximation of the amount of fluid crossing ΔS in a unit of time: $\Delta V \approx (F \cdot n)\Delta S$. The "$\cdot n$" comes into play since, in each situation, the height of the shaded region is not perpendicular to its base [the base Δs in (a), and the base ΔS in (b)].

To compute a unit normal to a surface S of the form $z = f(x, y)$ we first rewrite the equation as $z - f(x, y) = 0$, which is a level surface for the function:

$$g(x, y, z) = z - f(x, y)$$

Theorem 13.7, page 566, tells us that:

$$\nabla g = -f_x(x, y)\boldsymbol{i} - f_y(x, y)\boldsymbol{j} + \boldsymbol{k}$$

is a normal to the surface, with corresponding unit normal:

$$n = \frac{-f_x \boldsymbol{i} - f_y \boldsymbol{j} + \boldsymbol{k}}{\sqrt{(f_x)^2 + (f_y)^2 + 1}}$$

Since $\boldsymbol{k}$ is positive, it is the upward unit normal to the surface. That being the case, the downward unit normal is given by:

$$-n = \frac{f_x \boldsymbol{i} + f_y \boldsymbol{j} - \boldsymbol{k}}{\sqrt{(f_x)^2 + (f_y)^2 + 1}}$$

The next theorem addresses the flux issue across a surface S of the form $z = f(x, y)$.

THEOREM 14.17 Let the surface S be defined by a differentiable function $z = f(x, y)$ defined on a region D in the xy-plane.
If S is oriented by upward normals:

$$\iint\limits_S F \cdot n\, dS = \iint\limits_D F \cdot (-f_x \boldsymbol{i} - f_y \boldsymbol{j} + \boldsymbol{k})\, dA$$

If S is oriented by downward normals:

$$\iint\limits_S F \cdot n\, dS = \iint\limits_D F \cdot (f_x \boldsymbol{i} + f_y \boldsymbol{j} - \boldsymbol{k})\, dA$$

PROOF: Assume that S is oriented by upward normals. Turning to Theorem 14.1, page 592, we have:

$$\iint\limits_S F \cdot n\, dS = \iint\limits_D F \cdot \frac{-f_x \boldsymbol{i} - f_y \boldsymbol{j} + \boldsymbol{k}}{\sqrt{(f_x)^2 + (f_y)^2 + 1}} \sqrt{1 + (f_x)^2 + (f_y)^2}\, dA = \iint\limits_D F \cdot (-f_x \boldsymbol{i} - f_y \boldsymbol{j} + \boldsymbol{k})\, dA$$

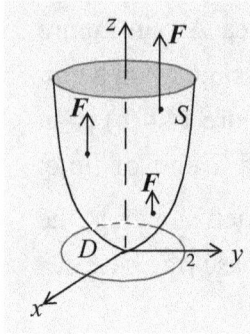

EXAMPLE 14.23 Determine the flux of the field $F(x, y, z) = zk$ in an upward direction through the portion of the paraboloid $z = x^2 + y^2$ lying below the plane $z = 4$.

SOLUTION: For $f(x, y) = x^2 + y^2$: $-f_x i - f_y j + k = -2xi - 2yj + k$ and $D = \{(x, y) | x^2 + y^2 \le 4\}$. Bringing us to:

$$\text{Flux}(F_S) = \iint_D F \cdot (-f_x i - f_y j + k) \, dy\, dx$$

$$= \iint_D zk \cdot (-2xi - 2yj + k) \, dy\, dx$$

$$= \iint_D z \, dy\, dx = \iint_D (x^2 + y^2) \, dy\, dx = \int_0^{2\pi} \int_0^2 r^2 r \, dr\, d\theta = 8\pi$$

Example 11.16, page 457

CHECK YOUR UNDERSTANDING 14.25

Answer: $\frac{3}{2}\pi$

Determine the flux of the field $F(x, y, z) = xi + yj + zk$ in an upward direction through the surface $z = 1 - (x^2 + y^2)$, $z \ge 0$.

PARAMETRIZED SURFACES

It is often convenient to parametrically describe a surface. As the surface S is a two-dimensional object, each coordinate of a point $(x, y, z) \in S$ needs to be specified in terms of **two parameters**; say u and v, yielding a parametrization of the form:

$$r(u, v) = x(u, v)i + y(u, v)j + z(u, v)k$$

> In previous parametrizations, $r(t)$ referred to a point on a curve (a one-dimensional object). Here, $r(u, v)$ refers to a point on a surface (a two-dimensional object).

Our next goal is to find the area $A(S)$ of a surface S that is the graph of a function with a given parametrization $r(u, v)$, where u and v range over a region D in the plane:

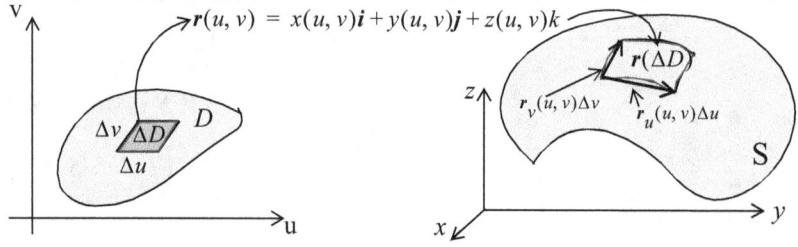

By stretching the reasoning process used at the beginning of the section, we conclude that the area of $r(\Delta D)$ can be approximated by that of the parallelogram determined by the vectors $r_u(u, v)\Delta u$ and $r_v(u, v)\Delta v$; namely, by:

$$\|r_u(u, v)\Delta u \times r_v(u, v)\Delta v\| = \|(r_u \times r_v)(u, v)\| \Delta u \Delta v$$

Bringing us to:

DEFINITION 14.12
SURFACE AREA
(Parametrization Form)

Let S be the surface parametrized by the differentiable function

$$r(u, v) = x(u, v)\boldsymbol{i} + y(u, v)\boldsymbol{j} + z(u, v)\boldsymbol{k}$$

defined on a region D in the uv-plane. The **surface area** of $S = r(D)$, denoted by $A(S)$, is given by:

$$A(S) = \iint_D \|r_u \times r_v\| \, du \, dv$$

EXAMPLE 14.24 Find the surface area S of a sphere of radius r.

SOLUTION: In spherical coordinates, the points on the sphere $S = \{(x, y, z) \mid x^2 + y^2 + z^2 = r^2\}$ take the form (see page 475):

$$x = r\sin\phi\cos\theta, \; y = r\sin\phi\sin\theta, \; z = r\cos\phi; \; 0 \le \phi \le \pi, \; 0 \le \theta \le 2\pi$$

Linking with the notation of Definition 14.12:

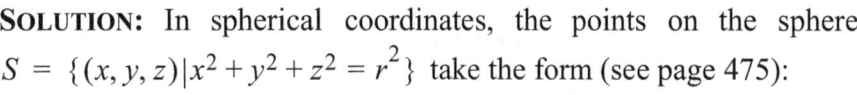

$$r(u, v) = x(u, v)\boldsymbol{i} + y(u, v)\boldsymbol{j} + z(u, v)\boldsymbol{k}$$

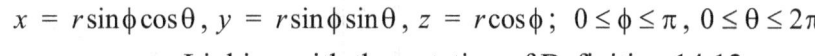

$$r(\phi, \theta) = (r\sin\phi\cos\theta)\boldsymbol{i} + (r\sin\phi\sin\theta)\boldsymbol{j} + (r\cos\phi)\boldsymbol{k}$$

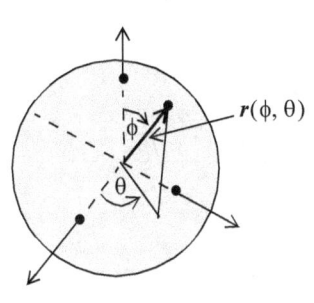

Then:

$$r_\phi \times r_\theta = \det \begin{bmatrix} \boldsymbol{i} & \boldsymbol{j} & \boldsymbol{k} \\ r\cos\phi\cos\theta & r\cos\phi\sin\theta & -r\sin\phi \\ -r\sin\phi\sin\theta & r\sin\phi\cos\theta & 0 \end{bmatrix}$$

$$= (r^2\sin^2\phi\cos\theta)\boldsymbol{i} - (r^2\sin^2\phi\sin\theta)\boldsymbol{j}$$
$$+ (r^2\cos\phi\sin\phi\cos^2\theta + r^2\cos\phi\sin\phi\sin^2\theta)\boldsymbol{k}$$

$$= (r^2\sin^2\phi\cos\theta)\boldsymbol{i} - (r^2\sin^2\phi\sin\theta)\boldsymbol{j} + (r^2\cos\phi\sin\phi)\boldsymbol{k}$$

And:

$$\|r_\phi \times r_\theta\| = \sqrt{r^4\sin^4\phi\cos^2\theta + r^4\sin^4\phi\sin^2\theta + r^4\cos^2\phi\sin^2\phi}$$

$$= r^2\sqrt{\sin^4\phi + \cos^2\phi\sin^2\phi} = r^2\sqrt{\sin^2\phi} = r^2\sin\phi$$

$\uparrow$ $\sin\phi \ge 0$ for $0 \le \phi \le \pi$

Thus:

$$A(S) = \iint_D \|r_\phi \times r_\theta\| \, d\phi \, d\theta = r^2 \int_0^{2\pi} \int_0^\pi \sin\phi \, d\phi \, d\theta$$

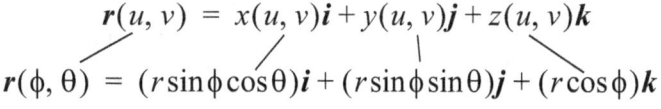

$$= r^2 \int_0^{2\pi} (-\cos\phi)\Big|_0^\pi \, d\theta = r^2 \int_0^{2\pi} 2 \, d\theta = 4\pi r^2$$

CHECK YOUR UNDERSTANDING 14.26

Solve Example 14.21(b) utilizing a cylindrical coordinates parametrization.

Answer: $\pi\sqrt{2}$

Proceeding as usual:

DEFINITION 14.13

SURFACE INTEGRAL
(Parametrization Form)

Let S be the surface parametrized by the differentiable function
$$r(u, v) = x(u, v)\mathbf{i} + y(u, v)\mathbf{j} + z(u, v)\mathbf{k}$$
defined on a region D in the uv-plane. The **surface integral** of g over S is given by:

$$\iint_S g(x, y, z)\,dS = \iint_D g[r(u, v)]\|r_u \times r_v\|\,du\,dv$$

EXAMPLE 14.25

Integrate the function $g(x, y, z) = x + zy$ over the surface S that is parametrized by:
$$r(u, v) = (u - v)\mathbf{i} + 2u\mathbf{j} - (3u + v)\mathbf{k} \quad 0 \le u \le 1, \; 0 \le v \le 2$$

SOLUTION:

$$\|r_u \times r_v\| = \left\|\det\begin{bmatrix} \mathbf{i} & \mathbf{j} & \mathbf{k} \\ 1 & 2 & -3 \\ -1 & 0 & -1 \end{bmatrix}\right\| = \|-2\mathbf{i} + 4\mathbf{j} + 2\mathbf{k}\| = \sqrt{24} = 2\sqrt{6}$$

$$\iint_S g(x, y, z)\,dS = \iint_D g[r(u, v)]\|r_u \times r_v\|\,du\,dv$$

$$= \iint_D g(u - v, 2u, -3u - v)2\sqrt{6}\,du\,dv$$

$$= 2\sqrt{6}\int_0^2\int_0^1 [(u - v) + (-3u - v)(2u)]\,du\,dv$$

$$= 2\sqrt{6}\int_0^2\int_0^1 (-6u^2 - 2uv + u - v)\,du\,dv$$

$$= 2\sqrt{6}\int_0^2 \left(-2u^3 - u^2v + \frac{u^2}{2} - vu\right)\Bigg|_0^1 dv$$

$$= 2\sqrt{6}\int_0^2 \left(-\frac{3}{2} - 2v\right)dv = 2\sqrt{6}\left(-\frac{3}{2}v - v^2\right)\Bigg|_0^2 = -14\sqrt{6}$$

CHECK YOUR UNDERSTANDING 14.27

Integrate the function $g(x, y, z) = xy^2z$ over the surface S that is parametrized by:
$$r(u, v) = u\mathbf{i} + uv\mathbf{j} - \mathbf{k} \quad 0 \le u \le 1, \; 0 \le v \le u$$

	EXERCISES	

Exercises 1-10. Find the area of the given surface.

1. That part of the surface $z = \sqrt{x^2 + y^2}$ that lies inside the cylinder $x^2 + y^2 = 4$.

2. That part of the plane $x + 2y + 3z + 4 = 0$ that lies within the cylinder $x^2 + y^2 = a^2$.

3. That part of the surface $z = x + \frac{2}{3}y^{3/2}$ that lies above $D = \{(x, y) | 1 \le x \le 4, 2 \le y \le 7\}$.

4. That part of the parabolic cylinder $z = y^2$ that lies above the triangle with vertices $(0, 0)$, $(0, 1)$, $(1, 1)$.

5. That part of the hyperbolic paraboloid $z = y^2 - x^2$ that lies between the cylinders $x^2 + y^2 = 1$ and $x^2 + y^2 = 4$.

6. That part of the hyperbolic paraboloid $z = xy$ that lies inside the cylinder $x^2 + y^2 = a^2$.

7. That part of the hemisphere $z = \sqrt{25 - x^2 - y^2}$ that lies above the disk $x^2 + y^2 \le 9$.

8. That portion of the sphere $x^2 + y^2 + z^2 = 9$ that lies above the plane $z = 2$.

9. That part of the surface $z = \frac{2}{3}(x^{3/2} + y^{3/2})$ that lies above $D = \{(x, y) | 2 \le x \le 4, 1 \le y \le 4\}$.

10. That part of the surface $z = e^{-y} + \sqrt{7}x$ that lies above $D = \{(x, y) | 0 \le x \le e^{-2y}, 0 \le y \le 3\}$

Exercises 11-18. Evaluate.

11. $\iint\limits_{S} (z + 3y - x^2)dS$ where S is the part of the surface $z = 2 - 3y + x^2$ that lies above the triangle with vertices $(0, 0), (2, 0), (2, -4)$.

12. $\iint\limits_{S} 32z\,dS$ where S is the part of the paraboloid $z = 2x^2 + 2y^2$ between the planes $z = 0$ and $z = 4$.

13. $\iint\limits_{S} y\,dS$ where S is the surface $z = x + y^2$ above $D = \{(x, y) | 0 \le x \le 1, 0 \le y \le 2\}$.

14. $\iint\limits_{S} 3y\,dS$ where S is the parabolic cylinder $z = \frac{1}{2}y^2$ above $D = \{(x, y) | 0 \le x \le 1, 0 \le y \le 1\}$.

15. $\iint\limits_S (x+z)dS$ where S is the part of the circular cylinder $y^2 + z^2 = 9$, in the first octant between the planes $x = 0$ and $x = 4$.

16. $\iint\limits_S xzdS$ where S is that part of the plane $x + y + z = 1$ that lies in the first octant. $\frac{\sqrt{3}}{24}$

17. $\iint\limits_S x^2 y^2 z^2 dS$ over the surface of the cone $x^2 + y^2 = z^2$ which lies between $z = 0$ and $z = 1$.

18. $\iint\limits_S x\sqrt{4 + y^2}dS$ where S is the surface of the parabolic cylinder $y^2 + 4z = 16$ cut by the planes $x = 0, x = 1$, and $z = 0$.

Exercises 19-22. Find the mass of the (thin) surface S with density δ.

19. S is the portion of the parabolic cylinder $y^2 = 9 - z$ between the planes $x = 0, x = 3, y = 0$, and $y = 3$, and $\delta(x, y, z) = y$.

20. S is the portion of the cone $z = \sqrt{x^2 + y^2}$ between the planes $z = 1$ and $z = 4$, and $\delta(x, y, z) = x^2 z$.

21. S is the triangle $(a, 0, 0), (0, a, 0), (0, 0, a)$, and $\delta(x, y, z) = kx^2$.

22. S is the surface $z = 1 - \frac{1}{2}(x^2 + y^2)$ that lies above $D = \{(x, y)|0 \le x \le 1, 0 \le y \le 1\}$, and $\delta(x, y, z) = xy$.

Exercises 23-26. Integrate the function g over the surface S with given parametrization $r(u, v)$.

23. $g(x, y, z) = xyz$, $r(u, v) = (u + v)i + vj + (u - v)k$ with $0 \le u \le 1, 0 \le v \le 2$.

24. $g(x, y, z) = x^2 + yz$, $r(u, v) = (u + v)i + uj + vk$ with $0 \le u \le 1, 0 \le v \le 2$.

25. $g(x, y, z) = \sqrt{1 + x^2 + y^2}$, $r(u, v) = (u\cos v)i + u\sin vj + vk$ with $0 \le u \le 1, 0 \le v \le \pi$.

26. $g(x, y, z) = \sqrt{zy - xy}$, $r(u, v) = vi + uj + (u + v)k$ with $3 \le u \le 5, 2 \le v \le 3$

Exercises 27-30. Calculate the flux of the vector field F in the upward direction through the surface S.

27. $F(x, y, z) = zk$, and S is the rectangular plate with corners at $(0, 0, 0), (1, 0, 0), (0, 1, 3)$, $(1, 1, 3)$.

28. $F(x, y, z) = xi + yj + zk$, and S is the paraboloid $z = x^2 + y^2 - 1$, for $-1 \le z \le 0$.

29. $F(x, y, z) = yi + xj + zk$, and S is the surface $z = 16 - x^2 - y^2$ that lies above the xy-plane.

30. $F(x, y, z) = yzi + xj - y^2k$, and S is the parabolic cylinder $z = x^2$ that lies above $D = \{(x, y) | 0 \leq x \leq 1, 0 \leq y \leq 4\}$

If D is the interior of a counterclockwise oriented simple closed curve in the plane, and if $F = Pi + Qj$ has continuous first-order partial derivatives on some open region in the plane containing D, then:

$$\oint_C F \cdot dr = \iint_D (\text{curl } F) \cdot k \, dA$$

§6. STOKE'S THEOREM

Green's theorem (margin) equates a line integral over a closed curve in the plane to a double integral over the region bounded by the curve. Stoke's theorem does pretty much the same thing, but in a higher dimension. Specifically:

THEOREM 14.18
STOKE'S THEOREM

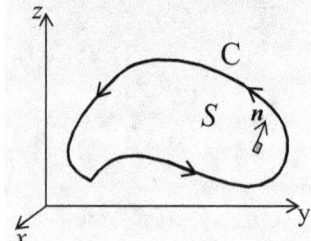

If S is an oriented surface that is bounded by a simple closed curve C with positive orientation (counterclockwise), and if $F = Pi + Qj + Rk$ has continuous first-order partial derivatives on some open region in three-space containing S, then:

$$\oint_C F \cdot dr = \iint_S \text{curl}(F) \cdot n \, dS$$

where the orientation of S is such that:

When walking around C in a counterclockwise direction with your head pointing in the direction of **n**, the surface will always be on your left.

In words: The line integral around the boundary curve of S of the tangential component of F is equal to the surface integral of the normal component of the curl of F.

A proof of the above result, in the special case where the surface S is the graph of a function, is offered in Appendix B, page B-3. At this point, we offer a geometrical argument that suggests the validity of the theorem:

Break up the surface S into positively oriented two-dimensional regions D_i. Being two-dimensional, Green's theorem holds on each D_i; which is to say:

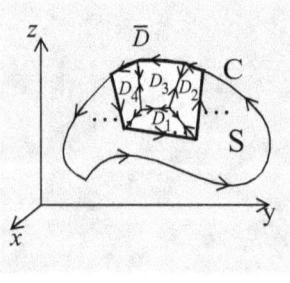

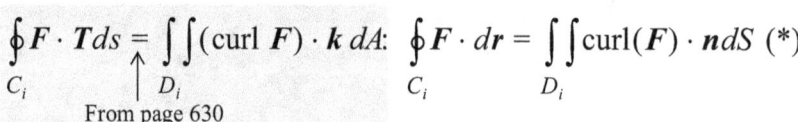

$$\oint_{C_i} F \cdot T \, ds = \iint_{D_i} (\text{curl } F) \cdot k \, dA: \quad \oint_{C_i} F \cdot dr = \iint_{D_i} \text{curl}(F) \cdot n \, dS \quad (*)$$

From page 630

where C_i denotes the closed curve bounding D_i.

Now consider the surface $\overline{D}$ composed of the four depicted regions D_1 through D_4 in the margin figure. From (*) we have:

$$\iint_{\overline{D}} \text{curl}(F) \cdot n \, dS = \oint_{C_1} F \cdot dr + \oint_{C_2} F \cdot dr + \oint_{C_3} F \cdot dr + \oint_{C_4} F \cdot dr$$

But if $\overline{C}$ is the boundary of $\overline{D}$, then:

$$\iint_{\overline{D}} \text{curl}(F) \cdot n \, dS = \oint_{\overline{C}} F \cdot dr$$

Why? Because, as is depicted in the figure, every inner-line segment of the four closed paths C_1 through C_4 is traversed twice, **but in opposite directions**. Expanding the region $\overline{D}$ to include all of the D_i's; which is to say, to all of S, we find that:

$$\iint_S \text{curl}(F) \cdot n \, dS = \oint_C F \cdot dr$$

Note: Just as Stoke's theorem represents an elevation of Green's theorem, so then is Green's theorem a compression of Stoke's theorem. To be more specific:

Green's theorem tells us that if D is a region in the xy-plane bounded by the curve C, then [see (*), page 630]:

$$\oint_C F \cdot dr = \iint_D (\text{curl } F) \cdot k \, dA$$

Applying Stokes theorem with S a (flat) region D in the xy-plane $z = 0$, and with $F = Pi + Qj + 0k$ we have:

$$\oint_C F \cdot dr = \iint_S \text{curl}(F) \cdot n \, dS = \iint_D (\text{curl } F) \cdot k \, dA$$

Stoke's theorem can be used in two directions.

One direction: Finding the value of a line integral using a surface integral. (Example 14.26 and 14.27).

Another direction: Finding the value of a surface integral using a line integral. (Example 14.28).

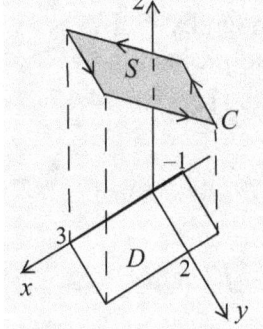

EXAMPLE 14.26 Let S be the graph of $f(x, y) = 2x - 6y + 25$ that lies above the region:

$$D = \{(x, y) \mid -1 \le x \le 3, 0 \le y \le 2\}$$

Find $\oint_C F \cdot dr$

where $F(x, y, z) = (x^2 i + xz j - y k)$, and where C is the boundary of S, oriented in the counterclockwise direction when viewed from above.

SOLUTION: Rather than involving four integrals, one for each of the line segments defining the boundary of S, we turn to Stoke's theorem.

Since: $\text{curl}(F) = \det \begin{bmatrix} i & j & k \\ \dfrac{\partial}{\partial x} & \dfrac{\partial}{\partial y} & \dfrac{\partial}{\partial z} \\ x^2 & xz & -y \end{bmatrix}$ 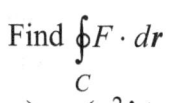 $= (-1-x)i - 0j + zk$

Employing Theorem 14.17, page 637, we then have:

$$\oint_C F \cdot dr = \iint_S \text{curl}(F) \cdot n \, dS = \iint_D [(-1-x)i + zk] \cdot (-f_x i - f_y j + k) \, dA$$

$$= \iint_D [(-1-x)i + (2x - 6y + 25)k] \cdot (-2i + 6j + k) \, dA$$

$$= \int_{-1}^{3} \int_0^2 [(2 + 2x) + (2x - 6y + 25)] \, dy \, dx$$

$$= \int_{-1}^{3} \int_0^2 (4x - 6y + 27) \, dy \, dx = \int_{-1}^{3} (4xy - 3y^2 + 27y)\Big|_0^2 \, dx$$

$$= \int_{-1}^{3} (8x + 42) \, dx = (4x^2 + 42x)\Big|_{-1}^{3}$$

$$= 36 + 126 - (-38) = 200$$

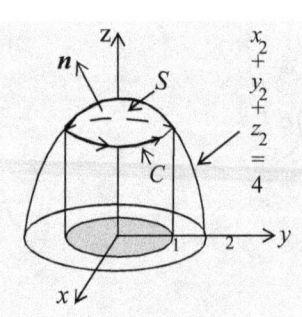

EXAMPLE 14.27 Use Stoke's theorem to evaluate $\oint_C F \cdot dr$ where $F(x, y, z) = 2z i + 4x j + 5y k$ and C is the curve of intersection of the plane $y + z = 4$ and the cylinder $x^2 + y^2 = 4$ oriented counterclockwise as viewed from above.

SOLUTION: For $F(x, y, z) = 2z i + 4x j + 5y k$ we have:

$$\text{curl}(F) = \det \begin{bmatrix} i & j & k \\ \dfrac{\partial}{\partial x} & \dfrac{\partial}{\partial y} & \dfrac{\partial}{\partial z} \\ 2z & 4x & 5y \end{bmatrix} = 5i + 2j + 4k$$

There are many surfaces with boundary C. The most convenient is the elliptical region S in the plane $y + z = 4$ bounded by C. If S is oriented upward, then C is positively oriented. The projection of S onto the xy-plane is the disk $x^2 + y^2 \leq 4$. Applying Theorem 14.17 of page 637, with $z = 4 - y$, we then have:

$$\oint_C F \cdot dr = \iint_S \text{curl}(F) \cdot n \, dS = \iint_D \text{curl}(F) \cdot (-f_x i - f_y j + k) \, dA$$

$$= \iint_D (5i + 2j + 4k) \cdot (j + k) \, dA$$

$$= \iint_D (2 + 4) \, dA = 8(\underset{\uparrow}{\pi \cdot 2^2}) = 32\pi$$
$$\text{Area of } D$$

EXAMPLE 14.28 Evaluate $\iint_S \text{curl}(F) \cdot n \, dS$ where

$$F(x, y, z) = xy i + xz j + yz k$$

and S is that part of the sphere $x^2 + y^2 + x^2 = 4$ that lies above the xy-plane, and within the cylinder $x^2 + y^2 = 1$.

SOLUTION: To find C, we solve the system of equations:

$$\left. \begin{array}{l} (1): x^2 + y^2 + z^2 = 4 \\ (2): x^2 + y^2 = 1 \end{array} \right\} \Rightarrow (1) - (2): z = \sqrt{3}$$

It follows that $C = \{(x, y, \sqrt{3}) \,|\, x^2 + y^2 = 1\}$. In polar-vector form:

$$r(t) = \cos t \, i + \sin t \, j + \sqrt{3} k, \quad 0 \leq t \leq 2\pi$$

with: $F(r(t)) = \cos t \sin t \, i + \sqrt{3} \cos t \, j + \sqrt{3} \sin t \, k$

and: $r'(t) = -\sin t \, i + \cos t \, j$

Applying Stoke's theorem:

$$\iint_S \text{curl}(F) \cdot n \, dS = \oint_C F \cdot dr = \int_0^{2\pi} (F(r(t)) \cdot r'(t) dt)$$

$$= \int_0^{2\pi} [\cos t \sin t i + \sqrt{3} \cos t j + \sqrt{3} \sin t k] \cdot (-\sin t i + \cos t j) dt$$

$$= \int_0^{2\pi} (-\cos t \sin^2 t + \sqrt{3} \cos^2 t) dt = -\int_0^{2\pi} \cos t \sin^2 t \, dt + \sqrt{3} \int_0^{2\pi} \frac{1 + \cos 2t}{2} dt$$

$$= \left[-\frac{1}{3} \sin^3 t \right]_0^{2\pi} + \frac{\sqrt{3}}{2} \left(t + \frac{\sin 2t}{2} \right) \Big|_0^{2\pi} = \sqrt{3}\pi$$

Note that the value of the above surface integral depended solely on *F*'s value on the boundary curve *C*. It follows that if we mold the spherical surface *S* of the previous example [see Figure 14.5(a)] into the surface *S'* depicted in Figure 14.5(b), we will still have:

$$\iint_{S'} \text{curl}(F) \cdot n \, dS = \iint_S \text{curl}(F) \cdot n \, dS = \oint_C F \cdot dr = \sqrt{3}\pi$$

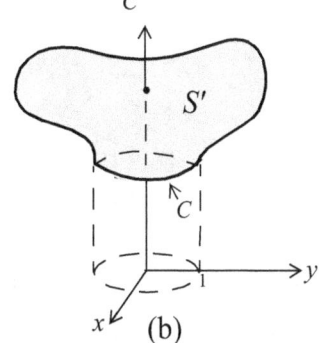

Figure 14.5

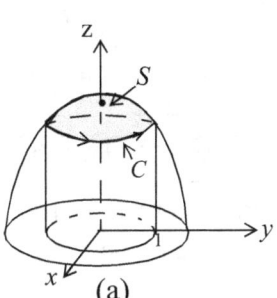

EXAMPLE 14.29   Verify Stoke's theorem where *S* is the portion of the paraboloid $z = 4 - x^2 - y^2$ lying above the *xy*-plane, and:

$$F(x, y, z) = xi + 2zj - 3yk$$

SOLUTION: We are to show (1) $\oint_C F \cdot dr$ and (2) $\iint_S \text{curl}(F) \cdot n \, dS$ have the same value. Turning to (1):

The positively oriented curve *C* can be parametrically represented by the equation:

$$x = 2\cos t, \; y = 2\sin t, \; z = 0, \quad 0 \le t \le 2\pi$$

$$r(t) = 2\cos t i + 2\sin t j + 0k, \quad 0 \le t \le 2\pi$$

Then: $\qquad r'(t) = -2\sin t i + 2\cos t j \quad ,$

$$F(x, y, z) = xi + 2zj - 3yk = 2\cos t i + (2 \cdot 0)j - 3(2\sin t)k$$

$$= 2\cos t i - 6\sin t k$$

and: $\oint_C F \cdot dr = \int_0^{2\pi} (2\cos t\, i - 6\sin t\, k) \cdot (-2\sin t\, i + 2\cos t\, j)\, dt$

$$= \int_0^{2\pi} -4\cos t \sin t\, dt = (-2\sin^2 t)\Big|_0^{2\pi} = 0$$

Turning to (2):

$$\text{curl}(F) = \det \begin{bmatrix} i & j & k \\ \dfrac{\partial}{\partial x} & \dfrac{\partial}{\partial y} & \dfrac{\partial}{\partial z} \\ x & 2z & -3y \end{bmatrix} = -5i + 0j + 0k$$

$\iint_S \text{curl}(F) \cdot n\, dS = \iint_D (-5i) \cdot (-f_x i - f_y j + k)\, dA \leftarrow$ Theorem 14.17, page 637

$f(x,y) = 4 - x^2 - y^2: = \iint_D (-5i) \cdot (2xi + 2yj + k)\, dA$

$$= -10\int_{-2}^{2} \int_{-\sqrt{4-y^2}}^{\sqrt{4-y^2}} x\, dy\, dx = 10\int_{-2}^{2} \left(\frac{x^2}{2}\right)\Big|_{-\sqrt{4-y^2}}^{\sqrt{4-y^2}} = 0$$

CHECK YOUR UNDERSTANDING 14.28

Verify Stoke's theorem where S is the portion of the paraboloid $z = x^2 + y^2$ lying below the plane $z = 1$ oriented upward, and $F(x, y, z) = y^2 i + x j + z^2 k$.

Answer: See page A-39.

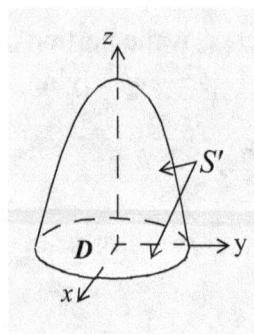

The closed the surface S' in the margin was obtained by taking the surface S of Example 14.29 and capping it below by the disk $D = \{(x, y) | x^2 + y^2 \le 4\}$.

We claim that $\iint_{S'} \text{curl}(F) \cdot n\, dS' = 0$ (*)

In an attempt to convince you that the above claim holds, we call your attention to the adjacent surface S_r, which is S' but with a hole at the bottom — a hole bounded by the circle C_r of radius r. Stokes theorem assures us that:

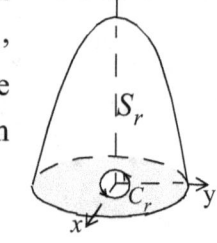

$$\iint_{S_r} \text{curl}(F) \cdot n\, dS_r = \oint_{C_r} F \cdot dr \text{ (**)}$$

We observe that as $r \to 0$: $\oint_{C_r} F \cdot dr \to 0$ and $S_r \to S'$.

Letting $r \to 0$ in (**), we arrive at (*).

In general: if $F = Pi + Qj + Rk$ has continuous first-order partial derivatives on some open region in three-space containing the **closed** surface S, then:

$$\iint\limits_{S} \text{curl}(F) \cdot n \, dS = 0$$

Answer: See page A-40.

CHECK YOUR UNDERSTANDING 14.29

Use Stoke's theorem to prove that if $\text{curl}(F) = 0$ in a region S, then F is path independent in S.

	EXERCISES	

Exercises 1-9. Use Stoke's Theorem to evaluate $\oint_C F \cdot dr$.

1. $F(x, y, z) = -3z\mathbf{i} + (x + y)\mathbf{j} + y\mathbf{k}$, S is the graph of $f(x, y) = 4x - 8y + 5$ that lies above the region: $D = \{(x, y)|(x - 1)^2 + 9(y - 3)^2 \le 36\}$, and C is the boundary of S, oriented in the counterclockwise direction when viewed from above.

2. $F(x, y, z) = z^2\mathbf{i} + y^2\mathbf{j} + x\mathbf{k}$ and C is the triangle with vertices $(1, 0, 0)$, $(0, 1, 0)$ and $(0, 0, 1)$, oriented in the counterclockwise direction when viewed from above.

3. $F(x, y, z) = y\mathbf{i} + xz^3\mathbf{j} - zy^3\mathbf{k}$ and C is the circle $x^2 + y^2 = 4$, $z = -3$, oriented in the counterclockwise direction when viewed from above.

4. $F(x, y, z) = (x^2 - y)\mathbf{i} + 4z\mathbf{j} + x^2\mathbf{k}$ and C is the curve of intersection of the cone $z = \sqrt{x^2 + y^2}$ and the plane $z = 2$, oriented in the counterclockwise direction when viewed from above.

5. $F(x, y, z) = -y^3\mathbf{i} + x^3\mathbf{j} - z^3\mathbf{k}$ and C is the intersection of the cylinder $x^2 + y^2 = 1$ and the plane $x + y + z = 1$ oriented in the counterclockwise direction when viewed from above.

6. $F(x, y, z) = x^2yz\mathbf{i} + yz^2\mathbf{j} + z^3e^{xy}\mathbf{k}$ and C is the intersection of the sphere $x^2 + y^2 + z^2 = 5$ and the plane $x = 1$ oriented in the counterclockwise direction when viewed from above.

7. $F(x, y, z) = x^2e^{5z}\mathbf{i} + x\cos y\mathbf{j} + 3y\mathbf{k}$ and C is the circle defined by the parametric equations $x = 0, y = 2 + 2\cos t, z = 2 + 2\sin t, 0 \le t \le 2\pi$.

8. $F(x, y, z) = (x^2 - y^2)\mathbf{i} + (y^2 - z^2)\mathbf{j} + (z^2 - x^2)\mathbf{k}$ and C is the boundary of the part of the plane $x + y + z = 2$ in the first octant, oriented in the counterclockwise direction when viewed from above.

9. $F(x, y, z) = xz\mathbf{i} + xy\mathbf{j} + 3xz\mathbf{k}$ and C is the boundary of the portion of the plane $2x + y + z = 2$ in the first octant, oriented in the counterclockwise direction when viewed from above.

Exercises 10-17. Use Stoke's Theorem to evaluate $\iint_S \text{curl}(F) \cdot \mathbf{n} dS$.

10. $F(x, y, z) = xz\mathbf{i} + yz\mathbf{j} + xy\mathbf{k}$ and S is that part of the sphere $x^2 + y^2 + z^2 = 4$ that lies above the xy-plane, and within the cylinder $x^2 + y^2 = 1$.

11. $F(x, y, z) = z^2\mathbf{i} - 3xy\mathbf{j} + x^3y^3\mathbf{k}$, and S is the part of $z = 5 - x^2 - y^2$ that lies above the plane $z = 1$. Assume that S is oriented upwards.

12. $F(x, y, z) = e^{xy}\cos z\mathbf{i} + x^2z\mathbf{j} + xy\mathbf{k}$, and S is the surface $x = \sqrt{1 - y^2 - z^2}$ oriented upwards.

13. $F(x, y, z) = z^2 i + 5xj$, and S is the square $0 \le x \le 1, 0 \le y \le 1, z = 1$.

14. $F(x, y, z) = x^2 z^2 i + y^2 z^2 j + xyzk$ and S is the part of the paraboloid $z = x^2 + y^2$ that lies inside the cylinder $x^2 + y^2 = 4$, oriented upward.

15. $F(x, y, z) = 2y \cos z i + e^x \sin z j + x e^y k$ and S is the part of the hemisphere $x^2 + y^2 + z^2 = 9$ above the plane $z = 0$, oriented upward.

16. $F(x, y, z) = x^2 yz^3 i + xyzj + \sin(xyz)k$ and S is the part of the cone $z^2 = x^2 + y^2$ between the planes $z = 0$ and $z = 3$, oriented upward.

17. $F(x, y, z) = -y^2 i + xj + z^2 k$ and S is the part of the plane $y + z = 2$ inside the cylinder $x^2 + y^2 = 1$, oriented upward.

Exercises 18-27. Verify Stoke's Theorem: $\oint\limits_C F \cdot dr = \iint\limits_S \text{curl}(F) \cdot n dS$

18. $F(x, y, z) = 2zi + 3xj + 5yk$ and S is that part of the paraboloid $z = 4 - x^2 - y^2$ for which $z \ge 0$, oriented upward.

19. $F(x, y, z) = yi + zj + xk$ and S is that part of the paraboloid $z = 1 - x^2 - y^2$ for which $z \ge 0$, oriented upward.

20. $F(x, y, z) = 3xi + \left(x + \dfrac{2x^3}{3} + 2xy^2\right)j + zk$ and S is that graph of the function $z = \sqrt{1 - 2(x^2 + y^2)}$ where $0 \le x^2 + y^2 \le \dfrac{1}{2}$, $z \ge 0$, oriented upward.

21. $F(x, y, z) = z^2 i - 2xj + y^3 k$ and S is the upper half of the sphere $x^2 + y^2 + z^2 = 1$, oriented upward.

22. $F(x, y, z) = 3yi + 4zj - 6xk$ and S is the paraboloid $z = 16 - x^2 - y^2$, $z \ge 0$, oriented upward.

23. $F(x, y, z) = 6xzi - x^2 j - 3y^2 k$ and S is the upper half of the sphere $x^2 + y^2 + z^2 = 1$, oriented upward.

24. $F(x, y, z) = x^4 i + xyj + z^4 k$ and S is the triangle $(2, 0, 0)$, $(0, 2, 0)$, $(0, 0, 2)$, oriented upward.

25. $F(x, y, z) = (x^2 + y^2)i + y^2 j + (x^2 + z^2)k$ and S is the triangle $(2, 0, 0)$, $(0, 2, 0)$, $(0, 0, 2)$ traversed counterclockwise.

26. $F(x, y, z) = xi + yj + xyzk$ and S is the part of the plane $2x + y + z = 2$ that lies in the first octant, oriented upward.

27. $F(x, y, z) = (z - y)i + (z + x)j - (x + y)k$ and S is the part of the paraboloid $z = 1 - x^2 - y^2$ that lies above the plane $z = 0$, oriented upward.

28. A particle moves along the line segments from the origin to the points $(1, 0, 0)$, $(1, 2, 1)$, $(0, 2, 1)$, and back to the origin under the influence of the force field $F(x, y, z) = z^2 i + 2xy j + 4y^2 k$. Determine the work done.

    Suggestion: See Theorem 14.2, page 628, and apply Stoke's theorem to the plane S containing the three points.

29. Let C be a simple closed smooth curve that lies in the plane $x + y + z = 1$. Show that $\oint_C z \, dx - 2x \, dy + 3y \, dz$ depends only on the area of the region enclosed by C and not on the shape of C or its location on the plane.

§7. THE DIVERGENCE THEOREM

A discussion on the surface integral of flux was initiated on page 636. Within that discussion you encountered:

Definition 14.11 (page 636): The **flux** of a three-dimensional vector field F across an oriented surface S is given by:

$$\text{Flux}(F_S) = \iint_S F \cdot n \, dS$$

And:

Theorem 14.17 (page 637): If S is the graph of a differentiable function $z = f(x, y)$ defined on a region D, then:

(if S is oriented by upward normals)

$$\iint_S F \cdot n \, dS = \iint_D F \cdot (-f_x i - f_y j + k) dA$$

(if S is oriented by downward normals)

$$\iint_S F \cdot n \, dS = \iint_D F \cdot (f_x i + f_y j - k) dA$$

Note that we cannot invoke the above theorem in the next example, as the spherical surface $S = \{(x, y, z) | x^2 + y^2 + z^2 = r^2\}$ is not of the form $z = f(x, y)$.

EXAMPLE 14.30 Find the outward flux of the vector field $F(x, y, z) = x i + y j + z k$ across the sphere $x^2 + y^2 + z^2 = r^2$.

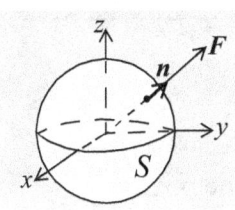

SOLUTION: Since both F and n are pointing radially away from the origin, they are parallel. Hence:

$$F \cdot n = \|F\| \|n\| \cos 0 = \|F\| \cdot 1 = \|F\| = \sqrt{x^2 + y^2 + z^2} = r$$

Thus: $\text{Flux}(F_S) = \iint_S F \cdot n \, dS = r \iint_S dS = r(4\pi r^2) = 4\pi r^3$

Example 14.24, page 639

While the above solution was easy, it is also atypical. Indeed, evaluating a surface integral $\iint_S F \cdot n \, dS$ when S is not generated by a function can prove to be an overwhelming if not impossible task.

Help is on the way:

One might anticipate that the vector form of Green's theorem, appearing on page 630:

$$\oint_C \mathbf{F} \cdot \mathbf{n} \, ds = \iint_D \text{div } \mathbf{F}(x, y) \, dA$$

$\swarrow$ $\nwarrow$
boundary curve of the plane region

extends to three space:

$$\iint_S \mathbf{F} \cdot \mathbf{n} \, dS = \iiint_E \text{div } \mathbf{F}(x, y, z) \, dV$$

$\swarrow$ $\nwarrow$
boundary surface of the solid region

Ans so it does:

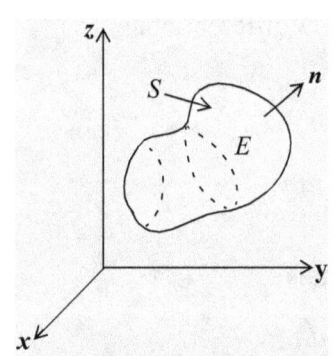

THEOREM 14.19

THE DIVERGENCE THEOREM

Let the surface S be oriented in the outward direction and let E be the solid region enclosed by S. If $\mathbf{F} = P\mathbf{i} + Q\mathbf{j} + R\mathbf{k}$ is continuously differentiable throughout E, then:

$$\iint_S \mathbf{F} \cdot \mathbf{n} \, dS = \iiint_E \text{div } (\mathbf{F}) \, dV$$

In words: The flux of $\mathbf{F}$ across the boundary S of a solid region E is equal to the triple integral of the divergence of $\mathbf{F}$ over E.

The above theorem, also known as Gauss' theorem, is undoubtedly the most far-reaching result in vector calculus; we do, after all, reside in three-dimensional space. A restricted proof of the theorem appears in Appendix B, page B-4.

CHECK YOUR UNDERSTANDING 14.30

Answer: $4\pi r^3$

Use the divergence theorem to solve Example 14.30.

EXAMPLE 14.31

Calculate the outward flux of:

$$\mathbf{F} = (x - y \sin z)\mathbf{i} + (y^2 - z^{\cos x})\mathbf{j} + yz\mathbf{k}$$

across the boundary S of the box:

$$E = \{(x, y, z) \mid 0 \leq x \leq 1, \, 0 \leq y \leq 2, \, 0 \leq z \leq 3\}$$

SOLUTION: The ugliness of the vector field $\mathbf{F}$, prevents us from attempting to compute $\iint_S \mathbf{F} \cdot \mathbf{n} \, dS$ directly. Fortunately, div$(\mathbf{F})$ is nice:

$$\text{div}(\mathbf{F}) = (x - y\sin z)_x + (y^2 - z^{\cos x})_y + (yz)_z = 1 + 2y + y = 1 + 3y$$

Turning to the divergence theorem we have:

$$\iint\limits_S F \cdot n\, dS = \iint\limits_E \int div(F)\, dV = \int_0^1 \int_0^2 \int_0^3 (1+3y)\, dz\, dy\, dx$$

$$= \int_0^1 \int_0^2 (z + 3yz)\big|_{z=0}^{z=3}\, dy\, dx$$

$$= \int_0^1 \int_0^2 (3 + 9y)\, dy\, dx$$

$$= \int_0^1 \left(3y + \frac{9y^2}{2}\right)\bigg|_{y=0}^{y=2}\, dx = \int_0^1 24\, dx = 24$$

CHECK YOUR UNDERSTANDING 14.31

Calculate the outward flux of:
$$F = (x\sin^2 y)i + (e^{x^2 z^3} - \tan x)j + (z\cos^2 y)k$$
across the boundary S of the cylindrical solid:
$$E = \{(x,y,z)\,|\,x^2 + y^2 = 9, 0 \le z \le 7\}$$

Answer: 63π

EXAMPLE 14.32   Calculate the outward flux of:
$$F = x^3 i + y^3 j + x^3 y^3 k$$
across the boundary S of the parabolic solid:
$$E = \{(x,y,z)\,|\,z = 4 - x^2 - y^2, z \ge 0\}$$

SOLUTION: We have:
$$div(F) = (x^3)_x + (y^3)_y + (x^3 y^3)_z = 3x^2 + 3y^2$$

Thus:
$$\iint\limits_S F \cdot n\, dS = \iint\limits_E \int div(F)\, dV = 3\iint\limits_E \int (x^2 + y^2)\, dV$$

Turning to cylindrical coordinates:
(see Figure 11.13, page 473)

$$= 3\int_0^{2\pi} \int_0^2 \int_0^{4-r^2} r^2 r\, dz\, dr\, d\theta$$

$$= 3\int_0^{2\pi} \int_0^2 (r^3 z)\big|_0^{4-r^2}\, dr\, d\theta$$

$$= 3\int_0^{2\pi} \int_0^2 (4r^3 - r^5)\, dr\, d\theta$$

$$= 3\int_0^{2\pi} \left(r^4 - \frac{r^6}{6}\right)\bigg|_0^2\, d\theta = 3\int_0^{2\pi} \frac{16}{3}\, d\theta = 32\pi$$

CHECK YOUR UNDERSTANDING 14.32

Calculate the outward flux of $F = 5xy^2 i + 5yz^2 j + 5x^2 z k$ across the boundary S of the sphere $E = (x,y,z)\,|\,x^2 + y^2 + z^2 \le 9$.

Answer: 972π

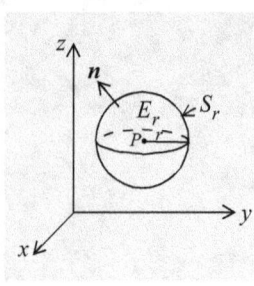

We conclude this brief section with a glimpse into the nature of divergence:

Let P be a point in space. Let E_r be the solid sphere centered at P of radius r with boundary S_r, and outward pointing normal $\boldsymbol{n}$. We then have:

$$\iint_{S_r} \boldsymbol{F} \cdot \boldsymbol{n} \, dS = \iiint_{E_r} \operatorname{div}(\boldsymbol{F}) dV \approx \iiint_{E_r} \operatorname{div}(\boldsymbol{F}(P)) dV$$

$$= \operatorname{div}(\boldsymbol{F}(P)) \iiint_{E_r} dV = \operatorname{div}(\boldsymbol{F}(P))\left(\frac{4}{3}\pi r^3\right)$$

Notice that the approximation $\operatorname{div}(\boldsymbol{F}(P)) \approx \dfrac{3}{4\pi r^3} \iint_{S_r} \boldsymbol{F} \cdot \boldsymbol{n} \, dS$

improves as r tends to 0. That said:

If $\operatorname{div}(\boldsymbol{F}) > 0$ at P, then so is $\displaystyle\iint_{S_r} \boldsymbol{F} \cdot \boldsymbol{n} \, dS$ for r small. I.e: Flow is directed away from P.
If $\operatorname{div}(\boldsymbol{F}) < 0$ at P, then so is $\displaystyle\iint_{S_r} \boldsymbol{F} \cdot \boldsymbol{n} \, dS$ for r small. I.e: Flow is directed toward P.
If $\operatorname{div}(\boldsymbol{F}) = 0$ at P, then $\displaystyle\iint_{S_r} \boldsymbol{F} \cdot \boldsymbol{n} \, dS \approx 0$. I.e: Flow toward P equals Flow from P.

| | **EXERCISES** | |

Exercises 1-12. Use the divergence theorem to evaluate $\iint\limits_{S} F \cdot n \, dS$.

1. $F(x, y, z) = x^3 i + y^3 j + z^3 k$ and S is the sphere $x^2 + y^2 + z^2 = a^2$.

2. $F(x, y, z) = (x^3 + y)i + (y^3 + z)j + (z^3 + x)k$ and S is the sphere $x^2 + y^2 + z^2 = 1$.

3. $F(x, y, z) = (x + y)i + z^2 j + x^2 k$ and S is the surface of the solid hemisphere:
   $$E = \{(x, y, z) | x^2 + y^2 + z^2 \leq 1, z \geq 0\} \text{ (with outward normal).}$$

4. $F(x, y, z) = xyi + yzj + xzk$ and S is the surface of the solid:
   $$E = \{(x, y, z) | 0 \leq x \leq 1, 0 \leq x \leq 1, 0 \leq z \leq 1 - x - y\} \text{ (with outward normal).}$$

5. $F(x, y, z) = ye^{z^2}i + y^2 j + e^{xy}k$ and S the surface of the solid:
   $$E = \{(x, y, z) | x^2 + y^2 \leq 9, 0 \leq z \leq y - 3\} \text{ (with outward normal).}$$

6. $F(x, y, z) = 3xy^2 i + xe^z j + z^3 k$ and S is the surface of the solid cylinder:
   $$E = \{(x, y, z) | y^2 + z^2 \leq 1, -1 \leq x \leq 2\} \text{ (with outward normal).}$$

7. $F(x, y, z) = y^2 i + xz^3 j + (z - 1)^2 k$ and S is the surface of the solid cylinder:
   $$E = \{(x, y, z) | x^2 + y^2 \leq 16, 1 \leq z \leq 5\} \text{ (with outward normal).}$$

8. $F(x, y, z) = 2xyi + 3ye^z j + x\sin(z)k$ and S is the surface of the solid unit cube:
   $$E = \{(x, y, z) | 0 \leq x \leq 1, 0 \leq y \leq 1, 0 \leq z \leq 1\} \text{ (with outward normal).}$$

9. $F(x, y, z) = [xz\sin(yz) + x^3]i + \cos(yz)j + [3zy^2 - e^{x^2 + y^2}]k$ and S is the surface of the solid:
   $$E = \{(x, y, z) | 0 \leq z \leq 4 - x^2 - y^2\} \text{ (with outward normal).}$$

10. $F(x, y, z) = xyi - \dfrac{y^2}{2}j + zk$ and S is the surface of the solid:
   $$E = \{(x, y, z) | z \leq 4 - x^2 - y^2, 1 \leq z \leq 4\} \cup \{(x, y, z) | x^2 + y^2 \leq 1, 0 \leq z \leq 1\}$$
   $$\text{(with outward normal).}$$

11. $F(x, y, z) = xyi + (y^2 + e^{xz^2})j + \sin(xy)k$ and S is the surface of the solid bounded by the parabolic cylinder $z = 1 - x^2$ and the planes $z = 0, y = 0$, and $y = 2 - z$ (with outward normal).

12. $F(x, y, z) = 2xzi - xyj - z^2 k$ and S is the surface, with outward normal, of the wedge cut from the first octant by the plane $z = 4 - y$ and the elliptical cylinder $4x^2 + y^2 = 16$.

13. Verify the divergence theorem for the vector field $F = x^2 i + y^2 j + z^2 k$ and S is the surface of the solid unit cube $E = \{(x, y, z)|0 \leq x \leq 1, 0 \leq y \leq 1, 0 \leq z \leq 1\}$.

   Note: You will need to compute six surface integrals to evaluate $\iint\limits_S F \cdot n \, dS$ directly.

14. Verify the divergence theorem for the vector field $F = xi + yj + zk$ and S is the surface of the solid $E = \{(x, y, z)|0 \leq z \leq 16 - x^2 - y^2\}$.

15. Verify the divergence theorem for the vector field $F = x^2 i + xzj + 3zk$ and S is the sphere $x^2 + y^2 + z^2 = 4$.

16. Show that for $F = xi + yj + zk$ and for S the surface of any solid E satisfying the conditions of the divergence theorem:

$$\text{Volume of E} = \frac{1}{3}\iint\limits_S F \cdot n \, dS$$

17. Show that the outward flux of a constant vector field $F = c$ across any closed surface satisfying the conditions of the divergence theorem is zero.

18. Verify that if the conditions of the divergence theorem are satisfied, then:

$$\iint\limits_S \text{curl } F \cdot n \, dS = 0$$

CHAPTER SUMMARY	
LINE INTEGRALS **For Scalar-Valued** **Functions**	For C a smooth curve C in the domain of a function $f(x, y)$ or $f(x, y, z)$: $$\int_C f ds = \lim_{\Delta s \to 0} \sum_a^b f \Delta s$$
THEOREM	If $x = x(t)$, $y = y(t)$, for $a \le t \le b$, then: $$\int_C f(x, y) ds = \int_a^b f[x(t), y(t)] \sqrt{\left(\frac{dx}{dt}\right)^2 + \left(\frac{dy}{dt}\right)^2} dt$$ If $x = x(t)$, $y = y(t)$, $z = z(t)$ for $a \le t \le b$, then: $$\int_C f(x, y, z) ds = \int_a^b f[x(t), y((t), z(t))] \sqrt{\left(\frac{dx}{dt}\right)^2 + \left(\frac{dy}{dt}\right)^2 + \left(\frac{dz}{dt}\right)^2} dt$$
LINE INTEGRALS **For Vector-Valued** **Functions**	Let $\boldsymbol{r}(t)$, $a \le t \le b$ be a parametrization of the smooth curve C, and let $\boldsymbol{F}$ be a continuous vector function defined on C. The **line integral** (or **path integral**) of $\boldsymbol{F}$ over C, is given by: $$\int_C \boldsymbol{F} \cdot \boldsymbol{T} ds = \int_C \boldsymbol{F} \cdot \boldsymbol{dr} = \int_a^b \boldsymbol{F}(\boldsymbol{r}(t)) \cdot \boldsymbol{r}'(t) dt$$
THEOREM	If C is a curve with parametrization $\boldsymbol{r}(t)$, $a \le t \le b$, then $-C$ denotes the curve with parametrization $\bar{\boldsymbol{r}}(t) = \boldsymbol{r}(a + b - t)$, $a \le t \le b$ and we have: $$\int_C \boldsymbol{F} \cdot \boldsymbol{dr} = -\int_{-C} \boldsymbol{F} \cdot \boldsymbol{dr}$$ (Note that $\bar{\boldsymbol{r}}(t)$ traces out C, but in the opposite direction of $\boldsymbol{r}(t)$)

Line integral of a vector-valued function $\boldsymbol{F}$ defined on a curve C with parametrization $\boldsymbol{r}(t)$ for $a \le t \le b$ can be represented in several forms:

$$\int_C \boldsymbol{F} \cdot \boldsymbol{T} ds \quad \text{or} \quad \int_C \boldsymbol{F} \cdot \boldsymbol{dr} \quad \text{or} \quad \int_a^b \boldsymbol{F} \cdot \frac{\boldsymbol{dr}}{dt} dt \quad \text{or} \quad \int_a^b \boldsymbol{F}(\boldsymbol{r}(t)) \cdot \boldsymbol{r}'(t) dt$$

For $\boldsymbol{F}(x, y) = P(x, y)\boldsymbol{i} + Q(x, y)\boldsymbol{j}$ and C in the plane:	For $\boldsymbol{F}(x, y, z) = P(x, y, z)\boldsymbol{i} + Q(x, y, z)\boldsymbol{j} + R(x, y, z)\boldsymbol{k}$ and C in three-space:
$$\int_a^b \left[P(x, y)\frac{dx}{dt} + Q(x, y)\frac{dy}{dt} \right] dt$$ or: $$\int_a^b P dx + Q dy$$	$$\int_a^b \left[P(x, y, z)\frac{dx}{dt} + Q(x, y, z)\frac{dy}{dt} + R(x, y, z)\frac{dz}{dt} \right] dt$$ or: $$\int_a^b P dx + Q dy + R dz$$

PATH-INDEPENDENT VECTOR FIELD	A vector field F is path-independent if for any two points p_0 and p_1 in the domain S of F, and any two smooth curves C and $\overline{C}$ in S from p_0 to p_1: $$\int_C F \cdot dr = \int_{\overline{C}} F \cdot dr$$
CONSERVATIVE FIELD	A vector field F is said to be **conservative** on a set S if there exists a scalar-valued function f, called a **potential function** for F, such that for every $p \in S$: $$F(p) = \nabla f(p)$$
curl(F)	Let $F(x,y) = P(x,y)i + Q(x,y)j$. The **curl of F**, denoted by curl(F) is the vector field: $$\text{curl}(F) = \left(\frac{\partial Q}{\partial x} - \frac{\partial P}{\partial y}\right)k$$ For $F(x,y,z) = P(x,y,z)i + Q(x,y,z)j + R(x,y,z)k$: $$\text{curl}(F) = \det\begin{bmatrix} i & j & k \\ \frac{\partial}{\partial x} & \frac{\partial}{\partial y} & \frac{\partial}{\partial z} \\ P & Q & R \end{bmatrix} = \left(\frac{\partial R}{\partial y} - \frac{\partial Q}{\partial z}\right)i + \left(\frac{\partial P}{\partial z} - \frac{\partial R}{\partial x}\right)j + \left(\frac{\partial Q}{\partial x} - \frac{\partial P}{\partial y}\right)k$$ Or: $\text{curl}(F) = \nabla \times F$ where $\nabla = \frac{\partial}{\partial x}i + \frac{\partial}{\partial y}j + \frac{\partial}{\partial z}k$

THEOREM
Let F be continuous on a simply connected open region S. The following are equivalent:
(i) $\oint_C F \cdot dr = 0$ for any smooth closed curve C in S.
(ii) F is a conservative vector field on S, i.e. $F = \nabla f$.
(iii) $F = \nabla f$ is path-independent on S with $\int_C F \cdot dr = f(p_1) - f(p_0)$ for any smooth C from p_0 to p_1.
(iv) For $F = P(x,y)i + Q(x,y)j$: $\frac{\partial P}{\partial y} = \frac{\partial Q}{\partial x}$. For $F = P(x,y,z)i + Q(x,y,z)j + R(x,y,z)k$: $\frac{\partial P}{\partial y} = \frac{\partial Q}{\partial x}, \ \frac{\partial P}{\partial z} = \frac{\partial R}{\partial x}, \ \frac{\partial Q}{\partial z} = \frac{\partial R}{\partial y}$. That is: $\text{curl}(F) = \mathbf{0}$

div(F)	Let $F(x, y) = P(x, y)i + Q(x, y)j$. The **divergence of** F, denoted by div(F) is given by: $$\text{div}(F) = \frac{\partial P}{\partial x} + \frac{\partial Q}{\partial y}$$ For $F(x, y, z) = P(x, y, z)i + Q(x, y, z)j + R(x, y, z)k$: $$\text{div}(F) = \frac{\partial P}{\partial x} + \frac{\partial Q}{\partial y} + \frac{\partial R}{\partial z}$$ Or: $\text{Div}(F) = \nabla \cdot F$ where $\nabla = \frac{\partial}{\partial x}i + \frac{\partial}{\partial y}j + \frac{\partial}{\partial z}k$

GREEN'S THEOREM

For $F \cdot T$	For $F \cdot n$
$$\oint_C F \cdot T ds = \oint_C P dx + Q dy = \iint_D \left(\frac{\partial Q}{\partial x} - \frac{\partial P}{\partial y}\right) dA$$ Vector Form	$$\oint_C F \cdot n ds = \oint_C P dy - Q dx = \iint_D \left(\frac{\partial P}{\partial x} + \frac{\partial Q}{\partial y}\right) dA$$ Vector Form
$$\oint_C F \cdot T ds = \iint_D (\text{curl } F) \cdot k\, dA = \iint_D (\nabla \times F) \cdot k\, dA$$	$$\oint_C F \cdot n ds = \iint_D (\text{div } F) dA = \iint_D (\nabla \cdot F) dA$$

SURFACE AREA (Function Form)	Let S be the surface $z = f(x, y)$, where f is a differentiable function defined on a region D. The **surface area** of S, denoted by $A(S)$, is given by: $$A(S) = \iint_D \sqrt{1 + [f_x(x, y)]^2 + [f_y(x, y)]^2}\, dA$$
(Parametrization Form)	Let S be the surface parametrized by the differentiable function $$r(u, v) = x(u, v)i + y(u, v)j + z(u, v)k$$ defined on a region D in the uv-plane. The **surface area** of $S = r(D)$, denoted by $A(S)$, is given by: $$A(S) = \iint_D \|r_u \times r_v\|\, du\, dv$$
SURFACE INTEGRAL (Function Form)	Let S be the surface $z = f(x, y)$, where f is a differentiable function, and let $g(x, y, z)$ be a continuous function on S. The **surface integral** of g over S, is given by: $$\iint_S g(x, y, z) dS = \iint_D g[x, y, f(x, y)]\sqrt{1 + [f_x(x, y)]^2 + [f_y(x, y)]^2}\, dA$$
(Parametrization Form)	Let S be the surface parametrized by the differentiable function $$r(u, v) = x(u, v)i + y(u, v)j + z(u, v)k$$ defined on a region D in the uv-plane. The **surface integral** of g over S is given by: $$\iint_S g(x, y, z) dS = \iint_D g[r(u, v)]\|r_u \times r_v\|\, du\, dv$$

FLUX ACROSS A SURFACE	The **flux** of a three-dimensional vector field F across an oriented surface S is given by: $$\text{Flux}(F_S) = \iint\limits_S F \cdot n \, dS$$
THEOREM	Let the surface S be defined by a differentiable function $z = f(x, y)$ defined on a region D in the xy-plane. If S is oriented by upward normals: $$\iint\limits_S F \cdot n \, dS = \iint\limits_D F \cdot (-f_x i - f_y j + k) \, dA$$ If S is oriented by downward normals: $$\iint\limits_S F \cdot n \, dS = \iint\limits_D F \cdot (f_x i + f_y j - k) \, dA$$
STOKE'S THEOREM 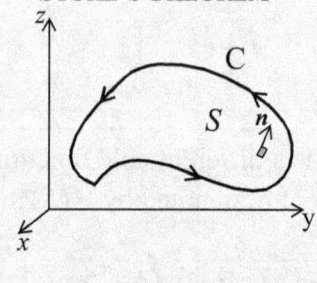	If S is an oriented surface that is bounded by a simple closed curve C with positive orientation (counterclockwise), and if $F = Pi + Qj + Rk$ has continuous first-order partial derivatives on some open region in three-space containing S, then: $$\oint\limits_C F \cdot dr = \iint\limits_S \text{curl}(F) \cdot n \, dS$$ where the orientation of S is such that: <blockquote>When walking around C in a counterclockwise direction with your head pointing in the direction of **n**, the surface will always be on your left.</blockquote>

In words: The line integral around the boundary curve of S of the tangential component of F is equal to the surface integral of the normal component of the curl of F.

DIVERGENCE THEOREM 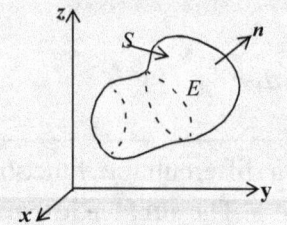	Let the surface S be oriented in the outward direction and let E be the solid region enclosed by S. If $F = Pi + Qj + Rk$ is continuously differentiable throughout E, then: $$\iint\limits_S F \cdot n \, dS = \iiint\limits_E \text{div}(F) \, dV$$

In words: The flux of F across the boundary S of a solid region E is equal to the triple integral of the divergence of F over E.

CHECK YOUR UNDERSTANDING SOLUTIONS

CHAPTER 11: FUNCTIONS OF SEVERAL VARIABLES

CYU 11.1 (a) The domain of $f(x) = \dfrac{1}{x^2 - 9} = \dfrac{1}{(x+3)(x-3)}$ consists of all real numbers for which the denominator is not zero, namely: $D_f = \{x \mid x \neq \pm 3\}$.

(b) The domain of $f(x) = \dfrac{1}{x^2 - y^2} = \dfrac{1}{(x+y)(x-y)}$ consists of all ordered pairs (x, y) for which the denominator is not zero, namely: $D_f = \{(x, y) \mid x \neq \pm y\}$.

(c) The domain of $f(x, y, z) = \dfrac{xy}{z+1}$ consists of all ordered 3-tuples for which the denominator is not zero, namely: $D_f = \{(x, y, z) \mid z \neq -1\}$.

CYU 11.2 Since as x and y tend to 0, both x^2 and y^2 also tend to 0, we can certainly anticipate that $\lim\limits_{(x, y) \to (0, 0)} (x^2 + y^2) = 0$. Kudos for our anticipation:

Let $\varepsilon > 0$ be given. We are to find $\delta > 0$ for which:

$$0 < \|(x, y) - (0, 0)\| < \delta \Rightarrow |(x^2 + y^2) - 0| < \varepsilon$$
$$0 < \|(x, y)\| < \delta \Rightarrow x^2 + y^2 < \varepsilon$$
$$\sqrt{x^2 + y^2} < \delta \Rightarrow x^2 + y^2 < \varepsilon$$
$$x^2 + y^2 < \delta^2 \Rightarrow x^2 + y^2 < \varepsilon$$

same

The above will certainly be satisfied for $\delta = \sqrt{\varepsilon}$.

CYU 11.3 (a) Let $\lim\limits_{(x, y) \to (x_0, y_0)} f(x, y) = L$. To prove that $\lim\limits_{(x, y) \to (x_0, y_0)} [cf(x, y)] = cL$. For given $\varepsilon > 0$ we are to find $\delta > 0$ such that:

$$0 < \|(x, y) - (x_0, y_0)\| < \delta \Rightarrow |cf(x, y) - cL| < \varepsilon$$
$$0 < \|(x, y) - (x_0, y_0)\| < \delta \Rightarrow |c||f(x, y) - L| < \varepsilon$$

In the event that $c = 0$, any $\delta > 0$ will surely work. If $c \neq 0$, then $\delta = \dfrac{\varepsilon}{|c|}$ will do the trick.

(b) $\lim\limits_{(x, y) \to (1, 1)} 5(x + y)\left(\dfrac{x^2 - y^2}{x - y}\right) = 5\left[\lim\limits_{(x, y) \to (1, 1)} (x + y)\right]\left[\lim\limits_{(x, y) \to (1, 1)} \left(\dfrac{x^2 - y^2}{x - y}\right)\right]$

Example 11.1: $= 5(2)(2) = 20$

CYU 11.4 (a) If f and g are continuous at (x_0, y_0), then:

$$\underset{(x,y)\to(x_0,y_0)}{\lim}[f(x,y)\cdot g(x,y)] \overset{\text{Theorem 11.1(c)}}{=} \underset{(x,y)\to(x_0,y_0)}{\lim}[f(x,y)]\underset{(x,y)\to(x_0,y_0)}{\lim}[g(x,y)]$$

since f and g are continuous at (x_0,y_0): $= f(x_0,y_0)g(x_0,y_0)$

(b) Since $f(y) = y^2$, $g(x) = \sin x + e^x$ and $h(r) = \sqrt{r}$ are continuous, so is $H(x,y) = \sqrt{y^2(\sin x + e^x)}$ (Theorem 11.3).

CYU 11.5 The projected traces of $z = f(x,y) = |x| - y$ for $z = 0, z = 2, z = 4$ onto the xy-plane, appearing in (a) below, are hoisted 0, 2, and 4 units up the z-axis in (b).

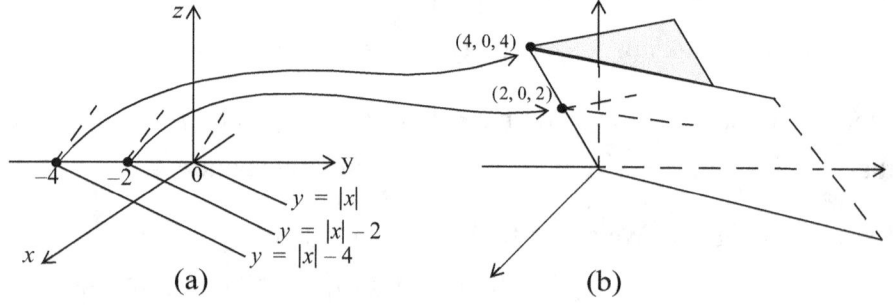

(a)   (b)

CYU 11.6 Since the variable z is missing in the equation $x^2 + \dfrac{y^2}{4} = 1$, the directrix resides in the xy-plane, and the rullings are parallel to the z-axis. A portion of the elliptical cylinder appears in the adjacent figure.

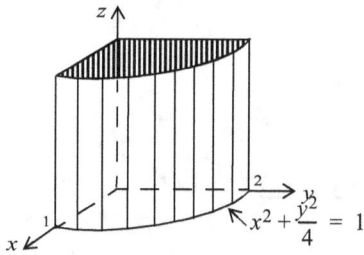

CYU 11.7 This is Example 11.6, with the roles of y and z reversed:

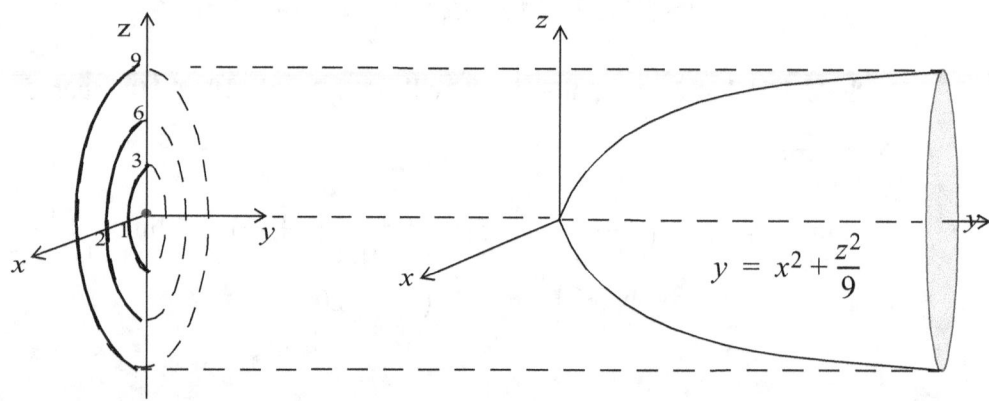

CYU 11.8 Portions of the cross sections of the ellipsoid on the planes $x = 0$, $y = 0$, and $z = 0$, along with that the planes $z = 1$, are depicted on the left side of the figure below. The associated ellipsoid appears at the right.

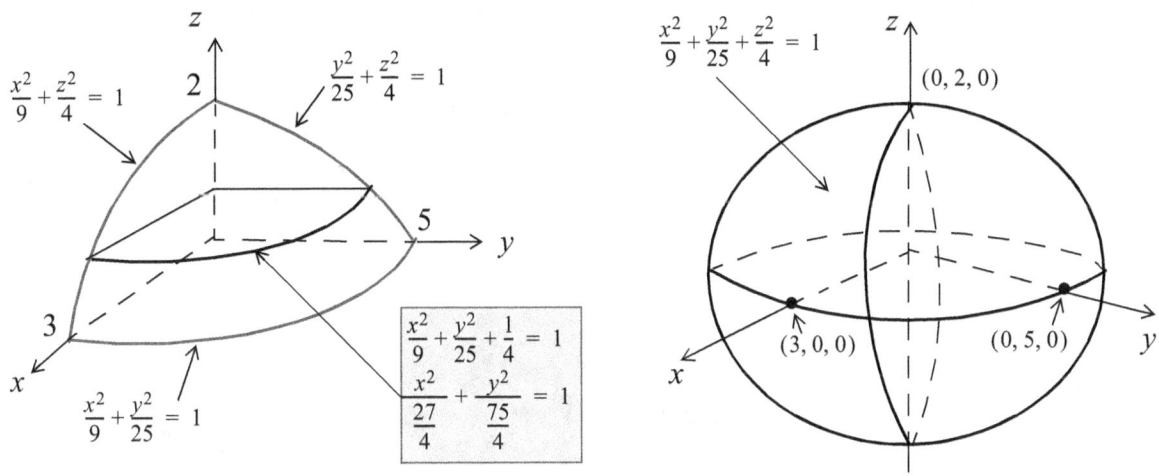

CYU 11.9 This is Example 11.8, with the roles of y and z reversed: $y^2 = x^2 + \dfrac{z^2}{9}$

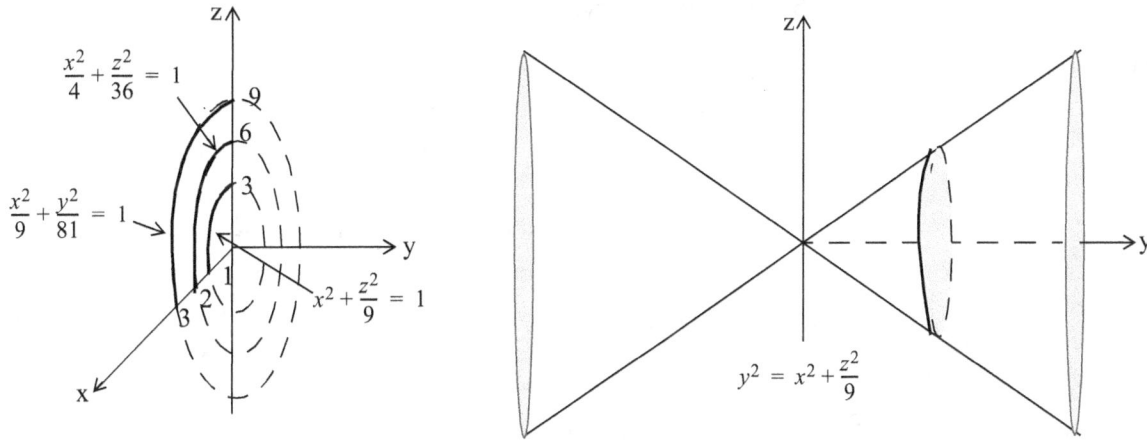

CYU 11.10 $V = \displaystyle\int_1^2 \left[\int_2^4 (2x + y + 3xy)\,dy \right] dx = \int_1^2 \left(2xy + \frac{1}{2}y^2 + \frac{3}{2}xy^2 \right) \Big|_2^4 dx$

$$= \int_1^2 [(8x + 8 + 24x) - (4x + 2 + 6x)]\,dx$$

$$= \int_1^2 (22x + 6)\,dx = (11x^2 + 6x)\Big|_1^2$$

$$= (44 + 12) - (11 + 6) = 39$$

$$V = \int_2^4 \left[\int_1^2 (2x + y + 3xy)dx \right] dy = \int_2^4 \left(x^2 + xy + \frac{3x^2 y}{2} \right) \Big|_1^2 dy$$

$$= \int_2^4 \left[(4 + 2y + 6y) - \left(1 + y + \frac{3}{2}y \right) \right] dy$$

$$= \int_2^4 \left(\frac{11}{2}y + 3 \right) dy = \left(\frac{11}{4}y^2 + 3y \right) \Big|_2^4$$

$$= (44 + 12) - (11 + 6) = 39$$

CYU 11.11

(a)

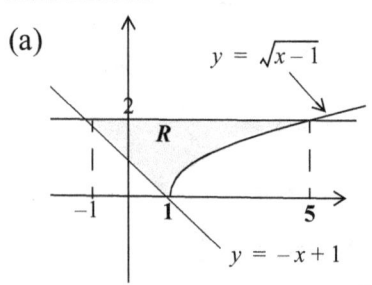

$$\iint_R 2xy \, dA = \int_{-1}^1 \int_{y = -x + 1}^{y = 2} 2xy \, dy \, dx + \int_1^5 \int_{y = \sqrt{x - 1}}^{y = 2} 2xy \, dy \, dx$$

$$(*) \quad \iint_R 2xy \, dA = \int_0^2 \int_{x = 1 - y}^{x = y^2 + 1} 2xy \, dx \, dy$$

(b) Choosing (*) we have:

$$\iint_R 2xy \, dA = \int_0^2 \int_{x = 1 - y}^{x = y^2 + 1} 2xy \, dx \, dy = \int_0^2 (x^2 y) \Big|_{1 - y}^{y^2 + 1} dy = \int_0^2 [y(y^2 + 1)^2 - y(1 - y)^2] dy$$

$$= \int_0^2 [y(y^4 + 2y^2 + 1) - y(1 - 2y + y^2)] dy$$

$$= \int_0^2 (y^5 + y^3 + 2y^2) dy = \left(\frac{y^6}{6} + \frac{y^4}{4} + \frac{2y^3}{3} \right) \Big|_0^2 = 20$$

CYU 11.12

$$V = \iint_R \left(-\frac{3}{2}y + 5 \right) - (-y + 2) dA = \int_0^2 \int_0^2 \left(-\frac{y}{2} + 3 \right) dy \, dx$$

$$= \int_0^2 \left(-\frac{y^2}{4} + 3y \right) \Big|_{y = 0}^{y = 2} dx = \int_0^2 5 \, dx = 10$$

CYU 11.13

$$\iint_R h(x, y) dA = \int_0^1 \left[\int_0^{x^2} h(x, y) dy \right] dx + \int_1^2 \left[\int_0^{-x + 2} h(x, y) dy \right] dx$$

Fixing x between 0 and 1, we go from $y = 0$ (the x-axis) to $y = x^2$ (hashed region). Fixing x between 1 and 2, we go from $y = 0$ to $y = -x + 2$.

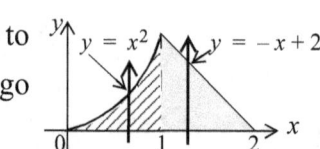

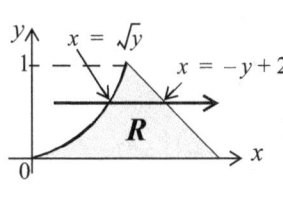

We can get by with one integral if we reverse the order of integration. Fixing y between 0 and 1 we go from the curve $x = \sqrt{y}$ to the curve $x = -y + 2$:

$$\iint_R h(x,y)\,dA = \int_0^1\left[\int_{\sqrt{y}}^{-y+2} h(x,y)\,dx\right]dy$$

CYU 11.14
$$M = \iint_R xy\,dA = \int_0^1\left[\int_0^{-x+1} xy\,dy\right]dx = \int_0^1\left(\frac{1}{2}xy^2\right)\Big|_0^{-x+1} dx$$

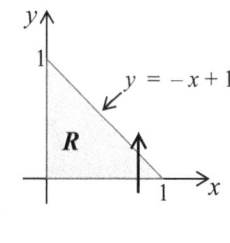

$$= \int_0^1\left(\frac{1}{2}x^3 - x^2 + \frac{1}{2}x\right)dx$$

$$= \left(\frac{x^4}{8} - \frac{x^3}{3} + \frac{x^2}{4}\right)\Big|_0^1 = \frac{1}{24}$$

CYU 11.15   We have: $M = \iint_R xy\,dA = \iint_R xy\,dA = \frac{1}{24}$ (see CYU 11.14). Moreover:

$$M_y = \iint_R x\delta(x,y)\,dA = \iint_R x^2 y\,dA = \int_0^1\left[\int_0^{-x+1} x^2 y\,dy\right]dx$$

$$= \int_0^1\left(x^2 \cdot \frac{y^2}{2}\right)\Big|_0^{-x+1} dx = \int_0^1 \frac{1}{2}x^2(-x+1)^2\,dx$$

$$= \int_0^1\left(\frac{1}{2}x^4 - x^3 + \frac{1}{2}x^2\right)dx = \frac{1}{60}$$

$$M_x = \iint_R y\delta(x,y)\,dA = \iint_R xy^2\,dA = \int_0^1\left[\int_0^{-x+1} xy^2\,dy\right]dx$$

$$= \int_0^1\left(x \cdot \frac{y^3}{3}\right)\Big|_0^{-x+1} dx = \int_0^1 \frac{1}{3}x(-x+1)^3\,dx$$

$$= \int_0^1\left(-\frac{1}{3}x^4 + x^3 - x^2 + \frac{1}{3}x\right)dx = \frac{1}{60}$$

Conclusion: $(\bar{x}, \bar{y}) = \left(\frac{1/60}{1/24}, \frac{1/60}{1/24}\right) = \left(\frac{2}{5}, \frac{2}{5}\right)$

CYU 11.16 (a) The paraboloid $z = 1 - x^2 - y^2$ intersects the plane $z = 0$ in the circle $x^2 + y^2 = 1$. It follows that the solid lies under the surface $z = f(x,y) = 1 - x^2 - y^2$ over the region $R = \{(x,y)\,|\,x^2 + y^2 \le 1\}$. Expressing the volume in polar form we have:

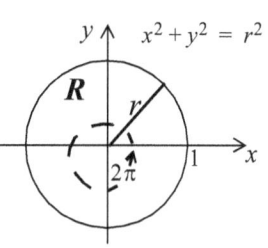

$$V = \iint_R (1 - x^2 - y^2)dA = \iint_R (1 - r^2)dA = \int_0^{2\pi}\left(\int_0^1 (1 - r^2)r\,dr\right)d\theta$$

$$= \int_0^{2\pi}\left(\frac{r^2}{2} - \frac{r^4}{4}\right)\bigg|_0^1 d\theta = \int_0^{2\pi}\frac{1}{4}d\theta = \frac{\pi}{2}$$

(b) $A = \int_0^{2\pi}\left(\int_0^{1-\cos\theta}(1 \cdot r\,dr)\right)d\theta = \int_0^{2\pi}\left(\frac{r^2}{2}\right)\bigg|_0^{1-\cos\theta}d\theta$

$$= \frac{1}{2}\int_0^{2\pi}(1 - \cos\theta)^2 d\theta$$

$$= \frac{1}{2}\int_0^{2\pi}(1 - 2\cos\theta + \cos^2\theta)d\theta$$

$$= \frac{1}{2}\int_0^{2\pi}\left(1 - 2\cos\theta + \frac{1 + \cos 2\theta}{2}\right)d\theta$$

$$= \frac{1}{2}\left(\theta - 2\sin\theta + \frac{\theta}{2} + \frac{\sin 2\theta}{4}\right)\bigg|_0^{2\pi} = \frac{3}{2}\pi$$

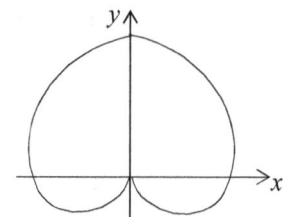

$r = 1 - \cos\theta$

CYU 11.17 $M = \int_0^{2\pi}\left(\int_0^{1+\sin\theta} r \cdot r\,dr\right)d\theta = \int_0^{2\pi}\left(\frac{r^3}{3}\right)\bigg|_0^{1+\sin\theta}d\theta$

$$= \frac{1}{3}\int_0^{2\pi}(1 + \sin\theta)^3\ d\theta$$

$$= \frac{1}{3}\int_{-\frac{\pi}{2}}^{\frac{\pi}{2}}(1 + 3\sin\theta + 3\sin^2\theta + \sin^3\theta)d\theta$$

$$= \frac{1}{3}\int_0^{2\pi}\left[1 + 3\sin\theta + \frac{3}{2}(1 - \cos 2\theta) + \sin\theta(1 - \cos^2\theta)\right]d\theta$$

$$= \frac{1}{3}\int_0^{2\pi}\left(\frac{5}{2} + 4\sin\theta - \frac{3}{2}\cos 2\theta - \sin\theta\cos^2\theta\right)d\theta$$

$$= \frac{1}{3}\left(\frac{5}{2}\theta - 4\cos\theta - \frac{3}{4}\sin 2\theta + \frac{1}{3}\cos^3\theta\right)\bigg|_0^{2\pi}$$

$$= \frac{1}{3}\left[5\pi - 4 + \frac{1}{3} - \left(-4 + \frac{1}{3}\right)\right] = \frac{5\pi}{3}$$

CYU 11.18 $M = \iint\limits_{R} \delta(x, y)\,dA = \int_0^{\frac{\pi}{2}}\left(\int_0^{\cos\theta} \sin\theta \cdot r\,dr\right)d\theta$

$$= \int_0^{\frac{\pi}{2}}\left(\left(\frac{r^2}{2}\sin\theta\right)\Big|_0^{\cos\theta}\right)d\theta = \frac{1}{2}\int_0^{\frac{\pi}{2}}\cos^2\theta\sin\theta\,d\theta = \frac{1}{2}\left(-\frac{1}{3}\cos^3\theta\right)\Big|_0^{\frac{\pi}{2}} = \frac{1}{6}$$

$M_y = \iint\limits_{R} x\delta(x, y)\,dA = \int_0^{\frac{\pi}{2}}\left(\int_0^{\cos\theta} r\cos\theta\cdot\sin\theta\cdot r\,dr\right)d\theta$

$$= \int_0^{\frac{\pi}{2}}\left(\frac{r^3}{3}\cos\theta\sin\theta\right)\Big|_0^{\cos\theta}d\theta = \frac{1}{3}\int_0^{\frac{\pi}{2}}\cos^3\theta\cos\theta\sin\theta\,d\theta$$

$$= \frac{1}{3}\int_0^{\frac{\pi}{2}}\cos^4\theta\sin\theta\,d\theta = \frac{1}{3}\left(-\frac{1}{5}\cos^5\theta\right)\Big|_0^{\frac{\pi}{2}} = \frac{1}{15}$$

$M_x = \iint\limits_{R} y\delta(x, y)\,dA = \int_0^{\frac{\pi}{2}}\left(\int_0^{\cos\theta} r\sin\theta\cdot\sin\theta\cdot r\,dr\right)d\theta$

$$= \int_0^{\frac{\pi}{2}}\left(\frac{r^3}{3}\sin^2\theta\right)\Big|_0^{\cos\theta}d\theta = \frac{1}{3}\int_0^{\frac{\pi}{2}}(\cos\theta)^3\sin^2\theta\,d\theta$$

$$= \frac{1}{3}\int_0^{\frac{\pi}{2}}\cos\theta(1 - \sin^2\theta)\sin^2\theta\,d\theta$$

$$= \frac{1}{3}\int_0^{\frac{\pi}{2}}(-\sin^4\theta\cos\theta + \sin^2\theta\cos\theta)\,d\theta$$

$$= \frac{1}{3}\left(-\frac{1}{5}\sin^5\theta + \frac{1}{3}\sin^3\theta\right)\Big|_0^{\frac{\pi}{2}} = \frac{2}{45}$$

Conclusion: $(\bar{x}, \bar{y}) = \left(\dfrac{M_y}{M}, \dfrac{M_x}{M}\right) = \left(\dfrac{1/15}{1/6}, \dfrac{2/45}{1/6}\right) = \left(\dfrac{6}{15}, \dfrac{12}{45}\right) = \left(\dfrac{2}{5}, \dfrac{4}{15}\right)$

CYU 11.19   Just as the area of a two-dimensional region R equals the volume of the solid of base R and height 1, which is to say: $\iint\limits_{R} 1\,dy\,dx$; so does the volume of a three-dimensional region W equal $\iiint\limits_{W} 1\,dz\,dy\,dx$. That being the case, we simply evaluate the integral of Example 11.12 with "1" replacing the "x":

$$V = \iiint x\,dz\,dy\,dx = \int_0^1\int_0^{1-x}\int_0^{1-x-y} 1\,dz\,dy\,dx = \int_0^1\int_0^{1-x} z\Big|_0^{1-x-y}\,dy\,dx$$

$$= \int_0^1\int_0^{1-x}(1-x-y)\,dy\,dx$$

$$= \int_0^1\left(y-xy-\frac{y^2}{2}\right)\Big|_0^{1-x}\,dx$$

$$= \int_0^1\left[(1-x)-x(1-x)-\frac{(1-x)^2}{2}\right]dx$$

$$= \int_0^1\left(\frac{x^2}{2}-x+\frac{1}{2}\right)dx$$

$$= \left(\frac{x^3}{6}-\frac{x^2}{2}+\frac{x}{2}\right)\Big|_0^1 = \frac{1}{6}-\frac{1}{2}+\frac{1}{2} = \frac{1}{6}$$

CYU 11.20 We show that $\iiint x\delta(x,y,z)\,dV = \iiint y\delta(x,y,z)\,dV = 0$:

$$\int_{-r}^{r}\int_{-\sqrt{r^2-x^2}}^{\sqrt{r^2-x^2}}\int_0^h x(kz)\,dz\,dy\,dx$$

$$= k\int_{-r}^{r}\int_{-\sqrt{r^2-x^2}}^{\sqrt{r^2-x^2}}\left(x\frac{z^2}{2}\right)\Big|_{z=0}^{z=h}\,dy\,dx$$

$$= k\frac{h^2}{2}\int_{-r}^{r}\int_{-\sqrt{r^2-x^2}}^{\sqrt{r^2-x^2}} x\,dy\,dx$$

$$= k\frac{h^2}{2}\int_{-r}^{r}\left[(xy)\Big|_{y=-\sqrt{r^2-x^2}}^{y=\sqrt{r^2-x^2}}\right]dx$$

$$= k\frac{h^2}{2}\int_{-r}^{r} 2x\sqrt{r^2-x^2}\,dx = -\frac{1}{2}kh^2\int_0^0\sqrt{u}\,du = 0$$

$$\uparrow$$

$$u = r^2-x^2,\, du = -2x\,dx$$
$$x = -r \Rightarrow u = 0,\, x = r \Rightarrow u = 0$$

$$\int_{-r}^{r}\int_{-\sqrt{r^2-x^2}}^{\sqrt{r^2-x^2}}\int_0^h y(kz)\,dz\,dy\,dx$$

$$= k\int_{-r}^{r}\int_{-\sqrt{r^2-x^2}}^{\sqrt{r^2-x^2}}\left(y\frac{z^2}{2}\right)\Big|_{z=0}^{z=h}\,dy\,dx$$

$$= k\frac{h^2}{2}\int_{-r}^{r}\int_{-\sqrt{r^2-x^2}}^{\sqrt{r^2-x^2}} y\,dy\,dx$$

$$= k\frac{h^2}{2}\int_{-r}^{r}\left[\left(\frac{y^2}{2}\right)\Big|_{y=-\sqrt{r^2-x^2}}^{y=\sqrt{r^2-x^2}}\right]dx$$

$$= \frac{kh^2}{4}\int_{-r}^{r}[(r^2-x^2)-(r^2-x^2)]\,dx = 0$$

CYU 11.21 Keep in mind that you have to go from surface to surface, then from curve to curve, and, finally, from point to point. Once you choose the surface to surface direction, then the limits of integration of the remaining two integrals can be observed by projecting W onto the coordinate plane perpendicular to the initially chosen direction.

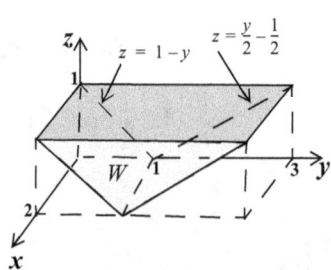

$$M = \int_0^1 \int_0^2 \int_{1-z}^{2z+1} z\,dy\,dx\,dz = \int_0^2 \int_0^1 \int_{1-z}^{2z+1} z\,dy\,dz\,dx$$

$$= \int_0^1 \int_{1-z}^{2z+1} \int_0^2 z\,dx\,dy\,dz$$

$$= \int_0^1 \int_{1-y}^1 \int_0^2 z\,dx\,dz\,dy + \int_1^3 \int_{\frac{y}{2}-\frac{1}{2}}^1 \int_0^2 z\,dx\,dz\,dy$$

$$= \int_0^1 \int_0^2 \int_{1-y}^1 z\,dz\,dx\,dy + \int_1^3 \int_0^2 \int_{\frac{y}{2}-\frac{1}{2}}^1 z\,dz\,dx\,dy$$

$$= \int_0^2 \int_0^1 \int_{1-y}^1 z\,dz\,dy\,dx + \int_0^2 \int_1^3 \int_{\frac{y}{2}-\frac{1}{2}}^1 z\,dz\,dy\,dx$$

CYU 11.22 (a) For $P = (r, \theta, z) = \left(4, \dfrac{\pi}{6}, -1\right)$:

$$x = r\cos\theta = 4\cos\frac{\pi}{6} = 4 \cdot \frac{\sqrt{3}}{2} = 2\sqrt{3}$$

$$y = r\sin\theta = 4\sin\frac{\pi}{6} = 4 \cdot \frac{1}{2} = 2$$

Thus $P = (2\sqrt{3}, 2, -1)$.

(b) For $P = (x, y, z) = (2, 2, 4)$: $r = \pm\sqrt{x^2+y^2} = \pm\sqrt{4+4} = \pm 2\sqrt{2}$

$$\theta = \tan^{-1}\frac{y}{x} = \tan^{-1}\frac{2}{2} = \tan^{-1}1 = \frac{\pi}{4} + n\pi$$

Thus $P = \left(2\sqrt{2}, \dfrac{\pi}{4}, 4\right)$

CYU 11.23 To determine the region of intersection of the paraboloid $z = x^2 + y^2$ and the sphere $x^2 + y^2 + z^2 = 20$ solve the system:

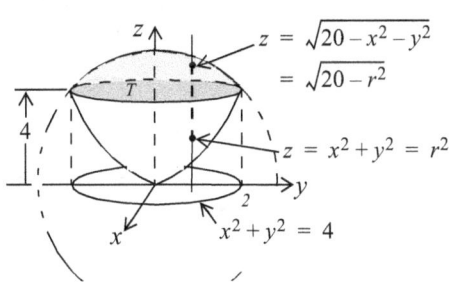

$$\left.\begin{array}{l} z = x^2 + y^2 \\ x^2 + y^2 + z^2 = 20 \end{array}\right\} \Rightarrow z + z^2 = 20 \Rightarrow z^2 + z - 20 = 0$$

$$(z+5)(z-4) = 0$$

Hence $z = 4$ and $x^2 + y^2 = 4$. We then have:

$$V = \iiint_W 1\,dV = \int_0^{2\pi} \int_0^2 \int_{r^2}^{\sqrt{20-r^2}} r\,dz\,dr\,d\theta = \int_0^{2\pi} \int_0^2 (rz)\Big|_{z=r^2}^{z=\sqrt{20-r^2}} dr\,d\theta$$

$$\int r\sqrt{20-r^2} = -\frac{1}{2}\int u^{1/2}du = -\frac{1}{3}u^{3/2} + C$$

$$u = 20 - r^2$$

$$du = -2r\,dr$$

$$= -\frac{1}{3}(20-r^2)^{3/2} + C$$

$$= \int_0^{2\pi} \int_0^2 (r\sqrt{20-r^2} - r \cdot r^2)\,dr\,d\theta$$

$$= \int_0^{2\pi} \left(-\frac{1}{3}(20-r^2)^{3/2} - \frac{r^4}{4}\right)\Big|_0^2 d\theta$$

$$= \int_0^{2\pi} \frac{1}{3}(40\sqrt{5} - 76)\,d\theta = \frac{2\pi}{3}(40\sqrt{5} - 76)$$

Also, for $\delta(x, y, z) = k$: $M = \iiint\limits_W \delta(x, y, z)dV = k\iiint\limits_W dV = \dfrac{2\pi k}{3}(40\sqrt{5} - 76)$

By symmetry, the x and y coordinates of the center of mass are 0. As for $\bar{z}$:

$$\bar{z} = \dfrac{1}{M}\iiint\limits_W z\delta(x, y, z)dV = \dfrac{1}{M}\int_0^{2\pi}\int_0^2\int_{r^2}^{\sqrt{20-r^2}} zkr\,dz\,dr\,d\theta$$

$$= \dfrac{k}{M}\int_0^{2\pi}\int_0^2\left(\dfrac{z^2}{2}\right)\bigg|_{r^2}^{\sqrt{20-r^2}} dr\,d\theta$$

$$= \dfrac{k}{2M}\int_0^{2\pi}\int_0^2 (20r - r^3 - r^5)dr\,d\theta = \dfrac{k}{2M}\int_0^{2\pi}\left(10r^2 - \dfrac{r^4}{4} - \dfrac{r^6}{6}\right)\bigg|_0^2 d\theta$$

$$= \dfrac{k}{2M}\int_0^{2\pi}\dfrac{76}{3}\bigg|_0^2 d\theta = \dfrac{76\pi k}{3M}$$

$$M = \dfrac{2\pi}{3}k(40\sqrt{5} - 76): \quad = \dfrac{38}{40\sqrt{5} - 76} \approx 1.4$$

CYU 11.24 (a) For $P(x, y, z) = (0, 2\sqrt{3}, -2)$: $\rho = \sqrt{0 + 12 + 4} = 4$

From $z = \rho\cos\phi$ we have: $-2 = 4\cos\phi \Rightarrow \cos\phi = -\dfrac{1}{2}$; as $0 \le \phi \le \pi$, $\phi = \dfrac{2\pi}{3}$.

From $x = \rho\sin\phi\cos\theta$ we have: $0 = 4\sin\dfrac{2\pi}{3}\cos\theta \Rightarrow \cos\theta = 0$; as $y > 0$, $\theta = \dfrac{\pi}{2}$.

Thus: $P(\rho, \phi, \theta) = \left(4, \dfrac{2\pi}{3}, \dfrac{\pi}{2}\right)$.

(b) For $P(\rho, \phi, \theta) = \left(2, \dfrac{2\pi}{3}, \dfrac{5\pi}{6}\right)$ (see Figure 13.9):

$$x = \rho\sin\phi\cos\theta = 2\sin\dfrac{2\pi}{3}\cos\dfrac{5\pi}{6} = 2\left(\dfrac{\sqrt{3}}{2}\right)\left(-\dfrac{\sqrt{3}}{2}\right) = -\dfrac{3}{2}$$

$$y = \rho\sin\phi\sin\theta = 2\sin\dfrac{2\pi}{3}\sin\dfrac{5\pi}{6} = 2\left(\dfrac{\sqrt{3}}{2}\right)\left(\dfrac{1}{2}\right) = \dfrac{\sqrt{3}}{2}$$

$$z = \rho\cos\phi = 2\cos\dfrac{2\pi}{3} = 2\left(-\dfrac{1}{2}\right) = -1$$

Thus: $P(x, y, z) = \left(-\dfrac{3}{2}, \dfrac{\sqrt{3}}{2}, -1\right)$

As for the cylindrical coordinates, $P(r, \theta, z)$, we have $\theta = \dfrac{5\pi}{6}$; and, from above: $z = -1$.

As for r: $r = \sqrt{x^2 + y^2} = \sqrt{\left(-\dfrac{3}{2}\right)^2 + \left(\dfrac{\sqrt{3}}{2}\right)^2} = \sqrt{3}$. Thus: $P(r, \theta, z) = \left(\sqrt{3}, \dfrac{5\pi}{6}, -1\right)$.

CYU 11.25 Here is the sphere's equation in spherical coordinates:

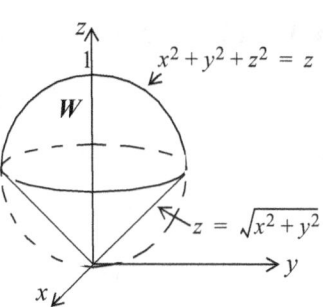

$$x^2 + y^2 + z^2 = z$$

$$\rho^2 = \rho\cos\phi \Rightarrow \rho = \cos\phi$$

And here is the cone's equation in spherical coordinates:

$$z = \sqrt{x^2 + y^2}$$

$$\rho\cos\phi = \sqrt{(\rho\sin\phi\cos\theta)^2 + (\rho\sin\phi\sin\theta)^2}$$

$$= \sqrt{\rho^2\sin^2\phi(\cos^2\theta + \sin^2\theta)} = \rho\sin\phi$$

$$\cos\phi = \sin\phi \Rightarrow \phi = \frac{\pi}{4}$$

So: $$W = \left\{ (\rho, \phi, \theta) \mid 0 \le \rho \le \cos\phi, 0 \le \phi \le \frac{\pi}{4}, 0 \le \theta \le 2\pi \right\};$$ and:

$$V = \iiint 1\,dV = \int_0^{2\pi}\int_0^{\frac{\pi}{4}}\int_0^{\cos\phi} \rho^2\sin\phi\,d\rho\,d\phi\,d\theta = \int_0^{2\pi}\int_0^{\frac{\pi}{4}}\left(\frac{\rho^3}{3}\sin\phi\right)\Bigg|_{\rho=0}^{\rho=\cos\phi} d\phi\,d\theta$$

$$= \frac{1}{3}\int_0^{2\pi}\int_0^{\frac{\pi}{4}} \cos^3\phi\sin\phi\,d\phi\,d\theta = \frac{1}{3}\int_0^{2\pi}\left(-\frac{1}{4}\cos^4\phi\right)\Bigg|_0^{\frac{\pi}{4}} d\theta = \frac{1}{12}\int_0^{2\pi}\left(\frac{3}{4}\right)d\theta = \frac{\pi}{8}$$

CHAPTER 12: VECTORS AND VECTOR-VALUED FUNCTIONS

CYU 12.1

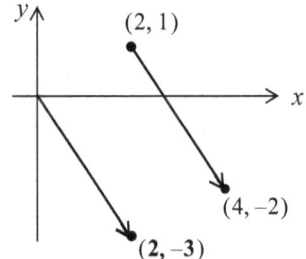

To move the initial point to the origin, we have to move that point two units to the left and one unit down. Doing the same to the terminal point $(4, -2)$ we end up with:

$$\langle a_1, a_2 \rangle = \langle 4 - 2, -2 - 1 \rangle = \langle 2, -3 \rangle$$

CYU 12.2 For $v = \langle 3, 2 \rangle, w = \langle -1, 1 \rangle, r = 2, s = 3$:

$$rv + sw = 2\langle 3, 2 \rangle + 3\langle -1, 1 \rangle = \langle 6, 4 \rangle + \langle -3, 3 \rangle = \langle 6 - 3, 4 + 3 \rangle = \langle 3, 7 \rangle$$

CYU 12.3 $\frac{1}{2}[\langle 2, -3, 0 \rangle - \langle 1, 0, -\sqrt{5} \rangle] - \langle 1, 2, 3 \rangle = \frac{1}{2}\langle 1, -3, \sqrt{5} \rangle - \langle 1, 2, 3 \rangle$

$$= \langle \frac{1}{2}, -\frac{3}{2}, -\frac{\sqrt{5}}{2} \rangle - \langle 1, 2, 3 \rangle$$

$$= \langle -\frac{1}{2}, -\frac{7}{2}, \frac{\sqrt{5} - 6}{2} \rangle$$

CYU 12.4 For $u = \langle u_1, u_2, u_3 \rangle$ and $v = \langle v_1, v_2, v_3 \rangle$:

$$u + v \equiv \langle u_1, u_2, u_3 \rangle + \langle v_1, v_2, v_3 \rangle \equiv \langle u_1 + v_1, u_2 + v_2, u_3 + v_3 \rangle$$

$$\textbf{PofR:} = \langle v_1 + u_1, v_2 + u_2, v_3 + u_3 \rangle \equiv v + u$$

For $v = \langle v_1, v_2, \ldots, v_n \rangle$, and scalars r and s:

$$(r + s)v = (r + s)\langle v_1, v_2, \ldots, v_n \rangle \equiv \langle (r+s)v_1, (r+s)v_2, \ldots, (r+s)v_n \rangle$$

$$\textbf{PofR:} = \langle rv_1 + sv_1, rv_2 + sv_2, \ldots, rv_n + sv_n \rangle$$

$$\equiv \langle rv_1, rv_2, \ldots, rv_n \rangle + \langle sv_1, sv_2, \ldots, sv_n \rangle$$

$$\equiv r\langle v_1, v_2, \ldots, v_n \rangle + s\langle v_1, v_2, \ldots, v_n \rangle = rv + sv$$

CYU 12.5 (a) $\|3\langle 2, 1, 0 \rangle - \langle 4, -2, 1 \rangle\| = \|\langle 6, 3, 0 \rangle - \langle 4, -2, 1 \rangle\|$

$$= \|\langle 2, 5, -1 \rangle\| = \sqrt{2^2 + 5^2 + (-1)^2} = \sqrt{30}$$

(b) $\|cv\| = \sqrt{c(v_1, v_2, v_3)} = \sqrt{(cv_1, cv_2, cv_3)} = \sqrt{c^2 v_1^2 + c^2 v_2^2 + c^2 v_3^2}$

$$= \sqrt{c^2}\sqrt{v_1^2 + v_2^2 + v_3^2} = |c|\|v\|$$

CYU 12.6 $3(2i + 4j - k) - 5(3i - k) = 6i + 12j - 3k - 15i + 5k = -9i + 12j + 2k$

Or, if you prefer: $3\langle 2, 4, -1 \rangle - 5\langle 3, 0, -1 \rangle = \langle 6, 12, -3 \rangle - \langle 15, 0, -5 \rangle = \langle -9, 12, 2 \rangle$

CYU 12.7 Here is the unit vector in the direction of F_1:

$$\frac{1}{\sqrt{3^2 + 4^2}}(-3i + 4j) = -\frac{3}{5}i + \frac{4}{5}j$$

Since we are not given a specific point on F_2, we do the best we can, and exhibit its unit vector in terms of the point $(1, a)$ in the figure: $\dfrac{1}{\sqrt{1^2 + a^2}}(i + aj)$.

From the equilibrium equation $F_1 + F_2 + F = 0$ we have:

since $\|F_2\| = 2\|F_1\|$

$$\frac{\|F_1\|}{5}(-3i + 4j) + \frac{2\|F_1\|}{\sqrt{1 + a^2}}(i + aj) - 100j = 0i + 0j$$

Equating the i-components:

$$-\frac{3}{5}\|F_1\| + \frac{2}{\sqrt{1 + a^2}}\|F_1\| = 0 \Rightarrow \frac{2}{\sqrt{1 + a^2}} = \frac{3}{5} \Rightarrow \sqrt{1 + a^2} = \frac{10}{3} \Rightarrow 1 + a^2 = \frac{100}{9}$$

$$a^2 = \frac{91}{9}$$

The figure indicates that a is positive: $a = \dfrac{\sqrt{91}}{3}$

Turning to the j-components:

$$\frac{4}{5}\|F_1\| + \frac{2a}{\sqrt{1+a^2}}\|F_1\| - 100 = 0 \Rightarrow \frac{4}{5}\|F_1\| + \frac{\frac{2\sqrt{91}}{3}}{\sqrt{1+\left(\frac{\sqrt{91}}{3}\right)^2}}\|F_1\| = 100$$

$$\|F_1\|\left(\frac{4}{5} + \frac{\sqrt{91}}{5}\right) = 100$$

$$\|F_1\| = \frac{500}{4+\sqrt{91}}$$

Conclusion: $F_2 = \dfrac{\|F_2\|}{\sqrt{1^2+a^2}}(i+aj) = \dfrac{2\|F_1\|}{\sqrt{1+\left(\frac{\sqrt{91}}{3}\right)^2}}\left(i+\frac{\sqrt{91}}{3}j\right)$

$$= \frac{2\left(\frac{500}{4+\sqrt{91}}\right)}{\frac{10}{3}}\left(i+\frac{\sqrt{91}}{3}j\right) = 4(\sqrt{91}-4)\left(i+\frac{\sqrt{91}}{3}j\right)$$

__CYU 12.8__ $u \cdot rv = \langle u_1, u_2, \ldots, u_n\rangle \cdot r\langle v_1, v_2, \ldots, v_n\rangle = \langle u_1, u_2, \ldots, u_n\rangle \cdot \langle rv_1, rv_2, \ldots, rv_n\rangle$

$$= u_1(rv_1) + u_2(rv_2) + \ldots + u_n(rv_n)$$

$$= r(u_1 v_1 + u_2 v_2 + \ldots + u_n v_n) = r(u \cdot v)$$

__CYU 12.9__ (a) $\|3i-4j+2k\| = \sqrt{(3i-4j+2k)\cdot(3i-4j+2k)} = \sqrt{3^2+4^2+2^2} = \sqrt{29}$

$\|\langle 5, 1\rangle - \langle 2, -3\rangle\|^2 = \|\langle 3, 4\rangle\|^2 = \langle 3, 4\rangle \cdot \langle 3, 4\rangle = 3(3)+4(4) = 25$

and $\|\langle 5, 1\rangle\|^2 - 2\langle 5, 1\rangle \cdot \langle 2, -3\rangle + \|\langle 2, -3\rangle\|^2$

$= \langle 5, 1\rangle \cdot \langle 5, 1\rangle - \langle 10, 2\rangle \cdot \langle 2, -3\rangle + \langle 2, -3\rangle \cdot \langle 2, -3\rangle$

(b) $= (25+1) - (20-6) + (4+9) = 26 - 14 + 13 = 25$

__CYU12.10__ $\theta = \cos^{-1}\left(\dfrac{(i+2j)\cdot(-i+3j)}{\sqrt{1+4}\sqrt{1+9}}\right) = \cos^{-1}\left(\dfrac{-1+6}{\sqrt{50}}\right) = \cos^{-1}\left(\dfrac{5}{5\sqrt{2}}\right) = \cos^{-1}\left(\dfrac{1}{\sqrt{2}}\right)$

$= 45°$

__CYU 12.11__ (a) Since $\langle 2, 3\rangle \cdot \langle 1, -4\rangle = 2 - 12 \neq 0$, the two vectors are not orthogonal.

(b) Since $(2i+3j)\cdot(-3i+2j) = -6+6 = 0$, the two vectors are orthogonal.

(c) Since $\langle 1, 2, 3\rangle \cdot \langle -1, -1, 1\rangle = -1 -2 +3 = 0$, the two vectors are orthogonal.

CYU 12.12 $\text{proj}_u v = \left(\dfrac{v \cdot u}{u \cdot u}\right)u = \left[\dfrac{\langle 3, 1\rangle \cdot \langle 0, 2\rangle}{\langle 0, 2\rangle \cdot \langle 0, 2\rangle}\right]\langle 0, 2\rangle = \dfrac{2}{4}\langle 0, 2\rangle = \langle 0, 1\rangle$

and $v - \text{proj}_u v = \langle 3, 1\rangle - \langle 0, 1\rangle = \langle 3, 0\rangle$

$\text{proj}_{\langle 0, 2\rangle} v = \langle 0, 1\rangle \rightarrow$ $v = \langle 3, 1\rangle$

$v - \text{proj}_{(0, 2)} v = \langle 0, 3\rangle$

CYU 12.13   (a) $\det \begin{bmatrix} 5 & -3 \\ 0 & 6 \end{bmatrix} = 5 \cdot 6 - (-3 \cdot 0) = 30$ (b) $\det \begin{bmatrix} 2 & -3 \\ 4 & 6 \end{bmatrix} = 2 \cdot 6 - (-3 \cdot 4) = 24$

(c) $\det \begin{bmatrix} 2 & 5 \\ 4 & 0 \end{bmatrix} = 2 \cdot 0 - (5 \cdot 4) = -20$

CYU 12.14

$\det \begin{bmatrix} i & j & k \\ 1 & 1 & -1 \\ 3 & 2 & \\ 0 & -3 & \sqrt{2} \end{bmatrix} = \det \begin{bmatrix} \frac{1}{2} & -1 \\ & \\ -3 & \sqrt{2} \end{bmatrix} i - \det \begin{bmatrix} \frac{1}{3} & -1 \\ & \\ 0 & \sqrt{2} \end{bmatrix} j + \det \begin{bmatrix} \frac{1}{3} & \frac{1}{2} \\ & \\ 0 & -3 \end{bmatrix} k$

$= \left(\dfrac{1}{2}\sqrt{2} - 3\right)i - \left(\dfrac{1}{3}\sqrt{2}\right)j + (-1)k = \dfrac{\sqrt{2} - 6}{2}i - \dfrac{\sqrt{2}}{3}j - k$

$(2, 3, 4) \times (3, 1, -2) = \det \begin{bmatrix} i & j & k \\ 2 & 3 & 4 \\ 3 & 1 & -2 \end{bmatrix}$

CYU 12.15

$= \det \begin{bmatrix} 3 & 4 \\ 1 & -2 \end{bmatrix} i - \det \begin{bmatrix} 2 & 4 \\ 3 & -2 \end{bmatrix} j + \det \begin{bmatrix} 2 & 3 \\ 3 & 1 \end{bmatrix} k$

$= (-6 - 4)i - (-4 - 12)j + (2 - 9)k = \langle -10, 16, -7\rangle$

CYU 12.16

$\langle 2, 3, 4\rangle \times \langle 3, 1, -2\rangle \cdot \langle 3, 1, -2\rangle = \langle -10, 16, -7\rangle \cdot \langle 3, 1, -2\rangle = -30 + 16 + 14 = 0$

$\uparrow$
CYU 12.15

CYU 12.17   (a) $j \times k = \det \begin{bmatrix} i & j & k \\ 0 & 1 & 0 \\ 0 & 0 & 1 \end{bmatrix} = i, \quad k \times j = \det \begin{bmatrix} i & j & k \\ 0 & 0 & 1 \\ 0 & 1 & 0 \end{bmatrix} = -i$

$i \times k = \det \begin{bmatrix} i & j & k \\ 1 & 0 & 0 \\ 0 & 0 & 1 \end{bmatrix} = -j, \quad k \times i = \det \begin{bmatrix} i & j & k \\ 0 & 0 & 1 \\ 1 & 0 & 0 \end{bmatrix} = j$

(b) $i \times (i \times j) = i \times k = -j$ and $(i \times i) \times j = 0 \times j = 0$

CYU 12.18 The area of triangle OAB is one half its base times it height:

$$\frac{1}{2}\|v\|h = \frac{1}{2}\|v\|\|u\|\sin\theta \underset{\underset{\text{Theorem 12.6}}{\uparrow}}{=} \frac{1}{2}\|u \times v\|$$

Flipping that triangle about and line segment joining A to B yields the depicted parallelogram. It follows that the parallelogram has area $\|u \times v\|$.

CYU 12.19 (a) The given slope of $-\frac{2}{3}$ (over 3, down 2) gives us the direction vector: $u = \langle 3, -2 \rangle$, and the given point $(5, 1)$ gives us the translation vector: $u = \langle 5, 1 \rangle$. Vector equation: $w = \langle 5, 1 \rangle + t\langle 3, -2 \rangle$. Parametric equations:

$$x = 5 + 3t, y = 1 - 2t$$

(b) A direction vector for any vertical line is j. Taking $\langle 3, 7 \rangle = 3i + 7j$ as our translation vector we have: $w = (3i + 7j) + tj$, and: $x = 3, y = 7 + t$.

CYU 12.20 Taking $v = \langle 0, 1, -2 \rangle - \langle 1, 2, 9 \rangle = -\langle 1, -1, -11 \rangle$ as the direction vector, and $u = \langle 1, 2, 9 \rangle$, as the translation vector we arrive at the vector equation:

$$w = \langle 1, 2, 9 \rangle + t\langle -1, -1, -11 \rangle = \langle 1 - t, 2 - t, 9 - 11t \rangle \,(*)$$

Which brings us to the parametric equations:

$$x = 1 - t, y = 2 - t, z = 9 - 11t$$

Note: If you chose $v = \langle 1, 2, 9 \rangle - \langle 0, 1, -2 \rangle$ as the direction vector and $\langle 0, 1, -2 \rangle$ as the translation vector you will end up with the vector equation: $w = \langle 0, 1, -2 \rangle + t\langle 1, 1, 11 \rangle = \langle t, 1 + t, -2 + 11t \rangle$ (**) , which certainly looks different than (*) above. But appearance can be deceiving, for if you replace t in (**) with $1 - t$ you will arrive at (*), and as t takes on all real numbers, so does $1 - t$.

CYU 12.21 (a) A direction vector for the line L passing through $(1, -2)$ and $(2, 4)$:

$$u = \langle 2, 4 \rangle - \langle 1, -2 \rangle = \langle 1, 6 \rangle$$

The vector from the point $(1, -2)$ on L to $P = (2, 5)$:

$$v = (2, 5) - (1, -2) = \langle 1, 7 \rangle$$

Applying Theorem 12.4: $\text{proj}_u v = \left(\dfrac{u \cdot v}{u \cdot w}\right)u = \left(\dfrac{\langle 1, 6 \rangle \cdot \langle 1, 7 \rangle}{\langle 1, 6 \rangle \cdot \langle 1, 6 \rangle}\right)\langle 1, 6 \rangle = \dfrac{43}{37}\langle 1, 6 \rangle$

Hence: $\left\|v - \text{proj}_u v\right\| = \left\|\langle 1, 7 \rangle - \langle \dfrac{43}{37}, \dfrac{258}{37} \rangle\right\| = \left\|\langle -\dfrac{6}{37}, \dfrac{1}{37} \rangle\right\| = \dfrac{\sqrt{6^2 + 1^2}}{37} = \dfrac{1}{\sqrt{37}}$.

(b) $u = \langle 1, 2, 2, 1 \rangle - \langle 1, 2, 0, 1 \rangle = \langle 0, 0, 2, 0 \rangle$. $v = \langle 1, 0, 1, 3 \rangle - \langle 1, 2, 0, 1 \rangle = \langle 0, -2, 1, 2 \rangle$.

$\text{proj}_u v = \left(\dfrac{u \cdot v}{u \cdot w}\right)u = \left(\dfrac{\langle 0, 0, 2, 0 \rangle \cdot \langle 0, -2, 1, 2 \rangle}{\langle 0, 0, 2, 0 \rangle \cdot \langle 0, 0, 2, 0 \rangle}\right)\langle 0, 0, 2, 0 \rangle = \dfrac{1}{2}\langle 0, 0, 2, 0 \rangle = \langle 0, 0, 1, 0 \rangle$

Hence: $\|v - \text{proj}_u v\| = \|\langle 0, -2, 1, 2 \rangle - \langle 0, 0, 1, 0 \rangle\| = \|\langle 0, -2, 0, 2 \rangle\| = \sqrt{8} = 2\sqrt{2}$.

CYU 12.22 A normal to the desired plane will have the same direction as that of the line passing through the two given points; namely: $n = \langle 0, 2, 1 \rangle - \langle 1, 1, 0 \rangle = \langle -1, 1, 1 \rangle$.

Vector equation: $\langle -1, 1, 1 \rangle \cdot \langle x - 1, y - 3, z + 2 \rangle = 0$.

Scalar equation: $-(x - 1) + (y - 3) + z + 2 = 0$

General equation: $-x + y + z = 0$

CYU 12.23 We chose $A = (0, 16, 0)$, $B = \left(0, 0, -\dfrac{16}{5}\right)$, and $C = \left(\dfrac{16}{9}, 0, 0\right)$. Then:

$$\overrightarrow{AB} = (0, 16, 0) - \left(0, 0, -\frac{16}{5}\right) = \langle 0, 16, \frac{16}{5} \rangle \text{ and } \overrightarrow{AC} = (0, 16, 0) - \left(\frac{16}{9}, 0, 0\right) = \langle -\frac{16}{9}, 16, 0 \rangle.$$

Here is a normal to the plane:

$$n = \det \begin{bmatrix} i & j & k \\ 0 & 16 & \frac{16}{5} \\ -\frac{16}{9} & 16 & 0 \end{bmatrix} = -\frac{16^2}{5}i + \frac{16^2}{45}j + \frac{16^2}{9}k = \langle -\frac{16^2}{5}, \frac{16^2}{45}, \frac{16^2}{9} \rangle.$$

Here is a "nicer" normal: $n = \dfrac{45}{16^2}\langle -\dfrac{16^2}{5}, \dfrac{16^2}{45}, \dfrac{16^2}{9} \rangle = \langle -9, 1, 5 \rangle$.

Choosing the point $A = (0, 16, 0)$ on the plane, we arrive at the vector and general equation of the plane: $\langle -9, 1, 5 \rangle \cdot \langle x, y - 16, z \rangle = 0$, and $9x + y - 5z = 16$.

CYU 12.24 (a) From the given equation $x + 2y + 2z = 13$, we see that $n = (1, 2, 2)$ is a normal to the plane, and that $Q = (13, 0, 0)$ is a point on the plane.

For $v = \overrightarrow{QP} = (2, -3, 4) - (13, 0, 0) = \langle -11, -3, 4 \rangle$, we calculate the length d of the vector $w = \text{proj}_n v$, as that is the distance between P and the plane:

$$d = \|\text{proj}_n v\| = \frac{|v \cdot n|}{\sqrt{n \cdot n}} = \frac{|\langle -11, -3, 4 \rangle \cdot \langle 1, 2, 2 \rangle|}{\sqrt{\langle 1, 2, 2 \rangle \cdot \langle 1, 2, 2 \rangle}} = \frac{9}{3} = 3$$

(b) We know that $n = (a, b, c)$ is a normal to the plane $ax + by + cz + d = 0$.

Let $Q = (x_1, y_1, z_1)$ be any point on the plane, and let:

$$v = \overrightarrow{QP} = (x_0, y_0, z_0) - (x_1, y_1, z_1) = \langle x_0 - x_1, y_0 - y_1, z_0 - z_1 \rangle$$

Following the procedure of (a) above we calculate the distance d between $P = (x_0, y_0, z_0)$ and the plane:

$$d = \frac{|\boldsymbol{v} \cdot \boldsymbol{n}|}{\sqrt{\boldsymbol{n} \cdot \boldsymbol{n}}} = \frac{|a(x_0 - x_1) + b(y_0 - y_1) + c(z_0 - z_1)|}{\sqrt{a^2 + b^2 + c^2}}$$

$$= \frac{|ax_0 + by_0 + cz_0 + (-ax_1 - by_1 - cz_1)|}{\sqrt{a^2 + b^2 + c^2}} = \frac{|ax_0 + by_0 + cz_0 + d|}{\sqrt{a^2 + b^2 + c^2}}$$

since $ax_1 + by_1 + cz_1 + d = 0$

CYU 12.25 $\boldsymbol{n_1} = \boldsymbol{i} + 2\boldsymbol{j} + \boldsymbol{k}$, and $\boldsymbol{n_2} = 3\boldsymbol{i} - 4\boldsymbol{j} - \boldsymbol{k}$ are normal vectors for the planes $x + 2y + z = 0$ and $3x - 4y - z = 1$, respectively. Consequently:

$$\theta = \cos^{-1}\left(\frac{(\boldsymbol{i} + 2\boldsymbol{j} + \boldsymbol{k}) \cdot (3\boldsymbol{i} - 4\boldsymbol{j} - \boldsymbol{k})}{\|\boldsymbol{i} + 2\boldsymbol{j} + \boldsymbol{k}\| \|(3\boldsymbol{i} - 4\boldsymbol{j} - \boldsymbol{k})\|}\right) = \cos^{-1}\left(\frac{3 - 8 - 1}{\sqrt{6}\sqrt{26}}\right)$$

$$= \cos^{-1}\left(\frac{-6}{\sqrt{156}}\right) \approx 119°$$

The line of intersection is parallel to the vector:

$$\boldsymbol{n_1} \times \boldsymbol{n_2} = \det\begin{bmatrix} \boldsymbol{i} & \boldsymbol{j} & \boldsymbol{k} \\ 1 & 2 & 1 \\ 3 & -4 & -1 \end{bmatrix} = 2\boldsymbol{i} + 4\boldsymbol{j} - 10\boldsymbol{k}$$

Setting $y = 0$ in the two given equations $x + 2y + z = 0$ and $3x - 4y - z = 1$, and solving for x and z we find that the line contains the point $(\frac{1}{4}, 0, -\frac{1}{4})$.

Consequently: $\boldsymbol{w} = (\frac{1}{4}\boldsymbol{i} - \frac{1}{4}\boldsymbol{k}) + t(2\boldsymbol{i} + 4\boldsymbol{j} - 10\boldsymbol{k})$ is a vector equation of the line of intersection. [As is $\boldsymbol{w} = (\frac{1}{4}\boldsymbol{i} - \frac{1}{4}\boldsymbol{k}) + t(\boldsymbol{i} + 2\boldsymbol{j} - 5\boldsymbol{k})$]

CYU 12.26 While $y = x^2$ represents a parabola in the plane, in $\mathfrak{R}^3$ it represents the **parabolic cylinder** with vector equation:

$$\boldsymbol{r}(t) = t\boldsymbol{i} + t^2\boldsymbol{j} + 2t\boldsymbol{k} = (t, t^2, 2t).$$

At time $t = 1$ the particle has position vector $(\boldsymbol{1}, \boldsymbol{1}, \boldsymbol{2})$, and occupies the point $(1, 1, 2)$. At time $t = 3$ it is at the point $(3, 9, 6)$. As time progresses from 1 to 3, the path remains on the parabolic cylinder as is suggested in the adjacent figure.

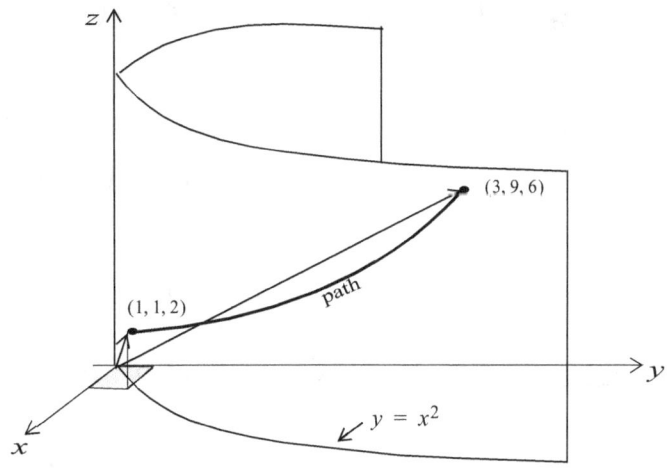

CYU 12.27 (a) For $\boldsymbol{r}(t) = (e^t)\boldsymbol{i} + \left(\frac{t}{t^2 + 1}\right)\boldsymbol{j} + (\sin t)\boldsymbol{k}$:

(i) $\lim\limits_{t\to 1} r(t) = \lim\limits_{t\to 1}(e^t)i + \lim\limits_{t\to 1}\left(\dfrac{t}{t^2+1}\right)j + \lim\limits_{t\to 1}(\sin t)k = ei + \dfrac{1}{2}j + (\sin 1)k$

(ii) $r'(t) = (e^t)'i + \left(\dfrac{t}{t^2+1}\right)'j + (\sin t)'k$

$= (e^t)i + \dfrac{(t^2+1)-t(2t)}{(t^2+1)^2}j + (\cos t)k = (e^t)i + \dfrac{-t^2+1}{(t^2+1)^2}j + (\cos t)k$

(iii) $\int r(t)dt = \left(\int e^t dt\right)i + \left(\int\dfrac{t}{t^2+1}dt\right)j + \left[\int(\sin t)dt\right]k$

$= e^t i + \left(\dfrac{1}{2}\int\dfrac{du}{u}\right)j - \cos t k = e^t i + \left(\dfrac{1}{2}\ln|u|\right)j - (\cos t)k + C$

$\qquad\qquad \begin{array}{l} u = t^2+1 \\ du = 2t\,dt \end{array}$

$\qquad\qquad\qquad\qquad\qquad = e^t i + \left[\dfrac{1}{2}\ln(t^2+1)\right]j - (\cos t)k + C$

(b) $r'(t) = \lim\limits_{h\to 0}\dfrac{1}{h}[r(t+h) - r(t)]$

$= \lim\limits_{h\to 0}\dfrac{1}{h}([f(t+h)i + g(t+h)j + h(t+h)k] - [f(t)i + g(t)j + h(t)k])$

$= \lim\limits_{h\to 0}\left(\dfrac{1}{h}[f(t+h) - f(t)]i + \dfrac{1}{h}[g(t+h) - g(t)]j + \dfrac{1}{h}[h(t+h) - h(t)]k\right)$

$= \lim\limits_{h\to 0}\dfrac{1}{h}[f(t+h) - f(t)]i + \lim\limits_{h\to 0}\dfrac{1}{h}[g(t+h) - g(t)]j + \lim\limits_{h\to 0}\dfrac{1}{h}[h(t+h) - h(t)]k$

$= f'(t)i + g'(t)j + h'(t)k$

CYU 12.28 $[u(t) + v(t)]' = ([u_1(t)i + u_2(t)j + u_3(t)k] + [v_1(t)i + v_2(t)j + v_3(t)k])'$

$= ([u_1(t) + v_1(t)]i + [u_2(t) + v_2(t)]j + [u_3(t) + v_3(t)]k)'$

$= [u_1(t) + v_1(t)]'i + [u_2(t) + v_2(t)]'j + [u_3(t) + v_3(t)]'k$

$= [u_1'(t) + v_1'(t)]i + [u_2'(t) + v_2'(t)]j + [u_3'(t) + v_3'(t)]k$

$= [u_1'(t)i + u_2'(t)j + u_3'(t)k] + [v_1'(t)i + v_2'(t)j + v_3'(t)k]$

$= u'(t) + v'(t)$

CYU 12.29 For $s(t) = (\cos t)i + (\sin t)j + tk$:

$v(t) = r'(t) = (-\sin t)i + (\cos t)j + k, \quad \|v(t)\| = \sqrt{\sin^2 t + \cos^2 t + 1^2} = \sqrt{2}$

$a(t) = v'(t) = (-\cos t)i - (\sin t)j$

CYU 12.30 Since $a(t) = i + 2j - k$:

$$v(t) = \int (i + 2j - k)dt = ti + 2tj - tk + \underbrace{(i - 12j + k)}_{\text{initial velocity}} = (t+1)i + (2t-12)j + (-t+1)k$$

Leading us to the speed function:

$$\|v(t)\| = \sqrt{(t+1)^2 + (2t-12)^2 + (-t+1)^2} = \sqrt{6t^2 - 48t + 146}$$

Differentiating the above real-valued function we have:

$$\|v(t)\|' = \frac{1}{2}(6t^2 - 48t + 146)^{-1/2}(6t^2 - 48t + 146)'$$

$$= \frac{12t - 48}{2\sqrt{6t^2 - 48t + 146}} = \frac{6t - 24}{\sqrt{6t^2 - 48t + 146}}$$

SIGN:

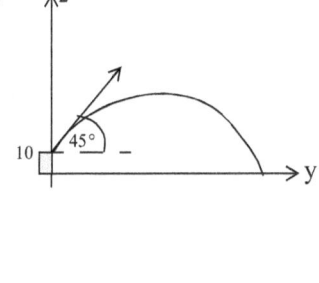

The above SIGN information tells us that the minimum speed occurs at $t = 4$. So:

$$\text{minimum speed} = \sqrt{6 \cdot 4^2 - 48 \cdot 4 + 146} = \sqrt{50} = 5\sqrt{2} \text{ ft/sec.}$$

We also see that the maximum speed occurs at either the endpoint 0 or the endpoint 5.

Calculating: $\|v(0)\| = \sqrt{146}$ and $\|v(5)\| = \sqrt{6 \cdot 5^2 - 48 \cdot 5 + 146} = \sqrt{56}$, we see

that during the specified interval the particle attains a maximum speed of $\sqrt{146}$ ft/sec.

CYU 12.31 Due to the force of gravity, the projectile will be subjected to a downward acceleration of 9.8 m/sec^2. Thus:

$$v(t) = \int -9.8k\,dt = (-9.8t)k + (175\cos 45°\, j + 175 \sin 45°\, k)$$

$$= (-9.8t)k + \left(\frac{175}{\sqrt{2}}j + \frac{175}{\sqrt{2}}k\right)$$

$$= \frac{175}{\sqrt{2}}j + \left(-9.8t + \frac{175}{\sqrt{2}}\right)k \quad (*)$$

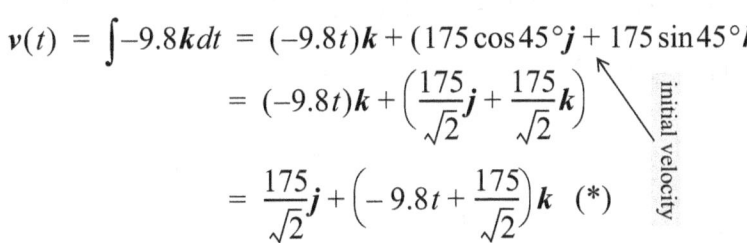

And: $s(t) = \int \left[\frac{175}{\sqrt{2}}j + \left(-9.8t + \frac{175}{\sqrt{2}}\right)k\right]dt = \left(\frac{175}{\sqrt{2}}t\right)j + \left(-4.9t^2 + \frac{175}{\sqrt{2}}t\right)k + (10k)$

$$= \left(\frac{175}{\sqrt{2}}t\right)j + \left(-4.9t^2 + \frac{175}{\sqrt{2}}t + 10\right)k$$

The projectile will hit the ground when the vertical component of $s(t)$ equals 0 (and $t > 0$):

$$-4.9t^2 + \frac{175}{\sqrt{2}}t + 10 = 0 \Rightarrow t = \frac{-\frac{175}{\sqrt{2}} \pm \sqrt{\left(\frac{175}{\sqrt{2}}\right)^2 + 40(4.9)}}{-2(4.9)} \approx 25.3 \text{ sec.}$$

Calculating the horizontal component of $s(25.3)$, we conclude that the projectile will hit the ground at a distance of $\frac{175}{\sqrt{2}}(25.3) \approx 3131$ meters from the base of the perch .

Turning to (*) we can calculate its approximate impact speed:

$$\|v(25.3)\| = \sqrt{\left(\frac{175}{\sqrt{2}}\right)^2 + \left(-9.8(25.3) + \frac{175}{\sqrt{2}}\right)^2} \approx 175 \text{ m/s}$$

CYU 12.32 $L = \int_1^2 \|r'(t)\|\,dt = \int_1^2 \sqrt{4 + t^4 + 4t^2}\,dt = \int_1^2 \sqrt{(t^2 + 2)^2}\,dt$

$$= \int_1^2 (t^2 + 2)\,dt = \left(\frac{t^3}{3} + 2t\right)\Bigg|_1^2$$

$$= \left(\frac{8}{3} + 4\right) - \left(\frac{1}{3} + 2\right) = \frac{13}{3}$$

CYU 12.33 (a) Noting that $P(1) = (2, 1, -2)$, we start at $t = 1$:

$$s = s(t) = \int_1^t \|r'(u)\|\,du = \int_1^t \sqrt{4 + 1 + 4}\,du = 3u\Big|_1^t = 3t - 3$$

(b) From $s = 3t - 3$, we have $t = \frac{s}{3} + 1$. Replacing t with $\frac{s}{3} + 1$ in

$$r = 2t\,i + t\,j - 2t\,k \quad \text{for } 1 \le t \le 3 \text{ we have:}$$

$$r = 2\left(\frac{s}{3} + 1\right)i + \left(\frac{s}{3} + 1\right)j - 2\left(\frac{s}{3} + 1\right)k \text{ for } 0 \le s \le 6.$$

(c) Arc Length using $r(t) = 2t\,i + t\,j - 2t\,k$ for $1 \le t \le 3$:

$$L = \int_1^3 \|r'(t)\|\,dt = \int_1^3 \sqrt{4 + 1 + 4}\,dt = 3t\Big|_1^3 = 9 - 3 = 6$$

Arc Length using $r(s) = \left(\frac{2s}{3} + 2\right)i + \left(\frac{s}{3} + 1\right)j - \left(\frac{2s}{3} + 2\right)k$ for $0 \le s \le 6$:

$$L = \int_0^6 \|r'(s)\|\,ds = \int_0^6 \sqrt{\left(\frac{2}{3}\right)^2 + \left(\frac{1}{3}\right)^2 + \left(\frac{2}{3}\right)^2}\,dt = 6$$

CYU 12.34 The solution is a tad tedious. The main thing is to understand the process:

$$T(t) = \frac{\langle 2t, t^2, -t\rangle'}{\|\langle 2t, t^2, -t'\rangle\|} = \frac{\langle 2, 2t, -1\rangle}{\sqrt{5 + 4t^2}} \text{ and: } T(1) = \frac{1}{3}(2, 2, -1)$$

$$N(t) = \frac{T'(t)}{\|T'(t)\|} = \frac{\left\langle \dfrac{2}{\sqrt{5 + 4t^2}}, \dfrac{2t}{\sqrt{5 + 4t^2}}, \dfrac{-1}{\sqrt{5 + 4t^2}}\right\rangle'}{\left\|\left\langle \dfrac{2}{\sqrt{5 + 4t^2}}, \dfrac{2t}{\sqrt{5 + 4t^2}}, \dfrac{-1}{\sqrt{5 + 4t^2}}\right\rangle'\right\|}$$

$$= \frac{\left\langle \dfrac{-8t}{(5 + 4t^2)^{3/2}}, \dfrac{10}{(5 + 4t^2)^{3/2}}, \dfrac{4t}{(5 + 4t^2)^{3/2}}\right\rangle}{\left\|\left\langle \dfrac{-8t}{(5 + 4t^2)^{3/2}}, \dfrac{10}{(5 + 4t^2)^{3/2}}, \dfrac{4t}{(5 + 4t^2)^{3/2}}\right\rangle\right\|}$$

and: $N(1) = \dfrac{\langle -\dfrac{8}{27}, \dfrac{10}{27}, \dfrac{4}{27} \rangle}{\sqrt{\left(-\dfrac{8}{27}\right)^2 + \left(\dfrac{10}{27}\right)^2 + \left(\dfrac{4}{27}\right)^2}} = \dfrac{\dfrac{1}{27}\langle -8, 10, 4 \rangle}{\dfrac{2\sqrt{5}}{9}} = \dfrac{\sqrt{5}}{15}\langle -4, 5, 2 \rangle$

Then: $B(1) = T(1) \times N(1) = \underset{\underset{\text{Theorem 12.7(c), page 509}}{\uparrow}}{\dfrac{\sqrt{5}}{45}} \det \begin{bmatrix} i & j & k \\ 2 & 2 & -1 \\ -4 & 5 & 2 \end{bmatrix} = \dfrac{\sqrt{5}}{45}\langle 9, 0, 18 \rangle = \dfrac{\sqrt{5}}{5}\langle 1, 0, 2 \rangle$

As for orthogonality:

$$T(1) \cdot N(1) = \dfrac{1}{3}\langle 2, 2, -1 \rangle \cdot \dfrac{\sqrt{5}}{15}\langle -4, 5, 2 \rangle = \dfrac{\sqrt{5}}{45}(-2 \cdot 4 + 2 \cdot 5 - 1 \cdot 2) = 0$$

$$T(1) \cdot B(1) = \dfrac{1}{3}\langle 2, 2, -1 \rangle \cdot \dfrac{\sqrt{5}}{5}\langle 1, 0, 2 \rangle = \dfrac{\sqrt{5}}{15}(2 \cdot 1 + 2 \cdot 0 - 1 \cdot 2) = 0$$

$$N(1) \cdot B(1) = \dfrac{\sqrt{5}}{15}\langle -4, 5, 2 \rangle \cdot \dfrac{\sqrt{5}}{5}\langle 1, 0, 2 \rangle = \dfrac{1}{15}(-4 \cdot 1 + 5 \cdot 0 + 2 \cdot 2) = 0$$

Since $T(1) = \dfrac{1}{3}\langle 2, 2, -1 \rangle$ is a normal to the normal plane, so then is $\langle 2, 2, -1 \rangle$, and since

$r(1) = \langle 2t, t^2, -t \rangle$, the point $(2, 1, -1)$ lies on the plane. Thus:

Normal Plane: $\langle 2, 2, -1 \rangle \cdot \langle x - 2, y - 1, z + 1 \rangle = 0$ or: $2x + 2y - z = 7$

(consider Example 12.12, page 517)

Since $B(1) = \dfrac{\sqrt{5}}{5}\langle 1, 0, 2 \rangle$ is a normal to the osculating plane, so then is $\langle 1, 0, 2 \rangle$. Thus:

Osculating Plane: $\langle 1, 0, 2 \rangle \cdot \langle x - 2, y - 1, z + 1 \rangle = 0$ or: $x + 2z = 0$

CYU 12.35   For $r(t) = ti + t^2 j + t^3 k$ we have:

$$r'(t) = i + 2tj + 3t^2 k$$
$$r''(t) = 2j + 6tk$$

$$r'(t) \times r''(t) = \det \begin{bmatrix} i & j & k \\ 1 & 2t & 3t^2 \\ 0 & 2 & 6t \end{bmatrix} = 6t^2 i - 6tj + 2k$$

Thus: $\kappa(t) = \dfrac{\|r'(t) \times r''(t)\|}{\|r'(t)\|^3} = \dfrac{\sqrt{36t^4 + 36t^2 + 4}}{(\sqrt{1 + 4t^2 + 9t^4})^3} = \dfrac{2\sqrt{9t^4 + 9t^2 + 1}}{(\sqrt{1 + 4t^2 + 9t^4})^3}$

Note that $(1, 1, 1)$ and $(2, 4, 8)$ are the terminal points of $r(1)$ and $r(2)$, respectively. Hence:

Curvature at $(1, 1, 1)$ = $\kappa(1)$ = $\dfrac{2\sqrt{19}}{14^{3/2}} \approx 0.17$ and

Curvature at $(2, 4, 8)$ = $\kappa(2)$ = $\dfrac{2\sqrt{181}}{161^{3/2}} \approx 0.01$

Roughly speaking: The curve is bending about 17 times faster at $(1, 1, 1)$ than it is at $(2, 4, 8)$.

CYU 12.36 The curve $y = f(x)$ is the curve traced out by the vector function
$r(t) = ti + f(t)j + 0k$. As such:

$$\kappa = \frac{\|r'(t) \times r''(t)\|}{\|r'(t)\|^3} = \frac{\|(i + f'(t)j + 0k) \times (0i + f''(t)j + 0k)\|}{\|i + f'(t)j + 0k\|^3}$$

$$= \frac{\left\| \det \begin{bmatrix} i & j & k \\ 1 & f'(t) & 0 \\ 0 & f''(t) & 0 \end{bmatrix} \right\|}{[\sqrt{1 + (f'(t))^2}]^3} = \frac{|f''(x)|}{[1 + (f'(x))^2]^{3/2}}$$

CYU 12.37 The parabola $f(x) = x^2$ is traced out by the vector function $r(t) = ti + t^2 j + 0k$,
with corresponding unit tangent vector $T(t) = \dfrac{r'(t)}{\|r'(t)\|} = \dfrac{i + 2tj}{\sqrt{1 + 4t^2}}$, for which:

$$T'(t) = \frac{\sqrt{1 + 4t^2}(2j) - (i + 2tj)\dfrac{4t}{\sqrt{1 + 4t^2}}}{1 + 4t^2} = \frac{(1 + 4t^2)(2j) - 4t(i + 2tj)}{(1 + 4t^2)^{3/2}} = \frac{-4ti + 2j}{(1 + 4t^2)^{3/2}}$$

In particular, the unit normals at $t = 0$ and at $t = 1$ (corresponding to the points
$(0, 0)$ and $(1, 1)$ on the curve, respectively) are:

$$N(0) = \frac{T'(0)}{\|T'(0)\|} = \frac{2j}{\|2j\|} = j \text{ and } N(1) = \frac{T'(1)}{\|T'(1)\|} = \frac{-4i + 2j}{\sqrt{20}} = \frac{-2i + j}{\sqrt{5}}$$

It follows that:

The radius of the circle of curvature at $(0, 0)$ is $r = \dfrac{1}{\kappa} = \dfrac{1}{2}$ and is centered at $\left(0, \dfrac{1}{2}\right)$ (vector position: $\dfrac{1}{2}\boldsymbol{j}$). Equation: $x^2 + \left(y - \dfrac{1}{2}\right)^2 = \dfrac{1}{4}$.

The radius of the circle of curvature at $(1, 1)$ is $r = \dfrac{5^{3/2}}{2}$ and is centered at the

endpoint of $\boldsymbol{i} + \boldsymbol{j} + \dfrac{5^{3/2}}{2}\left(\dfrac{-2\boldsymbol{i} + \boldsymbol{j}}{\sqrt{5}}\right) = -4\boldsymbol{i} + \dfrac{7}{2}\boldsymbol{j}$, which is to say, at $\left(-4, \dfrac{7}{2}\right)$.

$$\text{Equation: } (x + 4)^2 + \left(y - \dfrac{7}{2}\right)^2 = \dfrac{125}{4}$$

CHAPTER 13: DIFFERENTIATING FUNCTIONS OF SEVERAL VARIABLES

CYU 13.1 (a)
$$\dfrac{\partial}{\partial x}(x^2 y + e^{x + 3y}) \overset{\text{constant}}{=} y\dfrac{\partial}{\partial x}x^2 + e^{x + 3y}\dfrac{\partial}{\partial x}(x + 3y)$$
$$= y(2x) + e^{x + 3y}(1) = 2xy + e^{x + 3y}$$

(b)
$$\dfrac{\partial}{\partial y}(x^2 y + e^{x + 3y}) = x^2\dfrac{\partial y}{\partial y} + e^{x + 3y}\dfrac{\partial}{\partial y}(x + 3y)$$
$$= x^2 \cdot 1 + e^{x + 3y}(3) = x^2 + 3e^{x + 3y}$$

(c) Since $f_x(x, y) = 2xy + e^{x + 3y}$ [see (a)]: $f_x(2, 0) = e^2$.

(d) Since $\dfrac{\partial z}{\partial y} = x^2 + 3e^{x + 3y}$ [see (b)]: $\left.\dfrac{\partial z}{\partial y}\right|_{(2, 0)} = 4 + 3e^2$.

CYU 13.2 (a)
$$\dfrac{\partial^2 z}{\partial y^2} = \dfrac{\partial}{\partial y}\left[\dfrac{\partial}{\partial y}\left(\dfrac{x^2 + y}{5 + x - y}\right)\right] = \dfrac{\partial}{\partial y}\left[\dfrac{(5 + x - y) - (x^2 + y)(-1)}{(5 + x - y)^2}\right]$$
$$= \dfrac{\partial}{\partial y}\left[\dfrac{x^2 + x + 5}{(5 + x - y)^2}\right]$$
$$= \dfrac{0 - (x^2 + x + 5)2(5 + x - y)(-1)}{(5 + x - y)^3} = \dfrac{2(x^2 + x + 5)}{(5 + x - y)^3}$$

(b) $\dfrac{\partial^2 z}{\partial x^2} = \dfrac{\partial}{\partial x}\left[\dfrac{\partial}{\partial x}\left(\dfrac{x^2+y}{5+x-y}\right)\right] = \dfrac{\partial}{\partial x}\left[\dfrac{(5+x-y)2x-(x^2+y)}{(5+x-y)^2}\right]$

$$= \dfrac{\partial}{\partial x}\left[\dfrac{x^2-2xy+10x-y}{(5+x-y)^2}\right]$$

$$= \dfrac{(5+x-y)^2(2x-2y+10)-(x^2-2xy+10x-y)2(5+x-y)}{(5+x-y)^4}$$

$$= \dfrac{2(5+x-y)[-x^2-8x-y+2xy+10]}{(5+x-y)^4}$$

In particular: $\qquad \dfrac{\partial^2 z}{\partial x^2}\bigg|_{(0,\,4)} = \dfrac{2[-0^2-8(0)-4+2\cdot 0\cdot 4+10]}{(5+0-4)^3} = 12$

(c) $f_{yx} = \left[\left(\dfrac{x^2+y}{5+x-y}\right)_y\right]_x = \left[\dfrac{(5+x-y)-(x^2+y)(-1)}{(5+x-y)^2}\right]_x$

$$= \left[\dfrac{x^2+x+5}{(5+x-y)^2}\right]_x$$

$$= \dfrac{(5+x-y)^2(2x+1)-(x^2+x+5)\cdot 2(5+x-y)}{(5+x-y)^3}$$

$$= \dfrac{-2xy+9x-y-5}{(5+x-y)^3}$$

(d) $f_{yx}(2,\,3) = \dfrac{-2\cdot 2\cdot 3+9\cdot 2-3-5}{(5+2-3)^3} = \dfrac{-2}{4^3} = \dfrac{-1}{32}$

CYU 13.3 (a) For $z = f(x,y) = x^4y^3$:

(i) $\dfrac{\partial^3 z}{\partial x^3} = \dfrac{\partial^2}{\partial x^2}\left[\dfrac{\partial}{\partial x}(x^4y^3)\right] = \dfrac{\partial^2}{\partial x^2}(4x^3y^3) = \dfrac{\partial}{\partial x}\left[\dfrac{\partial}{\partial x}(4x^3y^3)\right]$

$$= \dfrac{\partial}{\partial x}(12x^2y^3) = 24xy^3$$

(ii) $z_{xyx} = [(x^4y^3)_x]_{yx} = [4x^3y^3]_{yx} = [(4x^3y^3)_y]_x = (12x^3y^2)_x = 36x^2y^2$

(iii) $z_{yxx} = [(x^4y^3)_y]_{xx} = (3x^4y^2)_{xx} = (12x^3y^2)_x = 36x^2y^2$

(b) For $w = f(x,y,z) = xy^2+z^3y-x^3yz$

(i) $\dfrac{\partial w}{\partial x} = \dfrac{\partial}{\partial x}(xy^2+z^3y-x^3yz) = y^2-3x^2yz$

(ii) $\dfrac{\partial^2 w}{\partial y^2} = \dfrac{\partial}{\partial y}\left[\dfrac{\partial}{\partial y}(xy^2 + z^3 y - x^3 yz)\right] = \dfrac{\partial}{\partial y}(2xy + z^3 - x^3 z) = 2x$

(iii) $w_{xxx} = [(xy^2 + z^3 y - x^3 yz)_x]_{xx} = [(y^2 - 3x^2 yz)_x]_x = (-6xyz)_x = -6yz$

(iv) $w_{yzx} = [(xy^2 + z^3 y - x^3 yz)_y]_{zx} = [(2xy + z^3 - x^3 z)_z]_x = (3z^2 - x^3)_x = -3x^2$

CYU 13.4 For $f(x, y) = x^2 \sin y$:

$$f_x(x, y) = 2x\sin y \quad \text{and} \quad f_y(xy) = x^2 \cos y$$

Theorem 11.3 assures us that both partial derivatives are continuous. It follows that f is differentiable (Theorem 13.2).

CYU 13.5 (a) For $z = (2x + e^y)^4$, $x = \sin t$, $y = t^2$:

$$\dfrac{dz}{dt} = \dfrac{\partial z}{\partial x}\dfrac{dx}{dt} + \dfrac{\partial z}{\partial y}\dfrac{dy}{dt} = 4(2x + e^y)^3 \cdot 2 \cdot \cos t + 4(2x + e^y)^3 e^y \cdot 2t$$

$$= 8(2x + e^y)^3(\cos t + t e^y)$$

$$= 8(2\sin t + e^{t^2})^3(\cos t + t e^{t^2})$$

Alternatively:

$$z = (2x + e^y)^4 = (2\sin t + e^{t^2})^4$$

$$\dfrac{dz}{dt} = 4(2\sin t + e^{t^2})^3 \dfrac{d}{dt}(2\sin t + e^{t^2})$$

$$= 4(2\sin t + e^{t^2})^3(2\cos t + 2t e^{t^2}) = 8(2\sin t + e^{t^2})^3(\cos t + t e^{t^2})$$

(b) For $z = e^{xy}, x = s + 3t, y = st^2$:

$$\dfrac{\partial z}{\partial t} = \dfrac{\partial z}{\partial x}\dfrac{\partial x}{\partial t} + \dfrac{\partial z}{\partial y}\dfrac{\partial y}{\partial t} = y e^{xy} \cdot 3 + x e^{xy} \cdot 2st = 3st^2 e^{(s+3t)st^2} + (s + 3t)e^{(s+3t)st^2} \cdot 2st$$

$$= ste^{(s^2 t^2 + 3st^3)}(3t + 2s + 6t)$$

$$= ste^{(s^2 t^2 + 3st^3)}(9t + 2s)$$

Alternatively:

$$z = e^{xy} = e^{(s+3t)st^2} = e^{s^2 t^2 + 3st^3}$$

and: $\dfrac{\partial}{\partial t}(e^{s^2 t^2 + 3st^3}) = e^{s^2 t^2 + 3st^3}\dfrac{\partial}{\partial t}(s^2 t^2 + 3st^3)$

$$= e^{s^2 t^2 + 3st^3}(2ts^2 + 9st^2) = ste^{(s^2 t^2 + 3st^3)}(9t + 2s)$$

CYU 13.6 (a) For $z = f(x, y) = e^{xy} + x\ln y$:

(i) $\dfrac{\partial^3 z}{\partial x^3} = [(e^{xy} + x\ln y)_x]_{xx} = [(y e^{xy} + \ln y)_x]_x = (y^2 e^{xy})_x = y^3 e^{xy}$

(ii) $z_{xyx} = [(e^{xy} + x \ln y)_x]_{yx} = [(ye^{xy} + \ln y)_y]_x = \left(yxe^{xy} + e^{xy} + \dfrac{1}{y}\right)_x$

$$= xy^2 e^{xy} + ye^{xy} + ye^{xy} = ye^{xy}(xy + 2)$$

(b) For $w = f(x, y, z) = xy^2 + z^3 y - xyz$:

(i) $\dfrac{\partial w}{\partial x} = y^2 - yz$

(ii) $\dfrac{\partial^2 w}{\partial y^2} = \dfrac{\partial}{\partial y}\left[\dfrac{\partial}{\partial y}(xy^2 + z^3 y - xyz)\right] = \dfrac{\partial}{\partial y}(2xy + z^3 - xz) = 2x$

(iii) $w_{xyz} = [(xy^2 + z^3 y - xyz)_x]_{yz} = [(y^2 - yz)_y]_z = (2y - z)_z = -1$

(c) Let $w = f(x, y, z) = zy^2 e^{3x}$. Since the partial derivatives $f_x = 3zy^2 e^{3x}$, $f_y = 2zye^{3x}$ and $f_z = y^2 e^{3x}$ are defined and continuous everywhere, f is differentiable.

(d) For $w = xy^2 z^3$, $x = t^2$, $y = \sin t$, $z = e^t$:

$$\dfrac{dw}{dt} = \dfrac{\partial w}{\partial x}\dfrac{dx}{dt} + \dfrac{\partial w}{\partial y}\dfrac{dy}{dt} + \dfrac{\partial w}{\partial z}\dfrac{dz}{dt} = y^2 z^3 \cdot 2t + 2xz^3 y \cos t + 3xy^2 z^2 e^t$$

$$= 2te^{3t}\sin^2 t + 2t^2 e^{3t}\sin t \cos t + 3t^2 e^{3t}\sin^2 t$$

$$= te^{3t}(\sin t)(2\sin t + 2t\cos t + 3t\sin t)$$

(e) For $w = xe^{yz}$, $x = rst^2$, $y = \sin t, z = \cos r^2$:

$$\dfrac{\partial w}{\partial t} = \dfrac{\partial w}{\partial x}\dfrac{\partial x}{\partial t} + \dfrac{\partial w}{\partial y}\dfrac{\partial y}{\partial t} + \dfrac{\partial w}{\partial z}\dfrac{\partial z}{\partial t} = e^{yz} \cdot 2rst + xze^{yz}\cos t + xye^{yz} \cdot 0$$

$$= 2rste^{\sin t \cos r^2} + rst^2 \cos r^2 e^{\sin t \cos r^2}\cos t$$

$$= rste^{\sin t \cos r^2}(2 + t\cos r^2 \cos t)$$

CYU 13.7 If $u = ai + bj$ is a unit vector making an angle θ with the positive x-axis then $a = \cos\theta$ and $b = \sin\theta$.

In particular, for $\theta = \dfrac{\pi}{6}$: $u = \left(\cos\dfrac{\pi}{6}\right)i + \left(\sin\dfrac{\pi}{6}\right)j = \dfrac{\sqrt{3}}{2}i + \dfrac{1}{2}j$.

Bringing us to: $D_u f\left(1, \dfrac{\pi}{2}\right) = \dfrac{\sqrt{3}}{2}f_x\left(1, \dfrac{\pi}{2}\right) + \dfrac{1}{2}f_y\left(1, \dfrac{\pi}{2}\right)$

Since $f_x = \dfrac{\partial}{\partial x}[x\sin xy] = xy\cos xy + \sin xy$, $f_y = \dfrac{\partial}{\partial y}[x\sin xy] = x^2 \cos xy$:

$$f_x\left(1, \dfrac{\pi}{2}\right) = 1 \quad \text{and} \quad f_y\left(1, \dfrac{\pi}{2}\right) = 0$$

Thus: $D_u f\left(1, \dfrac{\pi}{2}\right) = \dfrac{\sqrt{3}}{2}$.

CYU 13.8 Unit vector in the direction of $-i + j$: $u = \dfrac{1}{\sqrt{2}}(-i + j)$.

Gradient vector: $\nabla f(x, y) = f_x(x, y)i + f_y(x, y)j = e^{x+y^2}i + 2ye^{x+y^2}j$

In particular, $\nabla f(0, 1) = e\boldsymbol{i} + 2e\boldsymbol{j}$, and therefore:
$$D_{\boldsymbol{u}}f(0, 1) = \nabla f(0, 1) \cdot \boldsymbol{u} = (e\boldsymbol{i} + 2e\boldsymbol{j}) \cdot \frac{1}{\sqrt{2}}(-\boldsymbol{i} + \boldsymbol{j}) = -\frac{e}{\sqrt{2}} + \frac{2e}{\sqrt{2}} = \frac{e}{\sqrt{2}}$$

CYU 13.9 Gradient of $f(x, y) = x\sin y$ at (x, y):
$$\nabla f(x, y) = f_x(x, y)\boldsymbol{i} + f_y(x, y)\boldsymbol{j} = \sin y\,\boldsymbol{i} + x\cos y\,\boldsymbol{j}$$
In particular: $\nabla f(2, 0) = f_x(2, 0)\boldsymbol{i} + f_y(2, 0)\boldsymbol{j} = (\sin 0)\boldsymbol{i} + (2\cos 0)\boldsymbol{j} = 2\boldsymbol{j}$.
By Theorem 13.6, the greatest value of the directional derivative at (2,0) is $\|\nabla f(2, 0)\| = \|2\boldsymbol{j}\| = 2$, and the smallest value is $-\|\nabla f(2, 0)\| = -2$.

CYU 13.10 Gradient of $f(x, y, z) = \ln(x^2 + y^2 + z^2)$ at $(1, 1, 2)$:
$$\nabla f(x, y, z) = \frac{2x}{x^2 + y^2 + z^2}\boldsymbol{i} + \frac{2y}{x^2 + y^2 + z^2}\boldsymbol{j} + \frac{2z}{x^2 + y^2 + z^2}\boldsymbol{k}$$
In particular: $\nabla f(1, 1, 2) = \frac{2}{6}\boldsymbol{i} + \frac{2}{6}\boldsymbol{j} + \frac{4}{6}\boldsymbol{k} = \frac{1}{3}\boldsymbol{i} + \frac{1}{3}\boldsymbol{j} + \frac{2}{3}\boldsymbol{k}$.
By Theorem 13.6, the greatest value of the directional derivative at (1,1,2) is
$\|\nabla f(1, 1, 2)\| = \sqrt{\frac{1}{9} + \frac{1}{9} + \frac{4}{9}} = \frac{\sqrt{6}}{3}$, and the smallest value is $-\|\nabla f(1, 1, 2)\| = -\frac{\sqrt{6}}{3}$.

CYU 13.11 From $f(x, y) = 3y^2 - 2x^2 + x$ we have:
$$f_x(x, y) = -4x + 1 \text{ and } f_y(x, y) = 6y \Rightarrow f_x(2, -1) = -7 \text{ and } f_y(2, -1) = -6$$

Normal to the plane: $\boldsymbol{n} = -7\boldsymbol{i} - 6\boldsymbol{j} - \boldsymbol{k}$.
Consequently:
$$(-7, -6, -1) \cdot (x - 2, y + 1, z + 3) = 0 \Rightarrow -7(x - 2) - 6(y + 1) - 1(z + 3) = 0$$
$$\Rightarrow -7x - 6y - z = -5 \Rightarrow 7x + 6y + z = 5$$

CYU 13.12 Let the curve C_k, with position vector $\boldsymbol{r}(t) = x(t)\boldsymbol{i} + y(t)\boldsymbol{j}$, be the intersection of the surface $z = f(x, y)$ with the plane $z = k$.

Applying the Chain Rule (Theorem 13.4(a), page 555), to $f(x(t), y(t)) = k$ we have .

$$\frac{d}{dt}f(x(t), y(t)) = \frac{\partial f}{\partial x}\frac{dx}{dt} + \frac{\partial f}{\partial y}\frac{dy}{dt} = 0$$

Recalling that $\boldsymbol{i} \cdot \boldsymbol{i} = \boldsymbol{j} \cdot \boldsymbol{j} = 1$ and that $\boldsymbol{i} \cdot \boldsymbol{j} = 0$, we can express the above equation in vector form:
$$\left(\frac{\partial f}{\partial x}\boldsymbol{i} + \frac{\partial f}{\partial y}\boldsymbol{j}\right) \cdot \left(\frac{dx}{dt}\boldsymbol{i} + \frac{dy}{dt}\boldsymbol{j}\right) = 0$$

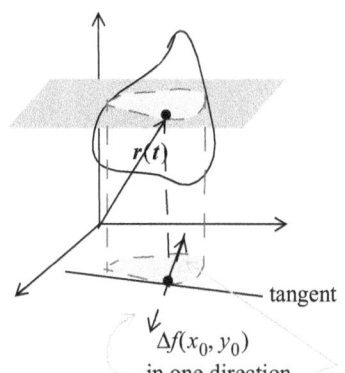

tangent

$\Delta f(x_0, y_0)$
in one direction
or the ohter

It follows that $\dfrac{\partial f}{\partial x}i + \dfrac{\partial f}{\partial y}j = \nabla f$ is orthogonal to the tangent vector $\dfrac{dx}{dt}i + \dfrac{dy}{dt}j = \dfrac{d\mathbf{r}}{dt}$ on the k-level curve at each point (x_0, y_0) on that curve.

CYU 13.13 Turning to $F(x, y, z) = x^2 + 4y^2 + z^2$, we have:
$$\nabla F(x, y, z) = 2xi + 8yj + 2zk$$
Using the normal $\nabla F(x, y, z) = 2i + 16j - 2k$ and the point $(1, 2, -1)$ we arrive at the equation of the tangent plane:
$$2(x - 1) + 16(y - 2) - 2(z + 1) = 0 \ \text{ or: } x + 8y - z = 18$$

CYU 13.14 Since $z = f(x, y) = \sqrt{x^2 + y^2}$:
$$f_x(x, y) = \frac{1}{2}(x^2 + y^2)^{-\frac{1}{2}}(2x) = \frac{x}{\sqrt{x^2 + y^2}}, \ \ f_y(x, y) = \frac{1}{2}(x^2 + y^2)^{-\frac{1}{2}}(2y) = \frac{y}{\sqrt{x^2 + y^2}}$$

In particular: $f_x(3, 4) = \dfrac{3}{\sqrt{3^2 + 4^2}} = \dfrac{3}{5}, f_y(3, 4) = \dfrac{4}{\sqrt{3^2 + 4^2}} = \dfrac{4}{5}$

Consequently:
$$\Delta z \approx dz = f_x(3, 4)\Delta x + f_y(3, 4)\Delta y = \frac{3}{5}(3.01 - 3) + \frac{4}{5}(3.98 - 4) = \frac{3}{5}(0.01) - \frac{4}{5}(0.02) \approx -0.01$$

CYU 13.15 For $f(x, y) = y^2 - x^2$, $f_x(x, y) = -2x$ and $f_y(x, y) = 2y$.

Since $f_x(0, 0) = f_y(0, 0) = 0$, $(0, 0)$ is a critical point of f.

Since there are both positive and negative function values in any open region containing $(0, 0)$:
$$f(0, y) = y^2 > 0 \text{ for all } y \neq 0, \text{ and } f(x, 0) = -x^2 < 0 \text{ for any } x \neq 0$$
neither a maximum nor minimum occurs at $(0, 0)$.

CYU 13.16 For $f(x, y) = x^2 + 3y^2 + 3xy - 6x - 3y$:
$$f_x(x, y) = 2x + 3y - 6 \text{ and } f_y(x, y) = 6y + 3x - 3 \text{ , and both exist for all } x, y.$$
Finding f's critical points:
$$\left.\begin{array}{l} 2x + 3y - 6 = 0 \\ 6y + 3x - 3 = 0 \end{array}\right\} \Rightarrow \left.\begin{array}{l} 2x + 3y = 6 \\ 3x + 6y = 3 \end{array}\right\} \Rightarrow \begin{array}{l} 4x + 6y = 12 \\ 3x + 6y = 3 \end{array}$$
$$\text{subtract: } x = 9 \text{ and } y = \frac{-3(9) + 3}{6} = -4$$

Substituting $y = -4$ in $2x + 3y = 6$, we can now say that the only critical point occurs at $(9, -4)$. To determine the nature of that critical point we turn to Theorem 13.9:
$$\text{From } f_{xx}(x, y) = 2, f_{yy}(x, y) = 6, \text{ and } f_{xy}(x, y) = 3, \text{ we have:}$$

$$D = f_{xx}(9, -4)f_{yy}(9, -4) - [f_{xy}(9, -4)]^2 = (2)(6) - 3^2 = 3 > 0 \text{ and}$$
$$f_{xx}(9, -4) = 2 > 0$$

Conclusion: A (local) minimum occurs at $(9, -4)$.

CYU 13.17 For x, y, and z positive, we are to minimize:
$$S = x^2 + y^2 + z^2 \text{ given (*) } x + y + z = 1$$
From (*): $z = 1 - x - y$. Hence:
$$S = x^2 + y^2 + (1 - x - y)^2 = 2x^2 + 2y^2 + 2xy - 2x - 2y + 1$$
Employing Theorem 13.8:
$$S_x = 4x + 2y - 2 = 0 \text{ and } S_y = 4y + 2x - 2 = 0$$
$$4x + 2y - 2 = 4y + 2x - 2$$
$$2x = 2y$$
$$x = y$$

Repeating the above argument, but now with the substitution $y = 1 - x - z$ leads to the conclusion that $x = z$. Consequently: $x = y = z$. Returning to (*), we have:
$$x = y = z = \frac{1}{3}$$
We leave it to you to verify, via Theorem 13.9, that S achieves its minimum value at the point $\left(\frac{1}{3}, \frac{1}{3}, \frac{1}{3}\right)$.

CYU 13.18 We know, from the solution of Example 13.14, that the absolute extremes of the function $f(x, y) = x^2 - 2y^3 - 3x + 2y$ must occur on the boundary of the domain D. Turning to the three boundary pieces we have:

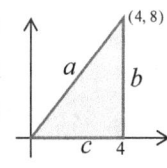

On line a:
$$g(x) = f(x, 2x) = x^2 - 2(2x)^3 - 3x + 2(2x) = -16x^3 + x^2 + x$$
Since $g'(x) = -48x^2 + 2x + 1 = (-6x + 1)(8x + 1)$, g has a critical point at $x = \frac{1}{6}$. It follows that f may assume an extreme value at $\left(\frac{1}{6}, \frac{1}{3}\right)$, $(0, 0)$ and $(4, 8)$.

On line b: $g(y) = f(4, y) = 16 - 2y^3 - 12 + 2y = -2y^3 + 2y + 4$
Since $g'(y) = -6y^2 + 2$, g has a critical point at $y = \frac{1}{\sqrt{3}}$. It follows that f may also assume its extreme values at $\left(4, \frac{1}{\sqrt{3}}\right)$ and $(4, 0)$.

On line c: $g(x) = f(x, 0) = x^2 - 3x$. Since $g'(x) = 2x - 3$, g has a critical point at $x = \frac{3}{2}$. It follows that f may assume an extreme value at $\left(\frac{3}{2}, 0\right)$.

Upon evaluating the function $f(x, y) = x^2 - 2y^3 - 3x + 2y$, at each of the above 6 boundary critical points, we found that:

$$f(0, 0) = 0 \qquad f(4, 0) = 4 \qquad f\left(4, \frac{1}{\sqrt{3}}\right) = \frac{36 + 4\sqrt{3}}{9} \approx 4.8$$

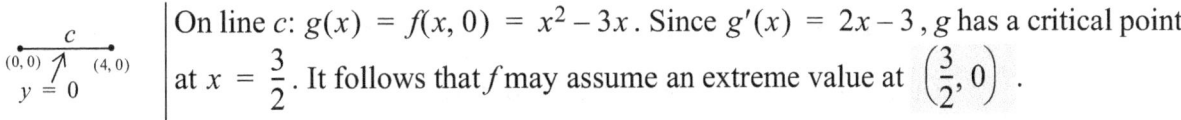

$$f\left(\frac{1}{6}, \frac{1}{3}\right) = \frac{13}{108} \qquad f\left(\frac{3}{2}, 0\right) = -\frac{9}{4} \qquad f(4, 8) = -1004$$

Conclusion: On the region D, f achieves its absolute maximum value of $\frac{36 + 4\sqrt{3}}{9}$ at $\left(4, \frac{1}{\sqrt{3}}\right)$, and its absolute minimum value of -1004 at $(4, 8)$.

CYU 13.19 For $f(x, y) = x^2 + 4y^2$ with constraint $g(x) = x^2 + y^2 = 1$ we have:

$$\nabla f(x, y) = \lambda \nabla g(x, y)$$
$$2x\mathbf{i} + 8y\mathbf{j} = \lambda(2x\mathbf{i} + 2y\mathbf{j})$$

Leading us to the following system of equations:

(1): $2x = \lambda 2x$ (2): $8y = \lambda 2y$ (3): $x^2 + y^2 = 1$

From (1): $x = 0$ or $\lambda = 1$.

   If $x = 0$, then, from (3): $y = \pm 1$; and (2) is also satisfied for $\lambda = 4$.

   If $\lambda = 1$, then, from (2): $y = 0$ and $x = \pm 1$ [from (3)].

Also, $\nabla g(x, y) = 2x\mathbf{i} + 2y\mathbf{j} = 0 \Rightarrow x = y = 0$, but $(0, 0)$ is not a candidate as it does not lie on the circle $x^2 + y^2 = 1$.

Evaluating $f(x, y) = x^2 + 4y^2$ at the four critical points
$$(0, 1), (0, -1), (1, 0), \text{ and } (-1, 0), \text{ we have:}$$
$$f(0, 1) = 4, \ f(0, -1) = 4, \ f(1, 0) = 1, \text{ and } f(-1, 0) = 1$$

Conclusion: The function $f(x, y) = x^2 + 4y^2$, when restricted to points on the circle $x^2 + y^2 = 1$, assumes a maximum value of 4 at $(0, 1)$ and $(0, -1)$, and a minimum value of 1 at $(1, 0)$ and $(-1, 0)$.

CYU 13.20 We are given $f(x, y, z) = xy + yz$ with constraints $x + 2y - 5 = 0$ and $x - 4z = 0$.
Setting $g(x, y, z) = x + 2y$ and $h(x, y, z) = x - 4z$, the vector equation:

$$\nabla f(x, y, z) = \lambda \nabla g(x, y, z) + \mu \nabla h(x, y, z)$$

becomes: $y\mathbf{i} + (x + z)\mathbf{j} + y\mathbf{k} = \lambda(\mathbf{i} + 2\mathbf{j} + 0\mathbf{k}) + \mu(\mathbf{i} + 0\mathbf{j} - 4\mathbf{k})$

Leading us to the following (linear) system of equations:

(1): $y = \lambda + \mu$ (2): $x + z = 2\lambda$ (3): $y = -4\mu$

     (4): $x + 2y - 5 = 0$ (5): $x - 4z = 0$

From (3): $\mu = -\frac{y}{4}$. Substituting in (1): $y = \lambda - \frac{y}{4} \Rightarrow \lambda = \frac{5y}{4}$

Substituting in (2): $x + z = 2\lambda \Rightarrow x + z = \frac{5y}{2} \Rightarrow 2x + 2z - 5y = 0$ (*)

Taking us to the following system of three equations in three unknowns:

(*): $2x - 5y + 2z = 0$

(4): $x + 2y = 5$ $\Rightarrow x = \frac{5}{2}, y = \frac{5}{4}, z = \frac{5}{8}$

(5): $x - 4z = 0$ steps ommited

We see that there is but one critical point for f; namely: $\left(\dfrac{5}{2}, \dfrac{5}{4}, \dfrac{5}{8}\right)$.

While we did what we were asked to do in CYU 13.20, let's do more:

We found but one critical point. Shouldn't there be at least two: one at which f assumes its minimum value and one for its maximum value? Not necessarily, for f is restricted to the region D consisting of those points satisfying both of the given constraints: $x + 2y - 5 = 0$ and $x - 4z = 0$.

That region turns out to be the plane $D = \left\{(x, y, z) \mid y = -\dfrac{1}{2}z + 5\right\}$ (just substitute $4z$ for x in $x + 2y - 5 = 0$). Since D is not bounded, there is no assurance that f will assume either a maximum or minimum value on D (see Theorem 13.4, page 502). To make matters worse, there is no second derivative (like Theorem 13.3) associated with the Lagrange method. So what can one do? In this comparably easy situation, we can do this:

$$x + 2y - 5 = 0 \Rightarrow y = -\frac{x}{2} + \frac{5}{2} \quad \text{and} \quad x - 4z = 0 \Rightarrow z = \frac{x}{4}$$

So: $\quad g(x) = f(x, y, z) = xy + yz = x\left(-\frac{x}{2} + \frac{5}{2}\right) + \left(-\frac{x}{2} + \frac{5}{2}\right)\frac{x}{4} = -\frac{5}{8}x^2 + \frac{25}{8}x$

Then: $\quad g'(x) = -\dfrac{5}{4}x + \dfrac{25}{8} = \dfrac{5}{8}(-2x + 5)$ SIGN: $\dfrac{\quad + \quad \overset{c}{\bullet} \quad - \quad}{\text{max}}$

Conclusion: On D, the function f assumes the (absolute) maximum value $f\left(\dfrac{5}{2}, \dfrac{5}{4}, \dfrac{5}{8}\right) = \dfrac{125}{32}$.

CHAPTER 14: VECTOR CALCULUS

CYU 14.1 $\displaystyle\int_C xy^2 ds = \int_0^{2\pi} \cos t \cdot \sin^2 t \sqrt{\left(\frac{dx}{dt}\right)^2 + \left(\frac{dy}{dt}\right)^2 + \left(\frac{dz}{dt}\right)^2}\, dt$

$$= \int_0^{2\pi} \cos t \cdot \sin^2 t \sqrt{\underbrace{(-\sin t)^2 + (\cos t)^2}_{=1} + (1)^2}\, dt = \sqrt{2}\int_0^{2\pi} \cos t \cdot \sin^2 t\, dt$$

$$= \sqrt{2}\left(\frac{1}{3}(\sin^3 t)\right)\Big|_0^{2\pi} = 0$$

CYU 14.2 Using the parametrization:

C_4: $r_4 = (1-t)(2, 2) + t(-1, 0) = (2 - 3t, 2 - 2t)$, $0 \le t \le 1$:

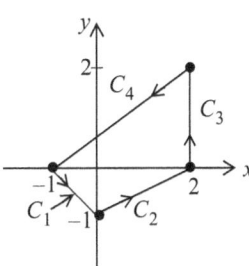

$$\int_{C_4} xy\, ds = \int_0^1 (2 - 3t)(2 - 2t)\sqrt{(-3)^2 + (-2)^2}\, dt$$

$$= \sqrt{13}\int_0^1 (4 - 10t + 6t^2)\, dt = \sqrt{13}(4t - 5t^2 + 2t^3)\Big|_0^1 = \sqrt{13}$$

We then have: $\displaystyle\int_C xy\,ds = \underbrace{\int_{C_1} xy\,ds + \int_{C_2} xy\,ds + \int_{C_3} xy\,ds + \int_{C_4} xy\,ds}_{\text{Example 14.3}} = \frac{1}{6}(\sqrt{2} - 2\sqrt{5} + 24) + \sqrt{13}$

CYU 14.3 $\displaystyle M = \int_C \delta(x,y)\,ds = \int_C (1+x+z)\,ds = \int_0^{6\pi} (1+\cos t + t)\sqrt{(-\sin t)^2 + (\cos t)^2 + 1}\,dt$

$$= \sqrt{2}\int_0^{6\pi} (1 + \cos t + t)\,dt$$

$$= \sqrt{2}\left(t + \sin t + \frac{t^2}{2}\right)\Big|_0^{6\pi} = \sqrt{2}(6\pi + 18\pi^2)$$

CYU 14.4 $\displaystyle W = \int_a^b F(r(t)) \cdot r'(t)\,dt = \int_0^1 F(t^2, t^3, t) \cdot (2t\mathbf{i} + 3t^2\mathbf{j} + \mathbf{k})\,dt$

$$= \int_0^1 (t^2\mathbf{i} + (t^2 + t^3)\mathbf{j} + t\mathbf{k}) \cdot (2t\mathbf{i} + 3t^2\mathbf{j} + \mathbf{k})\,dt$$

$$= \int_0^1 (2t^3 + 3t^4 + 3t^5 + t)\,dt = \left(\frac{t^4}{2} + \frac{3t^5}{5} + \frac{t^6}{2} + \frac{t^2}{2}\right)\Big|_0^1 = \frac{21}{10}$$

CYU 14.5 For $F(x,y,z) = y\mathbf{i} + 2x\mathbf{j} + 3z\mathbf{k}$ and $r(t) = (t, 2t, 2t)$, $0 \le t \le 1$ we have:

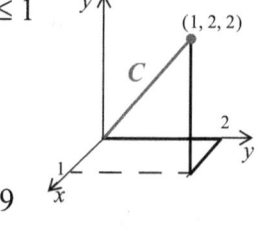

$\displaystyle\int_C F \cdot dr = \int_0^1 F(r(t)) \cdot r'(t)\,dt = \int_0^1 (2t\mathbf{i} + 2t\mathbf{j} + 6t\mathbf{k}) \cdot (\mathbf{i} + 2\mathbf{j} + 2\mathbf{k})\,dt$

$$= \int_0^1 (2 + 4 + 12)t\,dt = \int_0^1 18t\,dt = (9t^2)\Big|_0^1 = 9$$

CYU 14.6 We know that $\displaystyle\int_C F \cdot dr = 8$ (Example 14.6), and now show

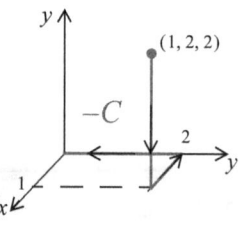

that $\displaystyle\int_{-C} F \cdot dr = -8$, where $F(x,y,z) = y\mathbf{i} + 2x\mathbf{j} + 3z\mathbf{k}$.

As noted above Theorem 14.3:

For a given parametrization $r(t)$, $a \le t \le b$ of a curve C, the parametrization $\bar{r}(t) = r(a+b-t)$, $a \le t \le b$, denoted by $-C$, traverses C in the opposite direction.

Firstly, $a = 0$ and $b = 1$ throughout $\Rightarrow a + b - t = 1 - t$. So:

$r_1(t) = \langle 0, 2t, 0\rangle \Rightarrow \bar{r}_1(t) = r_1(1-t) = \langle 0, 2(1-t), 0\rangle = \langle 0, 2 - 2t, 0\rangle$ for $0 \le t \le 1$

$r_2(t) = \langle t, 2, 0\rangle \Rightarrow \bar{r}_2(t) = r_2(1-t) = \langle 1 - t, 2, 0\rangle$ for $0 \le t \le 1$

$r_3(t) = \langle 1, 2, 2t\rangle \Rightarrow \bar{r}_3(t) = r_3(1-t) = \langle 1, 2, 2(1-t)\rangle = \langle 1, 2, 2 - 2t\rangle$ for $0 \le t \le 1$

Thus:

$$\int_{-C} F \cdot dr = \int_0^1 F(\bar{r}_1(t)) \cdot \bar{r}_1'(t)dt + \int_0^1 F(\bar{r}_2(t)) \cdot \bar{r}_2'(t)dt + \int_0^1 F(\bar{r}_3(t)) \cdot \bar{r}_3'(t)dt$$

$$= \int_0^1 (2 - 2t)i \cdot (-2j)dt + \int_0^1 [2i + 2(1 - t)j] \cdot (-i)dt + \int_0^1 [2i + 2j + 3(2 - 2t)k] \cdot (-2k)dt$$

$$= 0 + \int_0^1 (-2)dt - 6\int_0^1 (2 - 2t)dt = (-2t)\Big|_0^1 - 6(2t - t^2)\Big|_0^1 = -8 = -\int_C F \cdot dr$$

CYU 14.7 For $x(t) = t^2$, $y(t) = 5t$, and $z(t) = -2t^2$, $1 \le t \le 2$, we have: $\dfrac{dx}{dt} = 2t$, $\dfrac{dy}{dt} = 5$,

and $\dfrac{dz}{dt} = -4t$ Consequently:

$$\int_C xy\,dx + y\,dy + yz\,dz = \int_1^2 \left[t^2(5t)\frac{dx}{dt} + 5t\frac{dy}{dt} + (t^2)(-2t^2)\frac{dz}{dt} \right] dt$$

$$= \int_1^2 [5t^3(2t) + 5t(5) - 4t^4(-4t)]dt = \int_1^2 (10t^4 + 25t + 16t^5)dt$$

$$= \left(2t^5 + \frac{25t^2}{2} + \frac{8t^6}{3} \right)\Bigg|_1^2 = \frac{535}{2}$$

CYU 14.8 If C_1 and C_2 are two paths in S from P_1 to P_2, then $C = C_1 \cup (-C_2)$

is a closed path in S. As such $\oint_C F \cdot dr = 0$.

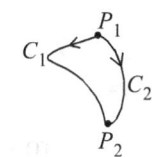

Since $\underset{C}{\oint F \cdot dr} = \underset{C_1}{\int F \cdot dr} \underset{\overset{\uparrow}{\text{Theorem 14.3}}}{-} \underset{C_2}{\int F \cdot dr}$, $\quad \underset{C_1}{\int F \cdot dr} = \underset{C_2}{\int F \cdot dr}$

CYU 14.9 If F is conservative with potential function f, then:

$$F = Pi + Qj + Rk = \nabla f(x, y, z) = f_x i + f_y j + f_z k$$

$$\Rightarrow P = f_x, \quad Q = f_y, \quad R = f_z$$

$$\Rightarrow \frac{\partial P}{\partial y} = f_{xy} = f_{yx} = \frac{\partial Q}{\partial x}, \qquad \frac{\partial P}{\partial z} = f_{xz} = f_{zx} = \frac{\partial R}{\partial x}, \qquad \frac{\partial Q}{\partial z} = f_{yz} = f_{zy} = \frac{\partial R}{\partial y}$$

Theorem 13.1, page 552

CYU 14.10 If, for given $F = P(x, y, z)i + Q(x, y, z)j + R(x, y, z)k$, any one of the equations

$\dfrac{\partial P}{\partial y} = \dfrac{\partial Q}{\partial x}$, $\dfrac{\partial P}{\partial z} = \dfrac{\partial R}{\partial x}$, $\dfrac{\partial Q}{\partial z} = \dfrac{\partial R}{\partial y}$ fails to hold in S, then F is not conservative in S.

While the first two of those equations do hold for $F(x, y, z) = xi + zj + yzk$, the third

does not: $\dfrac{\partial Q}{\partial z} = \dfrac{\partial z}{\partial z} = 1$ while $\dfrac{\partial R}{\partial y} = z$.

CYU 14.11 (a) Theorem 14.9(iv) holds for $F(x, y) = 2xy\mathbf{i} + x^2\mathbf{j}$: $(2xy)_y = (x^2)_x \boxed{= 2x}$.

For f such that $F = \nabla f$, we have: $(2xy)\mathbf{i} + (x^2)\mathbf{j} = \nabla f(x, y) = f_x(x, y)\mathbf{i} + f_y(x, y)\mathbf{j}$

Bringing us to: (1): $f_x(x, y) = 2xy$ and (2): $f_y(x, y) = x^2$.

Treating y as a **constant** in (1), we have: (3): $f(x, y) = \int 2xy\,dx = x^2y + g(y)$.

Taking the partial derivative of (3) with respect to y: $f_y(x, y) = x^2 + g'(y)$.

Replacing $f_y(x, y)$ with x^2 [see (2)]: $x^2 = x^2 + g'(y) \Rightarrow g'(y) = 0 \Rightarrow g(y) = C$.

Letting $C = 0$ we have an answer: $F(x, y) = 2xy\mathbf{i} + x^2\mathbf{j} = \nabla f = \nabla(x^2y)$.

(b) Theorem 14.9(iv) holds for $F(x, y, z) = (2xyz)\mathbf{i} + (x^2z)\mathbf{j} + (x^2y)\mathbf{k}$:

$$(2xyz)_y = (x^2z)_x = 2xz, (2xyz)_z = (x^2y)_x = 2xy, (x^2z)_z = (x^2y)_y = x^2$$

For f such that $F = \nabla f$ we have:

$$(2xyz)\mathbf{i} + (x^2z)\mathbf{j} + (x^2y)\mathbf{k} = \nabla f = f_x(x, y, z)\mathbf{i} + f_y(x, y, z)\mathbf{j} + f_z(x, y, z)\mathbf{k}$$

Bringing us to: (1): $f_x(x, y, z) = 2xyz$, (2): $f_y(x, y, z) = x^2z$, (3): $f_z(x, y, z) = x^2y$

Treating y and z as **constants** in (1), we have: (4): $f(x, y, z) = \int 2xyz\,dx = x^2yz + g(y, z)$.

Taking the partial derivative of (4) with respect to y: $f_y(x, y, z) = x^2z + [g(y, z)]_y$.

Replacing $f_y(x, y, z)$ with x^2z [see (2)]:

$$x^2z = x^2z + [g(y, z)]_y \Rightarrow [g(y, z)]_y = 0 \Rightarrow g(y, z) = h(z).$$

Returning to (4), we have: (5) $f(x, y, z) = x^2yz + h(z)$.

Taking the partial derivative with respect to z: $f_z(x, y, z) = x^2y + h'(z)$.

From (3): $x^2y = x^2y + h'(z) \Rightarrow h(z) = C$

Letting $C = 0$ and returning to (5) we have an answer: $F(x, y, z) = \nabla(x^2yz)$.

CYU 14.12 One approach: $F(x, y) = 2xy\mathbf{i} + x^2\mathbf{j}$ is a conservative field with potential function

$f(x, y) = x^2y$ (see CYU 14.11). Since $r(0) = \langle 0, 0 \rangle$ and $r\left(\dfrac{\pi}{4}\right) = \langle \dfrac{\pi}{4}\cos^2\dfrac{\pi}{4}, \dfrac{\pi^2}{16}\rangle$

the curve C starts at the point $(0, 0)$ and ends at the point $\left(\dfrac{\pi}{8}, \dfrac{\pi^2}{16}\right)$. Hence:

$$\int_C F \cdot dr = f\left(\dfrac{\pi}{8}, \dfrac{\pi^2}{16}\right) - f(0, 0) = \left(\dfrac{\pi}{8}\right)^2\left(\dfrac{\pi^2}{16}\right) - 0 = \dfrac{\pi^4}{1024}$$

Another approach: Choosing to integrate along the path $\bar{C} = C_1 \cup C_2$ with parametrizations:

$$r_1(t) = t\mathbf{i}, 0 \le t \le \dfrac{\pi}{8} \text{ for } C_1 \text{ and } r_2(t) = \dfrac{\pi}{8}\mathbf{i} + t\mathbf{j}, 0 \le t \le \dfrac{\pi^2}{16}$$

we have:

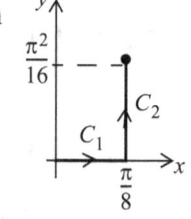

$$\int_C F \cdot dr = \int_{\bar{C}} F \cdot dr = \int_{C_1} F \cdot dr_1 + \int_{C_2} F \cdot dr_2$$

$$= \int_{C_1} [(2xyi + x^2j) \cdot i]dt + \int_{C_2} [(2xyi + x^2j) \cdot j]dt$$

$$= \int_{C_1} 2xy\,dt + \int_{C_2} x^2\,dt = 0 + \int_0^{\frac{\pi^2}{16}} \left(\frac{\pi}{8}\right)^2 dt = \left(\frac{\pi^2}{64}t\right)\Big|_0^{\frac{\pi^2}{16}} = \frac{\pi^4}{1024}$$

CYU 14.13 The force field $F(x, y, z) = 0i + 0j + (-mg)k$ is conservative:

$$F(x, y, z) = \nabla f(x, y, z), \text{ with } f(x, y, z) = -mgz.$$

It follows that for any path C from $(0, 0, 0)$ to (x_0, y_0, z_0):

$$W = \int_C F \cdot dr = f(x_0, y_0, z_0) - f(0, 0, 0) = -mgz_0$$

[Negative work indicates that the force field impedes movement along the curve (see page 594)]

CYU 14.14 Using the parametrization of a line segment from r_0 to r_1:

$$r(t) = r_0 + t\{r_1 - r_0\} = (1-t)r_0 + tr_1, \ 0 \le t \le 1$$

$$C_1: \ r = (1-t)\langle 0, 0\rangle + t\langle 1, 0\rangle = \langle t, 0\rangle, 0 \le t \le 1$$

We have: $C_2: \ r = (1-t)\langle 1, 0\rangle + t\langle 0, 1\rangle = \langle 1-t, t\rangle, 0 \le t \le 1$

$$C_3: \ r = (1-t)\langle 0, 1\rangle + t\langle 0, 0\rangle = \langle 0, 1-t\rangle, 0 \le t \le 1$$

Thus: $\oint_C F \cdot T\,ds = \oint_C (x^2i + xyj) \cdot dr = \int_{C_1} (x^2i + xyj) \cdot dr + \int_{C_2} (x^2i + xyj) \cdot dr + \int_{C_3} (x^2i + xyj) \cdot dr$

$$= \int_0^1 [(t^2i) \cdot i]dt + \int_0^1 \{[(1-t)^2i + t(1-t)j] \cdot (-i+j)\}dt + \int_0^1 0\,dt$$

$$= \int_0^1 [t^2 - (1-t)^2 + t(1-t)]dt = \int_0^1 (-t^2 + 3t - 1)dt = \left(-\frac{1}{3}t^3 + \frac{3}{2}t^2 - t\right)\Big|_0^1 = \frac{1}{6}$$

Using Green's Theorem, with $F(x, y) = Pi + Qj = x^2i + xyj$ and

$$\iint_D \left(\frac{\partial Q}{\partial x} - \frac{\partial P}{\partial y}\right)dA = \int_0^1\int_0^{-x+1} y\,dy\,dx = \int_0^1\left[\left(\frac{y^2}{2}\right)\Big|_0^{-x+1}\right]dx = \frac{1}{2}\int_0^1 (x^2 - 2x + 1)dx$$

$$= \frac{1}{2}\left(\frac{x^3}{3} - x^2 + x\right)\Big|_0^1 = \frac{1}{6}$$

CYU 14.15 From the adjacent figure we see that:

$$W = \iint\limits_{D}\left(\frac{\partial Q}{\partial x} - \frac{\partial P}{\partial y}\right)dA = \iint\limits_{D}(-2y-1)dA$$

$$= \int_{0}^{1}\int_{\frac{1}{2}(y-1)}^{-y+1}(-2y-1)dxdy$$

$$= \int_{0}^{1}(-2yx-x)\Big|_{\frac{y}{2}-\frac{1}{2}}^{-y+1}dy$$

$$= \int_{0}^{1}\left\{[-2y(-y+1)-(-y+1)]-\left[-2y\left(\frac{y}{2}-\frac{1}{2}\right)-\left(\frac{y}{2}-\frac{1}{2}\right)\right]\right\}dy$$

$$= \int_{0}^{1}\left(3y^2-\frac{3}{2}y-\frac{3}{2}\right)dy = \left(y^3-\frac{3}{4}y^2-\frac{3}{2}y\right)\Big|_{0}^{1} = 1-\frac{3}{4}-\frac{3}{2} = -\frac{5}{4}$$

(Figure: triangle region R with lines $y = 2x+1$, $y = -x+1$, with points at $-\frac{1}{2}$ and 1.)

CYU 14.16 $A = \dfrac{1}{2}\oint_{C}-y\,dx+x\,dy = \dfrac{1}{2}\int_{0}^{2\pi}[-\sin^3 t(3\cos^2 t)(-\sin t)+\cos^3 t(3\sin^2 t)\cos t]dt$

$$= \frac{3}{2}\int_{0}^{2\pi}[\sin^4 t(\cos^2 t)+\cos^4 t(\sin^2 t)]dt$$

$$= \frac{3}{2}\int_{0}^{2\pi}\sin^2 t\cdot\cos^2 t[\sin^2 t+\cos^2 t]dt$$

$$= \frac{3}{2}\int_{0}^{2\pi}\sin^2 t\cdot\cos^2 t\,dt = \frac{3}{2}\left(\frac{1}{8}t-\frac{1}{32}\sin 4t\right)\Big|_{0}^{2\pi} = \frac{3}{16}\cdot 2\pi = \frac{3}{8}\pi$$

Example 7.11(b), page 281

CYU 14.17 Using the parametrization of a line segment from r_0 to r_1:
$$r(t) = r_0+t\langle r_1-r_0\rangle = (1-t)r_0+tr_1,\ 0\le t\le 1$$

We have: $C_1:\ r = (1-t)\langle 0,0\rangle+t\langle 3,0\rangle = \langle 3t,0\rangle, 0\le t\le 1$

$\qquad\quad C_2:\ r = (1-t)\langle 3,0\rangle+t\langle 3,3\rangle = \langle 3,3t\rangle, 0\le t\le 1$

$\qquad\quad C_3:\ r = (1-t)\langle 3,3\rangle+t\langle 0,0\rangle = \langle 3-3t,3-3t\rangle, 0\le t\le 1$

(Figure: triangle with vertices showing C_1, C_2, C_3; axes marked 3 on x and y.)

$$\oint_{C}F\cdot n\,ds = \int_{C_1}\{[(y^2-x^2)i+(x^2+y^2)j]\cdot(-3j)\}dt + \int_{C_2}\{[(y^2-x^2)i+(x^2+y^2)j]\cdot(3i)\}dt$$

$$+ \int_{C_3}\{[(y^2-x^2)i+(x^2+y^2)j]\cdot(-3i+3j)\}dt$$

$$= \int_{C_1}-3(x^2+y^2)dt + \int_{C_2}3(y^2-x^2)dt + \int_{C_3}\underbrace{[-3(y^2-x^2)+3(x^2+y^2)]}_{6x^2}dt$$

$$= \int_{0}^{1}-3(9t^2)dt + \int_{0}^{1}3(9t^2-9)dt + \int_{0}^{1}6(3-3t)^2 dt$$

$$= \int_{0}^{1}(54t^2-108t+27)dt = (18t^3-54t^2+27t)\Big|_{0}^{1} = -9$$

CYU 14.18 For $F(x, y) = P(x, y)\mathbf{i} + Q(x, y)\mathbf{j} = (y^2 - x^2)\mathbf{i} + (x^2 + y^2)\mathbf{j}$ and

we have: $\displaystyle\oint_C F \cdot \mathbf{n}\, ds = \iint_D \left(\frac{\partial P}{\partial x} + \frac{\partial Q}{\partial y}\right) dA = \iint_D (-2x + 2y)\, dA = -2\int_0^3 \int_0^x (x - y)\, dy\, dx$

$$= -2\int_0^3 \left(xy - \frac{y^2}{2}\right)\Bigg|_0^x dx$$

$$= -2\int_0^3 \left(x^2 - \frac{x^2}{2}\right) dx$$

$$= -2\int_0^3 \frac{x^2}{2}\, dx = -\left(\frac{x^3}{3}\right)\Bigg|_0^3 = -9$$

(Compare with solution of CYU 14.17)

CYU 14.19 For $F(x, y, z) = \sin(xy)\mathbf{i} + e^{xz}\mathbf{j} + \ln(z^2 + 1)\mathbf{k}$ we have:

$$\text{curl}(F) = \nabla \times F = \det\begin{bmatrix} \mathbf{i} & \mathbf{j} & \mathbf{k} \\ \dfrac{\partial}{\partial x} & \dfrac{\partial}{\partial y} & \dfrac{\partial}{\partial z} \\ \sin(xy) & e^{xz} & \ln(z^2 + 1) \end{bmatrix}$$

$$= \left[\frac{\partial}{\partial y}\ln(z^2 + 1) - \frac{\partial}{\partial z}(e^{xz})\right]\mathbf{i} - \left[\frac{\partial}{\partial x}\ln(z^2 + 1) - \frac{\partial}{\partial z}\sin(xy)\right]\mathbf{j} + \left[\frac{\partial}{\partial x}(e^{xz}) - \frac{\partial}{\partial y}\sin(xy)\right]\mathbf{k}$$

$$= -xe^{xz}\mathbf{i} + (ze^{xz} - x\cos xy)\mathbf{k}$$

CYU 14.20 (a) For $F = P(x, y)\mathbf{i} + Q(x, y)\mathbf{j}$: $\dfrac{\partial P}{\partial y} = \dfrac{\partial Q}{\partial x} \Leftrightarrow \dfrac{\partial P}{\partial y} - \dfrac{\partial Q}{\partial x} = 0 \Leftrightarrow \text{curl}(F) = \mathbf{0}$.

For $F = P(x, y, z)\mathbf{i} + Q(x, y, z)\mathbf{j} + R(x, y, z)\mathbf{k}$:

$$\frac{\partial P}{\partial y} = \frac{\partial Q}{\partial x}, \; \frac{\partial P}{\partial z} = \frac{\partial R}{\partial x}, \; \frac{\partial Q}{\partial z} = \frac{\partial R}{\partial y} \Leftrightarrow \frac{\partial P}{\partial y} - \frac{\partial Q}{\partial x} = 0, \frac{\partial P}{\partial z} - \frac{\partial R}{\partial x} = 0, \frac{\partial Q}{\partial z} - \frac{\partial R}{\partial y} = 0 \Leftrightarrow \text{curl}(F) = \mathbf{0}$$

(b) In Example 14.9 we showed that $F = \dfrac{y}{(x^2 + y^2)}\mathbf{i} + \dfrac{-x}{(x^2 + y^2)}\mathbf{j}$ fails to be conservative

on the (not simply connected) open region $S = \{(x, y)|(x, y) \neq (0, 0)\}$, even though

$\text{curl}(F) = \mathbf{0}$ $\left(\dfrac{\partial P}{\partial y} = \dfrac{\partial Q}{\partial x}\right)$.

CYU 14.21 (a) From:

$$\text{curl}(F) = \nabla \times F = \det \begin{bmatrix} i & j & k \\ \dfrac{\partial}{\partial x} & \dfrac{\partial}{\partial y} & \dfrac{\partial}{\partial z} \\ xz & e^{xz} & -\sin z \end{bmatrix}$$

$$= [(-\sin z)_y - (e^{xz})_z]i - [(-\sin z)_x - (xz)_z]j + [(e^{xz})_x - (xz)_y]k$$

$$= -xe^{xz}i + xj + ze^{xz}k$$

we have: $\text{div}[\text{curl}(F)] = \text{div}(-xe^{xz}i + xj + ze^{xz}k) = (-xe^{xz})_x + (x)_y + (ze^{xz})_z$

$$= (-xze^{xz} - e^{xz}) + 0 + (zxe^{xz} + e^{xz}) = 0$$

(b) Since $\text{div}\,F$ is a scalar, and since the curl function acts on vectors, the expression $\text{curl}[\text{div}\,F]$ is not defined.

CYU 14.22 $\text{div}(\text{curl}[P(x, y)i + Q(x, y)j]) = \text{div}[0i + 0j + (Q_x - P_y)k] = (Q_x - P_y)_z = 0$

CYU 14.23 Since $z = \sqrt{x^2 + y^2}$, we have:

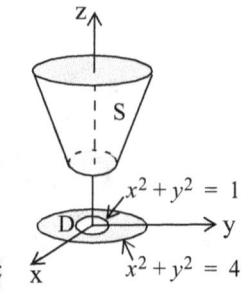

$$A(S) = \iint_D \sqrt{1 + \left[\frac{x}{\sqrt{x^2 + y^2}}\right]^2 + \left[\frac{y}{\sqrt{x^2 + y^2}}\right]^2}\, dA = \iint_D \sqrt{2}\, dA$$

$$= \sqrt{2}\iint_D dA = 3\sqrt{2}\pi$$

Area of D: $\pi 2^2 - \pi 1^2 = 3\pi$

$x^2 + y^2 = 1$

$x^2 + y^2 = 4$

CYU 14.24 $M = \iint_D \delta[x, y, f(x, y)]\sqrt{1 + [f_x(x, y)]^2 + [f_y(x, y)]^2}\, dA$

$$= \iint_D [x^2 + y^2 + (x^2 + y^2)^2]\sqrt{1 + (2x)^2 + (2y)^2}\, dA$$

Converting to polar coordinates (see page 405), we have:

$x^2 + y^2 = \dfrac{1}{4}$

$$M = \int_0^{2\pi} d\theta \int_0^{\frac{1}{2}} (r^2 + r^4)r\sqrt{1 + 4r^2}\, dr$$

Now: $\int_0^{\frac{1}{2}}(r^2+r^4)\sqrt{1+4r^2}\,r\,dr = \dfrac{1}{8}\int_1^2\left[\dfrac{u-1}{4}+\left(\dfrac{u-1}{4}\right)^2\right]u^{1/2}\,du$

$u = 1+4r^2 \Rightarrow r^2 = \dfrac{u-1}{4}$

$du = 8r\,dr$

$= \dfrac{1}{8\cdot 16}\int_1^2(u^{5/2}+2u^{3/2}-3u^{1/2})\,du$

$= \dfrac{1}{128}\left(\dfrac{2}{7}u^{7/2}+\dfrac{4}{5}u^{5/2}-2u^{3/2}\right)\Big|_1^2$

Steps omitted: $= \dfrac{1}{1120}(13\sqrt{2}+8)$

So: $M = \dfrac{1}{1120}(13\sqrt{2}+8)\int_0^{2\pi} d\theta = \dfrac{\pi}{560}(13\sqrt{2}+8)$

CYU 14.25 For $f(x,y) = 1-x^2-y^2$: $-f_x\mathbf{i}-f_y\mathbf{j}+\mathbf{k} = 2x\mathbf{i}+2y\mathbf{j}+\mathbf{k}$.

Thus: $\text{Flux}(\mathbf{F}_S) = \displaystyle\iint_D \mathbf{F}\cdot(-f_x\mathbf{i}-f_y\mathbf{j}+\mathbf{k})\,dA$

$= \displaystyle\iint_D (x\mathbf{i}+y\mathbf{j}+z\mathbf{k})\cdot(2x\mathbf{i}+2y\mathbf{j}+\mathbf{k})\,dA$

$= \displaystyle\iint_D (2x^2+2y^2+z)\,dA$

$= \displaystyle\iint_D [2x^2+2y^2+1-x^2-y^2]\,dA = \iint_D (x^2+y^2+1)\,dA$

$= \displaystyle\int_0^{2\pi}\int_0^1 (r^2+1)r\,dr\,d\theta$

$= \displaystyle\int_0^{2\pi}\int_0^1 (r^3+r)\,dr\,d\theta = \dfrac{3}{2}\pi$

CYU 14.26 Cylindrical coordinates (page 575) provide a direct parametrization for the surface:

$\mathbf{r}(r,\theta) = (r\cos\theta)\mathbf{i}+(r\sin\theta)\mathbf{j}+r\mathbf{k}$, $0\le r\le 1$, $0\le\theta\le 2\pi$

Linking with the notation of Definition 14.13:

$\mathbf{r}(u,v) = x(u,v)\mathbf{i}+y(u,v)\mathbf{j}+z(u,v)\mathbf{k}$

$\mathbf{r}(r,\theta) = (r\cos\theta)\mathbf{i}+(r\sin\theta)\mathbf{j}+r\mathbf{k}$

Grinding away: $r_r \times r_\theta = \det \begin{bmatrix} i & j & k \\ \cos\theta & \sin\theta & 1 \\ -r\sin\theta & r\cos\theta & 0 \end{bmatrix}$

$$= -r\cos\theta i - r\sin\theta j + (r\cos^2\theta + r\sin^2\theta)k$$

And so: $\|r_r \times r_\theta\| = \sqrt{r^2\cos^2\theta + r^2\sin^2\theta + r^2} = \sqrt{2r^2} = \sqrt{2}r$

Bringing us to:

$$A(S) = \iint_D \|(r_u \times r_v)\| \, dr \, d\theta = \sqrt{2}\int_0^{2\pi}\int_0^1 r \, dr \, d\theta$$

$$= \sqrt{2}\int_0^{2\pi} \left(\frac{r^2}{2}\right)\Big|_0^1 = \frac{\sqrt{2}}{2}(2\pi) = \pi\sqrt{2}$$

CYU 14.27 $\|r_u \times r_v\| = \left\|\det \begin{bmatrix} i & j & k \\ 1 & v & 0 \\ 0 & u & 0 \end{bmatrix}\right\| = \|uk\| = |u| = u$ (since $0 \le u \le 1$)

$$\iint_S g \, dS = \iint_D g[r(u,v)]\|r_u \times r_v\| \, du \, dv = \iint g(u, uv, -1)u \, du \, dv$$

$$= \int_0^1\int_0^u (-u^4v^2)\,dv\,du = -\int_0^1\left(u^4\frac{v^3}{3}\right)\Big|_0^u du$$

$$= -\frac{1}{3}\int_0^1 u^7 du = -\frac{1}{24}$$

CYU 14.28 The positively oriented curve C can be represented by:

$$x = \cos t, \ y = \sin t, \ z = 1, \quad 0 \le t \le 2\pi$$
$$r(t) = \cos t i + \sin t j + k, \quad 0 \le t \le 2\pi$$

Then: $r'(t) = -\sin t i + \cos t j$

$$F(x,y,z) = y^2 i + x j + z^2 k = \sin^2 t i + \cos t j + k$$

And: $\oint_C F \cdot dr = \int_0^{2\pi} (\sin^2 t i + \cos t j + k) \cdot (-\sin t i + \cos t j)\,dt$

$$= \int_0^{2\pi} (-\sin^3 t + \cos^2 t)\,dt = \int_0^{2\pi} -\sin^3 t\,dt + \int_0^{2\pi} \frac{1 + \cos 2t}{2}\,dt$$

Example 7.10(a), page 280: $= \left(\cos t - \frac{\cos^3 t}{3}\right)\Big|_0^{2\pi} + \left(\frac{t}{2} + \frac{\sin 2t}{4}\right)\Big|_0^{2\pi} = \pi$

For $\displaystyle\iint_S \mathrm{curl}(F)\cdot n\,dS$: $\mathrm{curl}(F) = \det\begin{bmatrix} i & j & k \\ \dfrac{\partial}{\partial x} & \dfrac{\partial}{\partial y} & \dfrac{\partial}{\partial z} \\ y^2 & x & z^2 \end{bmatrix} = (1-2y)k$. So:

$$\iint_S \mathrm{curl}(F)\cdot n\,dS = \iint_D (1-2y)k\cdot(-f_x i - f_y j + k)\,dA = \iint_D (1-2y)k\cdot(-2xi - 2yj + k)\,dA$$

$$= \int_{-1}^{1}\int_{-\sqrt{1-x^2}}^{\sqrt{1-x^2}} (1-2y)\,dA$$

using polar coordinates: $\displaystyle = \int_0^{2\pi}\int_0^1 (1 - 2\sin\theta)r\,dr\,d\theta$

$$= \int_0^{2\pi}\int_0^1 \left(\frac{r^2}{2} - r^2\sin\theta\right)\Bigg|_0^1 d\theta$$

$$= \int_0^{2\pi}\left(\frac{1}{2} - \sin\theta\right)d\theta$$

$$= \left(\frac{\theta}{2} + \cos\theta\right)\Bigg|_0^{2\pi} = \pi$$

CYU 14.29 If $\mathrm{curl}(F) = 0$, then $\displaystyle\oint_C F\cdot dr = \iint_S \mathrm{curl}(F)\cdot n\,dS = 0$ for every closed path in S. It then follows, from Theorem 14.4 (page 636), that F is path independent.

CYU 14.30 Since: $\mathrm{div}(F) = (x)_x + (y)_y + (z)_z = 1 + 1 + 1 = 3$:

$$\mathrm{Flux}(F_S) = \iint_S F\cdot n\,dS = \iiint_E \mathrm{div}\,F\,dV = 3\iiint_E dV = 3\left(\frac{4}{3}\pi r^3\right) = 4(\pi r^3)$$

$\uparrow$
volume of the sphere E of radius r
(Exercise 37, page 206)

CYU 14.31 Since: $\mathrm{div}(F) = (x\sin^2 y)_x + (e^{x^2 z^3} - \tan x)_y + (z\cos^2 y)_z = \sin^2 y + \cos^2 y = 1$:

$$\mathrm{Flux}(F_S) = \iint_S F\cdot n\,dS = \iiint_E \mathrm{div}\,F\,dV = \iiint_E dV = \pi\cdot 3^2\cdot 7 = 63\pi$$

$\uparrow$
volume of the cylinder E
of radius 3 and height 7

CYU 14.32 Since: $\mathrm{div}(F) = (5xy^2)_x i + (5yz^2)_y j + (5x^2 z)_z k = 5(y^2 + z^2 + x^2)$:

$$\text{Flux}(\boldsymbol{F}_S) = \iint_S \boldsymbol{F} \cdot \boldsymbol{n}\, dS = \iiint_E \text{div } \boldsymbol{F}\, dV = \iiint_E 5(x^2 + y^2 + z^2)\, dV$$

Turning to spherical coordinates:
(see Figure 13.12, page 580)

$$= \int_0^{2\pi} \int_0^{\pi} \int_0^3 5\rho^2 \rho^2 \sin\phi\, d\rho\, d\phi\, d\theta$$

$$= \int_0^{2\pi} \int_0^{\pi} (\rho^5 \sin\phi)\Big|_0^3\, d\phi\, d\theta = 3^5 \int_0^{2\pi} \int_0^{\pi} \sin\phi\, d\phi\, d\theta$$

$$= 3^5 \int_0^{2\pi} (-\cos\phi)\Big|_0^{\pi}\, d\theta$$

$$= 3^5 \int_0^{2\pi} 2\, d\theta = 972\pi$$

APPENDIX B
ADDITIONAL THEORETICAL DEVELOPMENT

THEOREM 11.3, PAGE 431

THEOREM 12.6 (a) If f and g are real-valued continuous functions of one variable, then:

(i) $H(x, y) = f(x) + g(y)$ and (ii) $K(x, y) = f(x)g(y)$

are continuous functions of two variables.

(b) If g *is* a real-valued continuous function of two variables, and if f is a real-valued continuous function of one variable, then:

$$H(x, y) = f[g(x, y)]$$

is a continuous function of two variables.

PROOF: (a) Let (x_0, y_0) be in the domain of H.

$$\lim_{(x,y) \to (x_0, y_0)} H(x, y) = \lim_{(x,y) \to (x_0, y_0)} [f(x) + g(y)] = \lim_{(x,y) \to (x_0, y_0)} f(x) + \lim_{(x,y) \to (x_0, y_0)} g(y)$$

$$= \lim_{x \to x_0} f(x) + \lim_{y \to y_0} g(y)$$

$$= f(x_0) + g(y_0) = H(x_0, y_0)$$

and: $$\lim_{(x,y) \to (x_0, y_0)} K(x, y) = \lim_{(x,y) \to (x_0, y_0)} [f(x)g(y)] = \lim_{(x,y) \to (x_0, y_0)} f(x) \cdot \lim_{(x,y) \to (x_0, y_0)} g(x)$$

$$= \lim_{x \to x_0} f(x) \cdot \lim_{y \to y_0} g(y)$$

$$= f(x_0)g(y_0) = K(x_0, y_0)$$

(b) (Compare with the proof of Theorem 2.5, page 58). Let (x_0, y_0) be in the domain of $H = g \circ f$, and let $\varepsilon > 0$ be given. Since f is continuous at $g(x_0, y_0)$, we can find a $\bar{\delta}$ such that $|y - g(x_0, y_0)| < \bar{\delta} \Rightarrow |f(y) - f[g(x_0, y_0)]| < \varepsilon$. Now, think of $\bar{\delta}$ as being an "ε-challenge" for the function g. By the continuity of g we can find a δ such that: $\|(x, y) - (x_0, y_0)\| < \delta \Rightarrow |g(x, y) - g(x_0, y_0)| < \bar{\delta}$. Putting this together, we have:

$$\|(x, y) - (x_0, y_0)\| < \delta \Rightarrow |g(x, y) - g(x_0, y_0)| < \bar{\delta}$$

$$\Rightarrow |f(g(x, y)) - f[g(x_0, y_0)]| = |H(x, y) - H(x_0, y_0)| < \varepsilon$$

THEOREM 13.2, PAGE 554

THEOREM 13.2 If the partial derivatives f_x and f_y exist and are **continuous** in an open region about (x_0, y_0), then f is differentiable at (x_0, y_0).

PROOF: We show that

$$\Delta z = f(x_0 + \Delta x, y_0 + \Delta y) - f(x_0, y_0) \quad (*)$$

can be expressed in the form

$$\Delta z = f_x(x_0, y_0)\Delta x + f_y(x_0, y_0)\Delta y + \varepsilon_1 \Delta x + \varepsilon_2 \Delta y \text{ where } \varepsilon_1, \varepsilon_2 \to 0 \text{ as } \Delta x, \Delta y \to 0$$

Adjoining a clever zero to Δz in (*) we obtain:

$$\Delta z = [f(x_0 + \Delta x, y_0 + \Delta y) - f(x_0, y_0 + \Delta y)] + [f(x_0, y_0 + \Delta y) - f(x_0, y_0)]$$

Applying the Mean Value Theorem to the single-variable function:

$$g(x) = f(x, y_0 + \Delta y) \quad (**)$$

on the interval $[x_0, x_0 + \Delta x]$, we have:

$$g'(c) = \frac{g(x_0 + \Delta x) - g(x_0)}{\Delta x} \quad \text{or: } g'(c)\Delta x = g(x_0 + \Delta x) - g(x_0)$$

for some c between x_0 and $x_0 + \Delta x$.

Noting that $g'(c) = f_x(c, y_0 + \Delta y)$ we can rewrite $g'(c)\Delta x = g(x_0 + \Delta x) - g(x_0)$ in the form:

$$f_x(c, y_0 + \Delta y)\Delta x = f(x_0 + \Delta x, y_0 + \Delta y) - f(x_0, y_0 + \Delta y)$$

$$f_x(c, y_0 + \Delta y)\Delta x + [f(x_0, y_0 + \Delta y) - f(x_0, y_0)]$$

We then have: $= f(x_0 + \Delta x, y_0 + \Delta y) - f(x_0, y_0 + \Delta y) + [f(x_0, y_0 + \Delta y) - f(x_0, y_0)]$

$$= f(x_0 + \Delta x, y_0 + \Delta y) - f(x_0, y_0) = \Delta z \text{ in } (*)$$

In a similar fashion, by applying the Mean Value Theorem to the single-valued function $h(y) = f(x_0, y)$ on the interval $[y_0, y_0 + \Delta x]$ we can arrive at:

$$\Delta z = f_x(c, y_0 + \Delta y)\Delta x + f_y(x_0, d)\Delta y, \text{ for some } y_0 < d < y_0 + \Delta y$$

Adding two more cleaver zeros we have:

$$\Delta z = f_x(x_0, y_0)\Delta x + [f_x(c, y_0 + \Delta y) - f_x(x_0, y_0)]\Delta x + f_y(x_0, y_0)\Delta y + [f_y(x_0, d) - f_y(x_0, y_0)]\Delta y$$

Or: $\Delta z = f_x(x_0, y_0)\Delta x + f_y(x_0, y_0)\Delta y + \varepsilon_1 \Delta x + \varepsilon_2 \Delta y$, where:

$$\varepsilon_1 = f_x(c, y_0 + \Delta y) - f_x(x_0, y_0) \text{ and } \varepsilon_2 = f_y(x_0, d) - f_y(x_0, y_0)$$

Finally, the continuity of f_x and f_y assures us that $\varepsilon_1, \varepsilon_2 \to 0$ as $\Delta x, \Delta y \to 0$.

THEOREM 13.5, PAGE 561

THEOREM 13.5 If $z = f(x, y)$ is differentiable, then for any unit vector $u = (a, b) = ai + bj$:
$$D_u f(x, y) = af_x(x, y) + bf_y(x, y)$$

PROOF: For any point (x_0, y_0) in the domain of f we define the single variable function:
$$g(h) = f(x_0 + ah, y_0 + bh)$$

We then have:
$$g'(0) = \lim_{h \to 0} \frac{g(h) - g(0)}{h} = \lim_{h \to 0} \frac{f(x_0 + ah, y_0 + bh) - f(x_0, y_0)}{h} = D_u f(x_0, y_0) \ (*)$$

On the other hand, we can write $g(h) = f(x, y)$ where $x = x_0 + ah$ and $y = y_0 + bh$, and be applying Theorem 12.8(a), page 473, with h replacing t, we have:
$$g'(h) = \frac{\partial f}{\partial x} \frac{dx}{dh} + \frac{\partial f}{\partial y} \frac{dy}{dh} = a\frac{\partial f}{\partial x} + b\frac{\partial f}{\partial y}$$

Substitution 0 for h, we then have $x = x_0$, $y = y_0$, and:
$$g'(0) = af_x(x_0, y_0) + bf_y(x_0, y_0)$$

Returning to (*) we have:
$$D_u f(x_0, y_0) = af_x(x_0, y_0) + bf_y(x_0, y_0)$$

THEOREM 14.18, PAGE 644

THEOREM 14.18

STOKE'S THEOREM

If S is an oriented surface that is bounded by a simple closed curve C with positive orientation (counterclockwise), and if $F = Pi + Qj + Rk$ has continuous first-order partial derivatives on some open region in three-space containing S, then:
$$\oint_C F \cdot dr = \iint_S \text{curl}(F) \cdot n \, dS$$

PROOF (When the surface S is the graph of a function):
We assume that S is the graph of a function f over a simple closed region whose boundary C_1 corresponds to the boundary C of S D. We also assume that f has continuous second-order partial derivative on D. Appealing to Theorem 14.17, page 637, with F replaced by curl(F) we have:

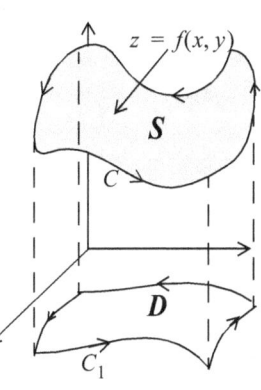

$$\iint_S \text{curl}(F) \cdot n \, dS = \iint_D -\left(\frac{\partial R}{\partial y} - \frac{\partial Q}{\partial z}\right)\frac{\partial z}{\partial x} - \left(\frac{\partial P}{\partial z} - \frac{\partial R}{\partial x}\right)\frac{\partial z}{\partial y} + \left(\frac{\partial Q}{\partial x} - \frac{\partial P}{\partial y}\right) dA$$

$$(*)$$

If $x = x(t)$, $y = y(t)$, $a \le t \le b$, is a parametrization of C_1, then

$$x = x(t), \, y = y(t), \, z = f(x(t), y(t)), \, a \le t \le b$$

is a parametrization of C. We then have:

$$\oint_C F \cdot dr = \int_a^b \left(P\frac{dx}{dt} + Q\frac{dy}{dt} + R\frac{dz}{dt} \right) dt$$

Chain Rule, page 555: $= \int_a^b \left[P\frac{dx}{dt} + Q\frac{dy}{dt} + R\left(\frac{\partial z}{\partial x}\frac{dx}{dt} + \frac{\partial z}{\partial y}\frac{dy}{dt}\right) \right] dt$

$$= \int_a^b \left[\left(P + R\frac{\partial z}{\partial x}\right)\frac{dx}{dt} + \left(Q + R\frac{\partial z}{\partial y}\right)\frac{dy}{dt} \right] dt$$

$$= \int_C \left(P + R\frac{\partial z}{\partial x}\right) dx + \left(Q + R\frac{\partial z}{\partial y}\right) dy$$

Green's Theorem page 617: $= \iint_D \frac{\partial}{\partial x}\left[Q + R\frac{\partial z}{\partial y}\right] - \frac{\partial}{\partial y}\left[P + R\frac{\partial z}{\partial x}\right] dA$

Chain Rule, page 555: $= \iint_D \left[\left(\frac{\partial Q}{\partial x} + \frac{\partial Q}{\partial z}\frac{\partial z}{\partial x} + \frac{\partial R}{\partial x}\frac{\partial z}{\partial y} + \frac{\partial R}{\partial z}\frac{\partial z}{\partial x}\frac{\partial z}{\partial y} + R\frac{\partial^2 z}{\partial x \delta y}\right) - \left(\frac{\partial P}{\partial y} + \frac{\partial P}{\partial z}\frac{\partial z}{\partial y} + \frac{\partial R}{\partial y}\frac{\partial z}{\partial x} + \frac{\partial R}{\partial z}\frac{\partial z}{\partial y}\frac{\partial z}{\partial x} + R\frac{\partial^2 z}{\partial y \delta x}\right) \right] dA$

Four of the terms in the above double integral cancel, and the remaining six terms can be rearranged to coincide with the double integral (*) at the bottom of the previous page.

THEOREM 14.19, PAGE 654

THEOREM 14.19

THE DIVERGENCE THEOREM

Let the surface S be oriented in the outward direction and let E be the solid region enclosed by S. If $F = Pi + Qj + Rk$ is continuously differentiable throughout E, then:

$$\iint_S F \cdot n \, dS = \iiint_E \text{div}(F) \, dV$$

PROOF: Restricted to the case where E lies between the graphs of two continuous functions of x and y, two continuous functions of y and z, and two continuous functions of x and z:

$$E = \{(x, y, z) | (x, y) \in D_z, z_1(x, y) \le z \le z_2(x, y)\}$$

$$= \{(x, y, z) | (y, z) \in D_x, x_1(y, z) \le x \le x_2(y, z)\}$$

$$= \{(x, y, z) | (x, z) \in D_y, y_1(x, z) \le y \le y_2(x, z)\}$$

Let $F = Pi + Qj + Rk$. Then $\text{div}(F) = P_x + Q_y + R_z$, and:

$$\iiint_E \text{div}(F) \, dV = \iiint_E P_x \, dV + \iiint_E Q_y \, dV + \iiint_E R_z \, dV$$

Moreover:

$$\iint\limits_{S} \boldsymbol{F} \cdot \boldsymbol{n}\, dS = \iint\limits_{S} (P\boldsymbol{i} + Q\boldsymbol{j} + R\boldsymbol{k}) \cdot \boldsymbol{n}\, dS = \iint\limits_{S} P\boldsymbol{i} \cdot \boldsymbol{n}\, dS + \iint\limits_{S} Q\boldsymbol{j} \cdot \boldsymbol{n}\, dS + \iint\limits_{S} R\boldsymbol{k} \cdot \boldsymbol{n}\, dS$$

To prove the theorem it suffices to establish the following three equations:

(1) $\displaystyle\iint\limits_{S} P\boldsymbol{i} \cdot \boldsymbol{n}\, dS = \iiint\limits_{E} P_x\, dV$ 　　(2) $\displaystyle\iint\limits_{S} Q\boldsymbol{j} \cdot \boldsymbol{n}\, dS = \iiint\limits_{E} Q_y\, dV$ 　　(3) $\displaystyle\iint\limits_{S} R\boldsymbol{k} \cdot \boldsymbol{n}\, dS = \iiint\limits_{E} R_z\, dV$

For (3), we take advantage or the fact that

$$E = \{(x, y, z) | (x, y) \in D_z, z_1(x, y) \le z \le z_2(x, y)\}$$

We then have:

$$\iiint\limits_{E} R_z\, dV = \iint\limits_{D_z} \left[\int_{z_1(x,y)}^{z_2(x,y)} R_z(x, y, z)\, dz \right] dA$$

$$= \iint\limits_{D_z} [R(x, y, z_2(x, y)) - R(x, y, z_1(x, y))]\, dA$$

(*)

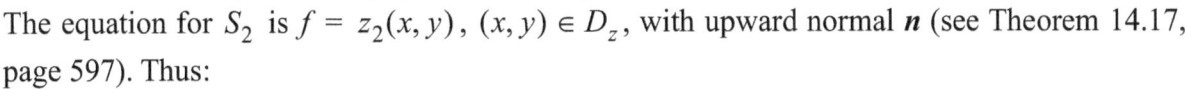

Moreover: $\displaystyle\iint\limits_{S} R\boldsymbol{k} \cdot \boldsymbol{n}\, dS = \iint\limits_{S_1} R\boldsymbol{k} \cdot \boldsymbol{n}\, dS + \iint\limits_{S_2} R\boldsymbol{k} \cdot \boldsymbol{n}\, dS$

(Note that $\boldsymbol{n}$ is perpendicular to $\boldsymbol{i}$ and $\boldsymbol{j}$)

The equation for S_2 is $f = z_2(x, y)$, $(x, y) \in D_z$, with upward normal $\boldsymbol{n}$ (see Theorem 14.17, page 597). Thus:

$$\iint\limits_{S_2} R\boldsymbol{k} \cdot \boldsymbol{n}\, dS = \iint\limits_{D_z} R\boldsymbol{k} \cdot (-f_x\boldsymbol{i} - f_y\boldsymbol{j} + \boldsymbol{k})\, dA = \iint\limits_{D_z} R(x, y, z_2(x, y))\, dA$$

On S_1 we have $g = z_1(x, y)$ with downward normal (see Theorem 14.17, page 597). Thus:

$$\iint\limits_{S_1} R\boldsymbol{k} \cdot \boldsymbol{n}\, dS = \iint\limits_{D_z} R\boldsymbol{k} \cdot (f_x\boldsymbol{i} + f_y\boldsymbol{j} - \boldsymbol{k})\, dA = \iint\limits_{D_z} -R(x, y, z_1(x, y))\, dA$$

Returning to (*) we have

$$\iiint\limits_{E} R_z\, dV = \iint\limits_{S_1} R\boldsymbol{k} \cdot \boldsymbol{n}\, dS + \iint\limits_{S_2} R\boldsymbol{k} \cdot \boldsymbol{n}\, dS = \iint\limits_{S} R\boldsymbol{k} \cdot \boldsymbol{n}\, dS$$

The same argument can be used to establish equations (1) and (2).

APPENDIX C
ANSWERS TO ODD EXERCISES
CHAPTER 11
FUNCTIONS OF SEVERAL VARIABLES

11.1 Limits and Continuity (page 427)

1. R^2 **3.** $\left\{(x, y)\,|\,xy \text{ is not an odd multiple of } \dfrac{\pi}{2}\right\}$ **5.** $\{(x, y, z)\,|\,x + y + z \neq 0\}$

7. For given $\varepsilon > 0$ we are to find $\delta > 0$ such that: $0 < \|(x, y) - (1, 1)\| < \delta \Rightarrow |(2x + y) - 3| < \varepsilon\,(*)$
Noting that $|(2x + y) - 3| = |2(x - 1) + (y - 1)| \le 2|x - 1| + |y - 1|$ we can conclude that (*) will be satisfied for any $\delta > 0$ for which $0 < \|(x, y) - (1, 1)\| < \delta \Rightarrow 2|x - 1| < \dfrac{\varepsilon}{2}$ and $|y - 1| < \dfrac{\varepsilon}{2}$, that is $|x - 1| < \dfrac{\varepsilon}{4}$ and $|y - 1| < \dfrac{\varepsilon}{2}$. It follows that $\delta = \dfrac{\varepsilon}{4}$ fits the bill.

9. For given $\varepsilon > 0$ we are to find $\delta > 0$ such that: $0 < \|(x, y) - (2, 1)\| < \delta \Rightarrow \left|\dfrac{x^2 - 4xy + 4y^2}{x - 2y}\right| < \varepsilon$ (*)

Noting that $\left|\dfrac{x^2 - 4xy + 4y^2}{x - 2y}\right| = \left|\dfrac{(x - 2y)^2}{x - 2y}\right| = |x - 2y| = |(x - 2) - 2(y - 1)| \le |x - 2| + 2|y - 1|$
we see that (*) will be satisfied for any $\delta > 0$ for which:
$$0 < \|(x, y) - (2, 1)\| < \delta \Rightarrow |x - 2| < \frac{\varepsilon}{2} \text{ and } |y - 1| < \frac{\varepsilon}{4}, \text{ and } \delta = \frac{\varepsilon}{4} \text{ fits that bill.}$$

11. $\dfrac{1}{\pi}$ **13.** 8   **15.** 1   **17.** 6   **19.** 1   **21.** $\{(x, \pm x)\,|\,x \in R\}$

23. $\{(0, y)\,|\,y \in R\}$ **25.** $\{(x, x, z)\,|\,x, z \in R\}$ **27.** $\varnothing$ **29.** No unique answer.

31. No unique answer.

33. Let $\varepsilon > 0$ be given. Then:
$$0 < \|(x, y) - (x_0, y_0)\| < \delta \Rightarrow |(f - g)(x, y) - (L - M)| < \varepsilon \Leftrightarrow |f(x, y) - g(x, y) - L + M| < \varepsilon$$
Since: $|f(x, y) - g(x, y) - L + M| = |[f(x, y) - L] - [g(x, y) - M]| \le |f(x, y) - L| + |g(x, y) - M|$
we choose $\delta_1 > 0$ and $\delta_2 > 0$ such that:
$$0 < \|(x, y) - (x_0, y_0)\| < \delta_1 \Rightarrow |f(x, y) - L| < \frac{\varepsilon}{2} \text{ and } 0 < \|(x, y) - (x_0, y_0)\| < \delta_2 \Rightarrow |g(x, y) - M| < \frac{\varepsilon}{2}.$$
It follows that for δ the smaller of δ_1 and δ_2:
$$0 < \|(x, y) - (x_0, y_0)\| < \delta \Rightarrow |(f - g)(x, y) - (L + M)| < \varepsilon$$

35.
$$\lim_{(x, y) \to (x_0, y_0)} (f + g)(x, y) = \lim_{(x, y) \to (x_0, y_0)} [f(x, y) + g(x, y)]$$
$$= \left[\lim_{(x, y) \to (x_0, y_0)} f(x, y)\right] + \left[\lim_{(x, y) \to (x_0, y_0)} g(x, y)\right]$$
$$= f(x_0, y_0) + g(x_0, y_0) = (f + g)(x_0, y_0)$$

37. $\displaystyle\lim_{(x,y)\to(x_0,y_0)}\left(\frac{f}{g}\right)(x,y) = \lim_{(x,y)\to(x_0,y_0)}\left[\frac{f(x,y)}{g(x,y)}\right] = \frac{\displaystyle\lim_{(x,y)\to(x_0,y_0)}f(x,y)}{\displaystyle\lim_{(x,y)\to(x_0,y_0)}g(x,y)}$

$$= \frac{f(x_0,y_0)}{g(x_0,y_0)} = \left(\frac{f}{g}\right)(x_0,y_0)$$

39. Consider the following continuous functions of **one** variable: $g(x) = x$ and $h(y) = 1$. By Theorem 11.4, the function $f(x,y) = g(x)h(y) = x\cdot 1 = x$ is a continuous function of two variables. Thus: $\displaystyle\lim_{(x,y)\to(x_0,y_0)}f(x,y) = f(x_0,y_0) = x_0$.

41. For given $\varepsilon > 0$, any δ will "work," and, for the simple reason that:
$$0 < \|(x,y)-(x_0,y_0)\| < \delta \Rightarrow |c-c| = 0 < \varepsilon$$

11.2 Graphing Functions of Two Variables (page 443)

1.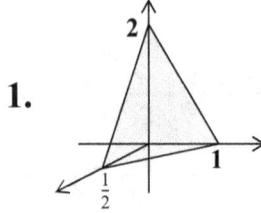

3. $4x + 5y - 2z = 0$

5.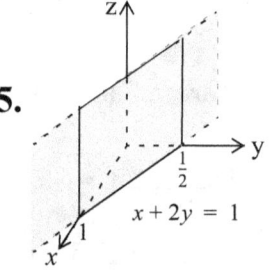

$x + 2y = 1$

7.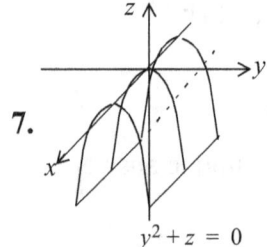

$y^2 + z = 0$

9. $x^2 - 2y^2 = 1$

11.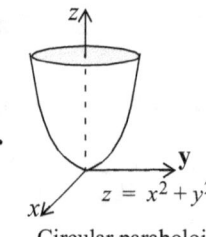

$z = x^2 + y^2$

Circular paraboloid

13.

Ellipsoid

$36x^2 + 9y^2 + 4z^2 = 36$

15.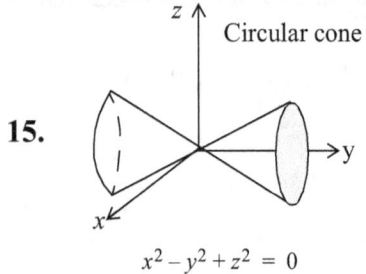

Circular cone

$x^2 - y^2 + z^2 = 0$

17. Hyperboloid of two sheets

$25x^2 - 4y^2 + 25z^2 + 100 = 0$

19.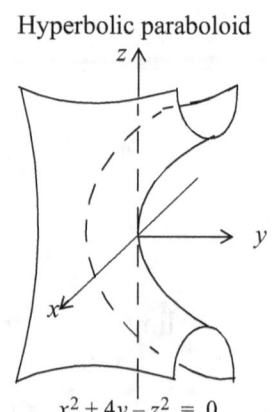

Hyperbolic paraboloid

$x^2 + 4y - z^2 = 0$

Hyperboloid or one sheet

21.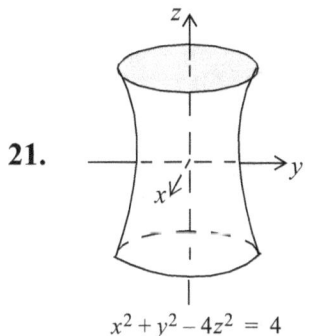

$x^2 + y^2 - 4z^2 = 4$

Hyperboloid of one sheet

23.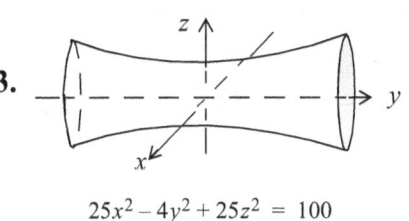

$25x^2 - 4y^2 + 25z^2 = 100$

Hyperbolic paraboloid

25.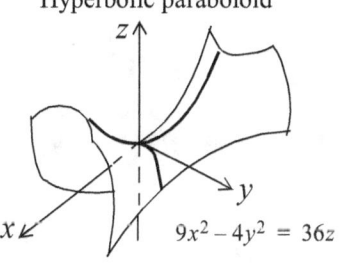

$9x^2 - 4y^2 = 36z$

11.3 Double Intgegrals (page 454)

1. (a) $\dfrac{7}{6}$ (b) $\dfrac{7}{6}$ **3.** (a) 2 (b) 2 **5.** (a) π (b) π **7.** (a) $\dfrac{96}{5}$ (b) $\dfrac{96}{5}$ **9.** $\dfrac{1}{6}$

11. $\dfrac{31}{8}$ **13.** $\dfrac{56}{15}$ **15.** $\dfrac{4}{5}$ **17.** $-\dfrac{23}{40}$ **19.** 8 **21.** $\dfrac{1}{4}(e^4 - 1)$ **23.** $\dfrac{13}{20}$

25. $\dfrac{\pi}{2}$ **27.** $\displaystyle\int_0^1\int_{-3}^2 h(x, y)\,dx\,dy$ **29.** $\displaystyle\int_0^1\int_{\sqrt{y}}^{y^{1/4}} h(x, y)\,dx\,dy$ **31.** $\displaystyle\int_{\frac{1}{e}}^{e}\int_{\ln y}^1 h(x, y)\,dx\,dy$

33. $\displaystyle\int_0^2\int_{\frac{x^2}{6}}^{\frac{x}{3}} h(x, y)\,dy\,dx$ **35.** $\displaystyle\int_{-3}^3\int_0^{\sqrt{9-x^2}} h(x, y)\,dy\,dx$

37. $\displaystyle\int_0^1\int_{-x+1}^1 h(x, y)\,dy\,dx + \int_1^2\int_{x-1}^1 h(x, y)\,dy\,dx$ **39.** $e^4 - 1$ **41.** $\dfrac{1}{2}(1 - \cos 1)$

43. $\displaystyle\int_0^1\int_y^3 h(x, y)\,dx\,dy$ **45.** $\displaystyle\int_1^4\int_x^{2x} h(x, y)\,dy\,dx$ **47.** $\dfrac{1}{16}$ **49.** 7 **51.** $\dfrac{128}{27}$ **53.** 1

55. $\left(\dfrac{12}{7}, \dfrac{23}{21}\right)$ **57.** $\left(\dfrac{3}{4}, \dfrac{3}{2}\right)$ **59.** $\left(\dfrac{190}{273}, \dfrac{6}{13}\right)$ **61.** $\left(\dfrac{e^2 + 1}{2(e^2 - 1)}, \dfrac{4(e^2 + e + 1)}{9(e + 1)}\right)$ **63.** $\left(\dfrac{\pi}{2}, \dfrac{16}{9\pi}\right)$

11.4 Double Integrals in Polar Coordinates (page 461)

1. 1 **3.** $\dfrac{\pi}{8}$ **5.** $\dfrac{(e - 1)\pi}{2}$ **7.** $\dfrac{1}{6}$ **9.** $\dfrac{11\sqrt{3} - 2\pi}{48}$ **11.** $\dfrac{2\pi}{5}$

13. 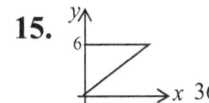 $\dfrac{16}{9}$

15. 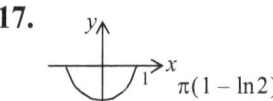 36

17. 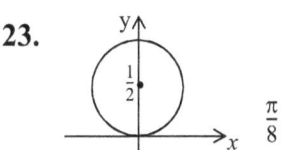 $\pi(1 - \ln 2)$

19. 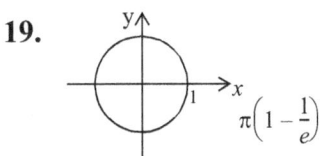 $\pi\left(1 - \dfrac{1}{e}\right)$

21. 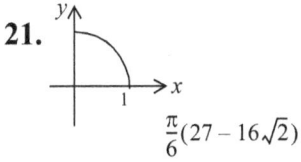 $\dfrac{\pi}{6}(27 - 16\sqrt{2})$

23. 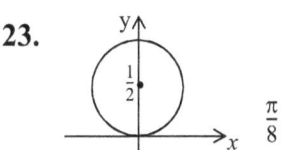 $\dfrac{\pi}{8}$

25. $\dfrac{3\pi}{4} - \dfrac{9\sqrt{3}}{16}$ **27.** $\dfrac{\pi}{2} - 1$ **29.** $\dfrac{3}{8}(2\pi + 3\sqrt{3})$ **31.** $2 - \dfrac{\pi}{4}$ **33.** $\dfrac{\pi}{6} + \dfrac{\sqrt{3}}{4}$

35. $\dfrac{4}{3}\pi a^3$ **37.** $\dfrac{16\pi}{3}$ **39.** $\dfrac{3\pi}{2}$ **41.** $\dfrac{2\pi}{3}(8 - 3\sqrt{3})$ **43.** $\dfrac{64\pi}{3}(8 - 3\sqrt{3})$ **45.** $\dfrac{5\pi}{3}$

47. $2(-\pi + 3\sqrt{3})$ **49.** $2a\left[\dfrac{2\pi}{3} + \sqrt{3} - 2\right]$ **51.** $\left(0, \dfrac{21}{20}\right)$ **53.** $\left(0, \dfrac{9\sqrt{3}}{6\sqrt{3} - 2\pi}\right)$ **55.** $\left(\dfrac{129}{200}, 0\right)$

11.5 Triple Integrals (page 469)

1. $-\dfrac{43}{420}$ **3.** $\dfrac{1}{24}$ **5.** 0 **7.** -30 **9.** $\dfrac{1}{2160}$ **11.** $\dfrac{1}{3}$ **13.** $\dfrac{\pi^3}{2}(1 - \cos 1)$

15. $\dfrac{1}{200}$ **17.** $\dfrac{1}{8}(e^4 - 13)$ **19.** 2 **21.** 36π **23.** $\dfrac{1}{3}$ **25.** $\dfrac{3\pi}{\sqrt{2}}$ **27.** a^5

29. $\dfrac{1}{4}$ **31.** $\dfrac{4k}{5}$ **33.** $\left(0, 0, \dfrac{4}{3}\right)$ **35.** $\left(\dfrac{5}{7}, 0, \dfrac{5}{14}\right)$ **37.** $\left(\dfrac{7a}{12}, \dfrac{7a}{12}, \dfrac{7a}{12}\right)$ **39.** $\left(\dfrac{31}{75}, \dfrac{31}{25}, \dfrac{124}{75}\right)$

41. $\displaystyle\int_0^4 \int_{-\sqrt{y}}^{\sqrt{y}} \int_0^{2-\frac{1}{2}y} k(x,y,z)\,dz\,dx\,dy,\ \int_{-2}^2 \int_{x^2}^4 \int_0^{2-\frac{1}{2}y} k(x,y,z)\,dz\,dy\,dx,\ \int_0^2 \int_0^{-2z+4} \int_{-\sqrt{y}}^{\sqrt{y}} k(x,y,z)\,dx\,dy\,dz$

$\displaystyle\int_0^4 \int_0^{2-\frac{1}{2}y} \int_{-\sqrt{y}}^{\sqrt{y}} k(x,y,z)\,dx\,dz\,dy,\ \int_0^2 \int_{-\sqrt{4-2z}}^{\sqrt{4-2z}} \int_{x^2}^{-2z+4} k(x,y,z)\,dy\,dx\,dz,$

$\displaystyle\int_{-2}^2 \int_0^{2-\frac{1}{2}x^2} \int_{x^2}^{-2z+4} k(x,y,z)\,dy\,dz\,dx$

43. $\displaystyle\int_0^4 \int_{-\sqrt{4-y}}^{\sqrt{4-y}} \int_{-\frac{1}{2}\sqrt{4-x^2-y}}^{\frac{1}{2}\sqrt{4-x^2-y}} k(x,y,z)\,dz\,dx\,dy,\ \int_{-2}^2 \int_0^{4-x^2} \int_{-\frac{1}{2}\sqrt{4-x^2-y}}^{\frac{1}{2}\sqrt{4-x^2-y}} k(x,y,z)\,dz\,dy\,dx$

$\displaystyle\int_{-1}^1 \int_0^{4-4z^2} \int_{-\sqrt{4-4z^2-y}}^{\sqrt{4-4z^2-y}} k(x,y,z)\,dx\,dy\,dz,\ \int_0^4 \int_{-\frac{1}{2}\sqrt{4-y}}^{\frac{1}{2}\sqrt{4-y}} \int_{-\sqrt{4-4z^2-y}}^{\sqrt{4-4z^2-y}} k(x,y,z)\,dx\,dz\,dy$

$\displaystyle\int_{-1}^1 \int_{-\sqrt{4-4z^2}}^{\sqrt{4-4z^2}} \int_0^{4-x^2-4z^2} k(x,y,z)\,dy\,dx\,dz,\ \int_{-2}^2 \int_{-\frac{1}{2}\sqrt{4-x^2}}^{\frac{1}{2}\sqrt{4-x^2}} \int_0^{4-x^2-4z^2} k(x,y,z)\,dy\,dz\,dx$

45. $\displaystyle\int_0^1 \int_0^x \int_0^y k(x,y,z)\,dz\,dy\,dx,\ \int_0^1 \int_z^1 \int_y^1 k(x,y,z)\,dx\,dy\,dz,\ \int_0^1 \int_0^y \int_y^1 k(x,y,z)\,dx\,dz\,dy$

$\displaystyle\int_0^1 \int_z^1 \int_z^x k(x,y,z)\,dy\,dx\,dz,\ \int_0^1 \int_0^x \int_z^x k(x,y,z)\,dy\,dz\,dx$

47. $\displaystyle\int_{-1}^0 \int_0^1 \int_0^{y^2} k(x,y,z)\,dz\,dx\,dy,\ \int_0^1 \int_{-1}^{-\sqrt{z}} \int_0^1 k(x,y,z)\,dx\,dy\,dz,\ \int_{-1}^0 \int_0^{y^2} \int_0^1 k(x,y,z)\,dx\,dz\,dy$

$\displaystyle\int_0^1 \int_0^1 \int_{-1}^{-\sqrt{z}} k(x,y,z)\,dy\,dx\,dz,\ \int_0^1 \int_0^1 \int_{-1}^{-\sqrt{z}} k(x,y,z)\,dy\,dz\,dx$

11.6 Cylindrical and Coordinates (page 479)

1. $(-\sqrt{3}, 1, 2)$ **3.** $(3\sqrt{2}, 3\sqrt{2}, 1)$ **5.** $\left(8, \frac{4\pi}{3}, -8\right)$ **7.** $\left(2, \frac{\pi}{2}, -2\right)$ **9.** $z = r^2$

11. $r = 2\sin\theta$ **13.** $\frac{2\pi}{3}$ **15.** $\frac{\pi}{3}$ **17.** $\frac{\pi}{6}(8 - 3\sqrt{3})$ **19.** $\frac{81\pi}{2}$ **21.** $\frac{16\pi}{3}$

23. 20π **25.** $\frac{16}{9}(3\pi - 4)$ **27.** $16\pi k$ **29.** (a) $\frac{\pi k}{6}$ (b) $\frac{\pi k}{4}$ **31.** $\left(0, 0, \frac{2a}{3}\right)$

33. $\left(0, 0, \frac{3}{8}\right)$ **35.** $(0, 0, 15)$ **37.** $(\sqrt{6}, \sqrt{6}, 2)$ **39.** $(-7, 0, 0)$ **41.** $\left(4, \frac{\pi}{6}, \frac{\pi}{3}\right)$

43. $\left(\sqrt{2}, \frac{3\pi}{4}, \frac{3\pi}{2}\right)$ **45.** $\phi = \frac{\pi}{4}$ or $\phi = \frac{3\pi}{4}$ **47.** $\rho^2(\sin^2\phi\cos^2\phi + \cos^2\phi) = 9$ **49.** π

51. $\frac{\pi^2}{64}$ **53.** $\frac{\pi a^5}{10}$ **55.** $\frac{1}{15}(4\sqrt{2} - 5)$ **57.** $\frac{4}{3}\pi a^3$ **59.** πa^3 **61.** $9\pi(2 - \sqrt{2})$

63. $\pi k a^4$ **65.** $\frac{\pi}{4}(2 - \sqrt{2})$ **67.** $4\pi - \pi^2$ **69.** $\left(0, 0, \frac{3}{4}h\right)$ **71.** $\left(0, 0, \frac{18 + 9\sqrt{2}}{16}\right)$

CHAPTER 12
VECTORS AND VECTOR-VALUED FUNCTIONS

12.1 Vectors in the Plane and Beyond (page 497)

1. 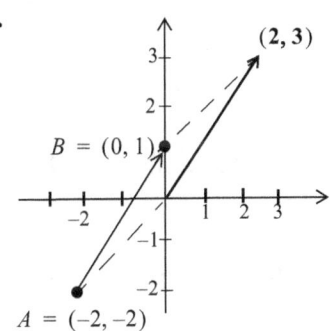 **3.** 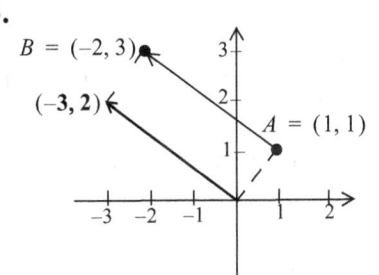 **5.**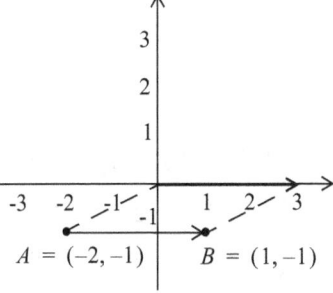

7. $(2, 0, -2)$ **9.** $(-9, -1, 11)$ **11.** $(13, -13)$ **13.** $(-3, -1, -1)$

15. $4i - 28j$ **17.** $-7i + j - 6k$ **19.** $\left(\frac{5}{\sqrt{29}}, \frac{2}{\sqrt{29}}\right)$, $v = \sqrt{29}\left(\frac{5}{\sqrt{29}}, \frac{2}{\sqrt{29}}\right)$

21. $\frac{2}{\sqrt{21}}i - \frac{4}{\sqrt{21}}j + \frac{1}{\sqrt{21}}k$, $v = \sqrt{21}\left(\frac{2}{\sqrt{21}}i - \frac{4}{\sqrt{21}}j + \frac{1}{\sqrt{21}}k\right)$

23. $\frac{3\sqrt{2}}{\sqrt{19}}i - \frac{1}{\sqrt{19}}j$, $v = \frac{\sqrt{19}}{3}\left(\frac{3\sqrt{2}}{\sqrt{19}}i - \frac{1}{\sqrt{19}}j\right)$ **25.** (a) $r = -14, s = 10$ (b) $r = -\frac{7}{5}, s = \frac{1}{10}$

(c) $r = \frac{5}{7}, s = \frac{1}{14}$ **27.** $r = -\frac{19}{6}, s = \frac{17}{6}, t = \frac{11}{6}$ **29.** $\left(\frac{5}{\sqrt{2}}, -\frac{5}{\sqrt{2}}\right)$

31. $F_1 = (5\sqrt{3} - 5)(-\sqrt{3}i + j)$ $F_2 = (15 - 5\sqrt{3})(i + j)$

33. $F_1 = 5(-\sqrt{3}i + j)$, $F_2 = (5\sqrt{3} - 5)(i + j)$

35. $F_1 = \dfrac{20 - 5\sqrt{2}}{4}(-2i - j)$, $F_2 = \dfrac{15\sqrt{2} - 20}{4}(-j)$

37. $\dfrac{15\sqrt{3}}{2}$ and $\dfrac{15}{2}$ lb   **39.** ≈ 460.06 km/hr.   **41.** $\approx 15.1°$, 290.1 km/hr.

43. (a)

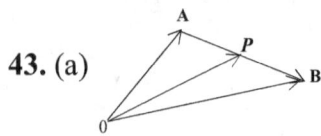

$\overrightarrow{AP} = \frac{1}{2}(\overrightarrow{AB}) = \frac{1}{2}(B - A) \Rightarrow P = A + \frac{1}{2}(B - A)$

$$= A + \frac{1}{2}B - \frac{1}{2}A = \frac{1}{2}A + \frac{1}{2}B$$

(b)

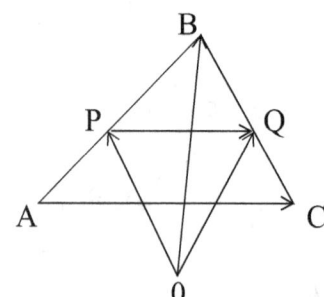

Let P and Q be the midpoints of AB and BC, respectively.

From (a): $P = \frac{1}{2}A + \frac{1}{2}B$ and $Q = \frac{1}{2}B + \frac{1}{2}C$

Thus: $\overrightarrow{PQ} = Q - P = \left(\frac{1}{2}B + \frac{1}{2}C\right) - \left(\frac{1}{2}A + \frac{1}{2}B\right)$

$$= \frac{1}{2}(C - A) = \frac{1}{2}\overrightarrow{AC}$$

It follows that $\overrightarrow{PQ}$ is half the length of $\overrightarrow{AC}$ and is parallel to $\overrightarrow{AC}$

45.

$P = A + \overrightarrow{AP} = A + \left(\dfrac{r}{r + s}\right)\overrightarrow{AB} = A + \left(\dfrac{r}{r + s}\right)(B - A)$

$$= \left(1 - \dfrac{r}{r + s}\right)A + \left(\dfrac{r}{r + s}\right)B = \left(\dfrac{r + s - r}{r + s}\right)A + \left(\dfrac{r}{r + s}\right)B$$

$$= \left(\dfrac{s}{r + s}\right)A + \left(\dfrac{r}{r + s}\right)B$$

47. Let $v = (v_1, v_2)$. Clearly if $r = 0$ or $v = 0$, then $rv = 0$. Conversely, if $rv = 0$, then $rv = r(v_1, v_2) = (rv_1, rv_2) = 0 \Rightarrow rv_1 = 0$ and $rv_2 = 0$. From $rv_1 = 0$: $r = 0$ or $v_1 = 0$. From $rv_2 = 0$: $r = 0$ or $v_2 = 0$. It follows that $r = 0$ or $v_1 = v_2 = 0$, and that therefore: $r = 0$ or $v = 0$.

49. $(u + v) + w = [(u_1, u_2, \ldots, u_n) + (v_1, v_2, \ldots, v_n)] + (w_1, w_2, \ldots, w_n)$

$$= (u_1 + v_1, u_2 + v_2, \ldots, u_n + v_n) + (w_1, w_2, \ldots, w_n)$$

$$= [(u_1 + v_1) + w_1, (u_2 + v_2) + w_2, \ldots, (u_n + v_n) + w_n]$$

$$= [u_1 + (v_1 + w_1), u_2 + (v_2 + w_2), \ldots, u_n + (v_n + w_n)]$$

$$= (u_1, u_2, \ldots, u_n) + [(v_1 + w_1), (v_2 + w_2), \ldots, (v_n + w_n)] = u + (v + w)$$

51. $(r + s)v = (r + s)(v_1, v_2, \ldots, v_n) = [(r + s)v_1, (r + s)v_2, \ldots, (r + s)v_n]$

$$= (rv_1 + sv_1, rv_2 + sv_2, \ldots, rv_n + sv_n)$$

$$= (rv_1, rv_2, \ldots, rv_n) + (sv_1, sv_2, \ldots, sv_n) = rv + sv$$

12.2 Dot and Cross Products (Page 510)

1. -26 **3.** -7 **5.** 60 **7.** $90°$ **9.** $\cos^{-1}\left(\dfrac{-5}{\sqrt{17}\sqrt{13}}\right) \approx 110°$ **11.** $90°$

13. $60°$ **15.** $135°$ **17.** $\cos^{-1}\left(\dfrac{5}{\sqrt{1015}}\right) \approx 81°$ **19.** orthogonal   **21.** neither

23. $(2, 1) = \left(\dfrac{12}{13}, -\dfrac{8}{13}\right) + \left(\dfrac{14}{13} + \dfrac{21}{13}\right)$ **25.** $3j + 4k = \left(\dfrac{3}{2}i + \dfrac{3}{2}j\right) + \left(-\dfrac{3}{2}i + \dfrac{3}{2}j + 4k\right)$

27. $\left(\dfrac{14}{3}, \dfrac{28}{3}, -\dfrac{14}{3}\right) + \left(\dfrac{10}{3}, -\dfrac{16}{3}, -\dfrac{22}{3}\right)$ **29.** $\left(-\dfrac{2}{\sqrt{3}}i - \dfrac{2}{\sqrt{3}}j + \dfrac{2}{\sqrt{3}}k\right) + \left(\dfrac{8}{\sqrt{3}}i + \dfrac{2}{\sqrt{3}}j + \dfrac{10}{\sqrt{3}}k\right)$

31. $(-25, -20, 5)$ **33.** $40i - 3j + 42k$ **35.** 55

37. (a) Expand both $(u - v) \cdot (u - v) = (u_1 - v_1, u_2 - v_2) \cdot (u_1 - v_1, u_2 - v_2)$ and

$u \cdot (u - v) - v \cdot (u - v)$ and observe that each is equal to $u_1^2 - 2u_1 v_1 + v_1^2 + u_2^2 - 2u_2 v_2 + v_2^2$.

(b) Expand both $(u - v) \cdot (u - v) = (u_1 - v_1, u_2 - v_2, u_3 - v_3) \cdot (u_1 - v_1, u_2 - v_2, u_3 - v_3)$

and $u \cdot (u - v) - v \cdot (u - v)$ directly and observe that each is equal to:

$$u_1^2 - 2u_1 v_1 + v_1^2 + u_2^2 - 2u_2 v_2 + v_2^2 + u_3^2 - 2u_3 v_3 + v_3^2$$

(c) Expand both

$(u - v) \cdot (u - v) = (u_1 - v_1, u_2 - v_2, \ldots, u_n - v_n) \cdot (u_1 - v_1, u_2 - v_2, \ldots, u_n - v_n)$ and

$u \cdot (u - v) - v \cdot (u - v)$ directly and observe that each is equal to:

$$u_1^2 - 2u_1 v_1 + v_1^2 + u_2^2 - 2u_2 v_2 + v_2^2 + \ldots + u_n^2 - 2u_n v_n + v_n^2$$

39. (a)

$$\left| \frac{u \cdot v}{\|u\| \|v\|} \right| = \left| \frac{u_1 v_1 + u_2 v_2}{\sqrt{u_1^2 + u_2^2}\sqrt{v_1^2 + v_2^2}} \right| \le 1 \Leftrightarrow |u_1 v_1 + u_2 v_2| \le \sqrt{u_1^2 + u_2^2}\sqrt{v_1^2 + v_2^2}$$

$$\Leftrightarrow (u_1 v_1 + u_2 v_2)^2 \le (u_1^2 + u_2^2)(v_1^2 + v_2^2)$$

$$\Leftrightarrow u_1^2 v_1^2 + 2u_1 v_1 u_2 v_2 + u_2^2 v_2^2 \le u_1^2 v_1^2 + u_1^2 v_2^2 + u_2^2 v_1^2 + u_2^2 v_2^2$$

$$\Leftrightarrow u_1^2 v_2^2 - 2u_1 v_1 u_2 v_2 + u_2^2 v_1^2 \ge 0$$

True, since $u_1^2 v_2^2 - 2u_1 v_1 u_2 v_2 + u_2^2 v_1^2 = (u_1 v_2 + u_2 v_1)^2$

(b) Omitted here. A solution appears in the Student Solutions Manual.

41. (a) First: $(u + v) \times w = (u_1 + v_1, u_2 + v_2, u_3 + v_3) \times (w_1, w_2, w_3)$

$$= \det \begin{bmatrix} i & j & k \\ u_1 + v_1 & u_2 + v_2 & u_3 + v_3 \\ w_1 & w_2 & w_3 \end{bmatrix}$$

$$= \boxed{(u_2 w_3 + v_2 w_3 - u_3 w_2 - v_3 w_2)i - (u_1 w_3 + v_1 w_3 - u_3 w_1 - v_3 w_1)j + (u_1 w_2 + v_1 w_2 - u_2 w_1 - v_2 w_1)k} \quad (*)$$

Then show that $u \times w + v \times w = \det \begin{bmatrix} i & j & k \\ u_1 & u_2 & u_3 \\ w_1 & w_2 & w_3 \end{bmatrix} + \det \begin{bmatrix} i & j & k \\ v_1 & v_2 & v_3 \\ w_1 & w_2 & w_3 \end{bmatrix}$ equals (*)

(b) Expand $cv \times w = \det \begin{bmatrix} i & j & k \\ cv_1 & cv_2 & cv_3 \\ w_1 & w_2 & w_3 \end{bmatrix}$, $v \times cw$, and $c(v \times w)$, and observe that each can

be expressed in the form $c[(v_2 w_3 - v_3 w_2)i - (v_1 w_3 - v_3 w_1)j + (v_1 w_2 - v_2 w_1)k]$

(c) $u \times v = \det \begin{bmatrix} i & j & k \\ u_1 & u_2 & u_3 \\ v_1 & v_2 & v_3 \end{bmatrix} = \det \begin{bmatrix} u_2 & u_3 \\ v_2 & v_3 \end{bmatrix} i - \det \begin{bmatrix} u_1 & u_3 \\ v_1 & v_3 \end{bmatrix} j + \det \begin{bmatrix} u_1 & u_2 \\ v_1 & v_2 \end{bmatrix} k$

$= -\det \begin{bmatrix} v_2 & v_3 \\ u_2 & u_3 \end{bmatrix} i + \det \begin{bmatrix} v_1 & v_3 \\ u_1 & u_3 \end{bmatrix} j - \det \begin{bmatrix} v_1 & v_2 \\ u_1 & u_2 \end{bmatrix} k = -\det \begin{bmatrix} i & j & k \\ v_1 & v_2 & v_3 \\ u_1 & u_2 & u_3 \end{bmatrix} = -(v \times u)$

$\det \begin{bmatrix} a & b \\ c & d \end{bmatrix} = ad - bc = -(bc - ad) = -\det \begin{bmatrix} c & d \\ a & b \end{bmatrix}$

43.

$ai + bj$ Diagonals are orhogonal $\Leftrightarrow (ai + bj) \cdot (ai - bj) = 0$

$\Leftrightarrow a^2 - b^2 = 0$

$\Leftrightarrow a = b$ (since both a and b are positive)

45. Omitted here. A solution appears in the Student Solutions Manual.

47. 20 foot-pounds   **49.** $10{,}000\sqrt{3}$ foot-pounds

51. $\tau = \overrightarrow{OP} \times F \Rightarrow \|\tau\| = \|\overrightarrow{OP} \times F\| = \|\overrightarrow{OP}\|\|F\| \sin\theta$ (Theorem 12.6)

53. (a) 11.25 foot-pounds   (b) 22.5 foot-pounds   (c) $\dfrac{45}{2\sqrt{2}} \approx 15.9$ foot-pounds

55. Show, directly, that both $u \cdot (v \times w)$ and $(u \times v) \cdot w$ can be expressed in the form:

$$u_1 v_2 w_3 - u_1 v_3 w_2 - u_2 v_1 w_3 + u_2 v_3 w_1 + u_3 v_1 w_2 - u_3 v_2 w_1$$

12.3 Lines and Planes (Page 520)

1. $w = (1, 3) + t(1, 3)$; $x = 1 + t, y = 3 + 3t$

3. $w = (0, 1, -1) + t(1, 1, 1)$; $x = t, y = 1 + t, z = -1 + t$

5. First, we show that for any given t there is a t' such that

$$(1, 1) + t[(1, 1) - (1, 0)] = (1, 0) + t'[(1, 0) - (1, 1)] \quad (*)$$

i.e. $(1, 1) + (0, t) = (1, 0) + (0, -t')$ i.e. $(1, 1 + t) = (1, -t') \Rightarrow t' = -(1 + t)$

From the above we also see that for any given t' there is a t for which (*) is satisfied; namely:

$t = -(1 + t')$.

7. (a) $\dfrac{3\sqrt{5}}{5}$ (b) $\dfrac{\sqrt{5}}{3}$ (c) $\dfrac{\sqrt{66}}{6}$

9. $(1, 4, 2) \cdot (x - 2, y, z - 3) = 0, (x - 2) + 4y + 2(z - 3) = 0, x + 4y + 2z = 8$

11. $(3, 2, -1) \cdot (x - 1, y - 1, z - 2) = 0, 3(x - 1) + 2(y - 1) - (z - 2) = 0, 3x + 2y - z = 3$

13. $7x + y + 3z = 9$ **15.** $x + z = 1$ **17.** $\dfrac{\sqrt{6}}{2}$ **19.** $\dfrac{4\sqrt{21}}{7}$ **21.** $\dfrac{8\sqrt{22}}{11}$ **23.** $a = \dfrac{3 \pm \sqrt{11}}{4}$

25. No unique answers.   **27.** $\cos^{-1}\left(\dfrac{1}{3}\right) \approx 70.5°$ **29.** $w = (0, 1, 2) + t(3, -1, -2)$

31. $5x + 2y + 3z = 3$ **33.** $2x - 7y - 3z = -13$ **35.** $\dfrac{95}{\sqrt{1817}}$

12.4 Vector-Valued Functions (Page 531)

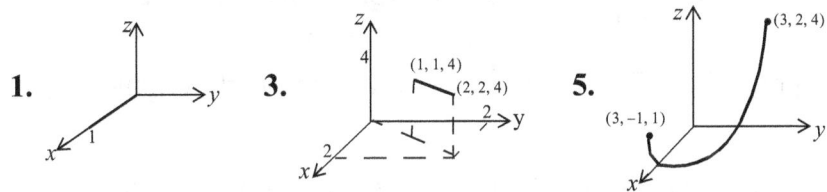

1.   **3.**   **5.**

7. $2i + 5j, \ 2i + 10tj, \ t^2 i + \dfrac{5}{3}t^3 j + C$ **9.** $(\sin 1)i - j, \ (\cos t)i - 2tj, \ (-\cos t)i - \dfrac{t^3}{3}j + C$

11. $i + j + k, \ i, \ \dfrac{t^2}{2}i + tj + tk + C$ **13.** $i - 3j, \ i + \dfrac{1 - \ln t}{t^2}k, \ \dfrac{t^2}{2}i - 3tj + \dfrac{(\ln t)^2}{2}k + C$

15. $i - j + (\tan 1)k, \ i - 2tj + (\sec^2 t)k, \dfrac{t^2}{2}i - \dfrac{t^3}{3}j + (\ln|\sec t|)k + C$ **17.** $x = 2 + 2t, y = 5 + 10t$

19. $x = t, y = -2, z = 1$ **21.** $v(1) = i + 2j, \ a(1) = 0, \ \|v(1)\| = \sqrt{5}$

23. $v(0) = i + j, \ a(0) = -k, \ \|v(0)\| = \sqrt{2}$ **25.** $\sqrt{\dfrac{755}{6}}$ (at $t = \dfrac{11}{6}$), $\sqrt{230}$ (at $t = 6$)

27. (a) Proof of Theorem 12.8(a):

$$[cu(t)]' = \{c[u_1(t)i + u_2(t)j + u_3(t)k]\}' = [cu_1(t)i + cu_2(t)j + cu_3(t)k]'$$
$$= [cu_1(t)]'i + [cu_2(t)]'j + [cu_3(t)]'k = cu_1'(t)i + cu_2'(t)j + cu_3(t)'k$$
$$= c[u_1'(t)i + u_2'(t)j + u_3(t)'k] = cu'(t)$$

(b) Proof of Theorem 12.8(e):

$$[u(t) - v(t)] = [u_1(t)i + u_2(t)j + u_3(t)k] - [v_1(t)i + v_2(t)j + v_3(t)k]$$
$$= [u_1(t) - v_1(t)]i + [u_2(t) - v_2(t)]j + [u_3(t) - v_3(t)]k$$
$$[u(t) - v(t)]' = [u_1'(t) - v_1'(t)]i + [u_2'(t) - v_2'(t)]j + [u_3'(t) - v_3'(t)]k$$
$$= [u_1'(t)i + u_2'(t)j + u_3'(t)k] - [v_1'(t)i + v_2'(t)j + v_3'(t)k] = u'(t) - v'(t)$$

29. $\{u[f(t)]\}' = \{u_1[f(t)]i + u_2[f(t)]j + u_3[f(t)]k\}'$

$$= \{u_1[f(t)]\}'i + \{u_2[f(t)]\}'j + \{u_3[f(t)]\}'k$$
$$= u_1'[f(t)]f'(t)i + u_2'[f(t)]f'(t)j + u_3'[f(t)]f'(t)k$$
$$= f'(t)\{u_1'[f(t)]i + u_2'[f(t)]j + u_3'[f(t)]k\} = f'(t)u'[f(t)]$$

31. Since $\dfrac{d}{dt}([\int f(t)dt]i + [\int g(t)dt]j + [\int h(t)dt]k) = \dfrac{d}{dt}[\int f(t)dt]i + \dfrac{d}{dt}[\int g(t)dt]j + \dfrac{d}{dt}[\int h(t)dt]k$:

$$= f(t)i + g(t)j + h(t)k = r(t)$$

$$\int r(t)dt = [\int f(t)dt]i + [\int g(t)dt]j + [\int h(t)dt]k + C$$

33. ≈ 500.6 m/sec.   35. Max height: 625/4 ft, Range: 625 ft, Impact speed: $100\sqrt{2}$ ft/sec

37. $40\sqrt{3}$ ft   39. ≈ 95 ft/sec   41. $\dfrac{4(45)}{\sqrt{21}} \approx 39$ ft/sec   43. $\sqrt{\dfrac{3200}{\sqrt{3}}} \approx 43$ ft/sec

45. Let α be the angle at which a projectile is fired at a speed of v_0 ft/sec from the origin, so $r_0 = 0$. Then:

$$v(t) = \int(-32k)dt = -32k + (v_0\cos\alpha)j + (v_0\sin\alpha)k = (v_0\cos\alpha)j + (-32t + v_0\sin\alpha)k$$
$$\text{initial velocity}$$

And: $r(t) = \int v(t)dt = [(v_0\cos\alpha)t]j + [-16t^2 + (v_0\sin\alpha)t]k$ (as $r_0 = 0$)

Maximum range occurs when the vertical component of $r(t)$ is zero, so:

$$-16t^2 + (v_0\sin\alpha)t = 0 \Rightarrow 16t = v_0\sin\alpha \Rightarrow t = \frac{v_0\sin\alpha}{16}$$

The range at that time is the horizontal component of $r(t)$:

$$(v_0\cos\alpha)\left(\frac{v_0\sin\alpha}{16}\right) = \frac{v_0^2\sin\alpha\cos\alpha}{16} = \frac{v_0^2 \cdot 2\sin\alpha\cos\alpha}{32} = \frac{v_0^2\sin2\alpha}{32}$$

Maximum value of the range occurs when $\sin2\alpha = 1 \Rightarrow 2\alpha = 90° \Rightarrow \alpha = 45°$.

47. The range, d, when a a projectile is fired at an angle α at an initials peed v_0 ft/sec is

$$d = \frac{v_0^2\sin2\alpha}{32}. \text{ Doubling the initial speed results in the range } \frac{(2v_0)^2\sin2\alpha}{32} = 4d.$$

49. Since acceleration $r''(t)$ is zero, the velocity $v(t) = r'(t)$ must be constant; say $r'(t) = b\boldsymbol{j} + c\boldsymbol{k}$. It follows that:

$$r(t) = \int r'(t)dt = bt\boldsymbol{j} + ct\boldsymbol{k} + \boldsymbol{r_0} = \boldsymbol{r_0} + t(b\boldsymbol{j} + c\boldsymbol{k}) \text{ (vector equation of a line).}$$

51. Letting g denote the force of gravity we have: $\boldsymbol{a}(t) = r''(t) = -g\boldsymbol{k}$. Integrating twice we have:
$v(t) = -gt\boldsymbol{k} + \boldsymbol{v_0} = -gt\boldsymbol{k} + (a\boldsymbol{j} + b\boldsymbol{k})$ and:

$$r(t) = \frac{-gt^2}{2}\boldsymbol{k} + (at\boldsymbol{j} + bt\boldsymbol{k}) + \boldsymbol{r_0} = \frac{-gt^2}{2}\boldsymbol{k} + (at\boldsymbol{j} + bt\boldsymbol{k}) + (c\boldsymbol{j} + d\boldsymbol{k})$$

$$= (at + c)\boldsymbol{j} + \left(\frac{-gt^2}{2} + bt + d\right)\boldsymbol{k}$$

Letting $r(t) = x\boldsymbol{j} + y\boldsymbol{k}$ we have:

$$x = at + c \Rightarrow t = \frac{x-c}{a} \text{ (note that since the trajectory is non-vertical, } a \neq 0\text{)}$$

$$y = \frac{-gt^2}{2} + bt + d \longrightarrow y = -\frac{1}{2}g\left(\frac{x-c}{a}\right)^2 + b\left(\frac{x-c}{a}\right) + d$$

since y is a quadratic function of x,
the graph is a parabola.

12.5 Arc Length and Curvature (Page 543)

1. $20\sqrt{29}$ **3.** $2\pi|a|$ **5.** $2\pi\sqrt{a^2 + b^2}$ **7.** $\ln(1 + \sqrt{2})$ **9.** $x = \frac{3}{5}s - 2$ $y = \frac{4}{5}s + 3$

11. $x = \frac{1}{3}[(3s + 1)^{2/3} - 1]^{3/2}, y = \frac{1}{2}[(3s + 1)^{2/3} - 1]$ **13.** $x = \sin\left(\frac{s}{\sqrt{2}}\right), y = \cos\left(\frac{s}{\sqrt{2}}\right), z = \frac{s}{\sqrt{2}}$

15. $T(t) = \left(\frac{3}{5}\cos t, -\frac{3}{5}\sin t, \frac{4}{5}\right), N(t) = (-\sin t, -\cos t, 0), B(t) = \left(\frac{4}{5}\cos t, -\frac{4}{5}\sin t, -\frac{3}{5}\right), \kappa = \frac{3}{25}$

17. $T(t) = \frac{1}{\sqrt{2}}(\cos t - \sin t)\boldsymbol{i} + \frac{1}{\sqrt{2}}(\cos t + \sin t)\boldsymbol{j}$,

$N(t) = \frac{1}{\sqrt{2}}(-\cos t - \sin t)\boldsymbol{i} + \frac{1}{\sqrt{2}}(-\sin t + \cos t)\boldsymbol{j}, B(t) = \boldsymbol{k}, \kappa = \frac{1}{e^t\sqrt{2}}$

19. $T(t) = \left(\frac{t}{\sqrt{t^2 + 1}}, \frac{1}{\sqrt{t^2 + 1}}, 0\right), N(t) = \left(\frac{1}{\sqrt{t^2 + 1}}, -\frac{t}{\sqrt{t^2 + 1}}, 0\right), B(t) = (0, 0, -1), \kappa = \frac{1}{t(t^2 + 1)^{3/2}}$

21. $-6x + y = \pi, x + 6y = 6\pi$ **23.** $-\frac{x}{\sqrt{2}} + \frac{y}{\sqrt{2}} + z = \frac{\pi}{4}, \frac{x}{\sqrt{2}} - \frac{y}{\sqrt{2}} + z = \frac{\pi}{4}$

25. $\left(x - \frac{\pi}{4}\right)^2 + \left(y - \frac{3}{4}\right)^2 = \frac{1}{16}$

27. Applying CYU 11.36 to $f(x) = ax^2$, with $f'(x) = 2ax$ and $f''(x) = 2a$, we have:

$$\kappa = \frac{|f''(x)|}{[1 + (f'(x))^2]^{3/2}} = \frac{|2a|}{(1 + 4a^2x^2)^{3/2}}$$

κ is greatest when the above denominator $(1 + 4a^2x^2)^{3/2}$ is smallest; which is to say at $x = 0$, which is where the vertex of the parabola is located.

29. We show that $\overline{B(t)} = \det \begin{bmatrix} i & j & k \\ f'(t) & g'(t) & h'(t) \\ f''(t) & g''(t) & h''(t) \end{bmatrix}$ is perpendicular to both $T(t)$ and $N(t)$; and, as

such, will be parallel to $T(t) \times N(t) = B(t)$. We first show that

$\overline{B} = (g'h'' - g''h')i - (f'h'' - f''h')j + (f'g'' - f''g')k$ is perpendicular to $T = \dfrac{(f', g', h')}{\sqrt{f'^2 + g'^2 + h'^2}}$:

$$\overline{B} \cdot T = \frac{1}{\sqrt{f'^2 + g'^2 + h'^2}}[f'(g'h'' - g''h') - g'(f'h'' - f''h') + h'(f'g'' - f''g')] = 0$$

To show that $\overline{B}$ is also perpendicular to $N = \dfrac{T'}{\|T'\|}$, we first determine T' :

$$T' = \left[\frac{(f', g', h')}{\sqrt{f'^2 + g'^2 + h'^2}}\right]' = \left(\frac{1}{\sqrt{f'^2 + g'^2 + h'^2}}\right)'(f', g', h') + \left(\frac{1}{\sqrt{f'^2 + g'^2 + h'^2}}\right)(f'', g'', h'')$$

$$= C(f', g', h') + D(f'', g'', h'') \text{ where } C = \left(\frac{1}{\sqrt{f'^2 + g'^2 + h'^2}}\right)' \text{ and } D = \left(\frac{1}{\sqrt{f'^2 + g'^2 + h'^2}}\right)$$

Then:

$$\overline{B} \cdot N = \overline{B} \cdot \frac{T'}{\|T'\|} = \frac{1}{\|T'\|}(\overline{B} \cdot T') = \frac{1}{\|T'\|}\{\overline{B} \cdot [C(f', g', h') + D(f'', g'', h'')]\}$$

$$= \frac{1}{\|T'\|}\{C[\overline{B} \cdot (f', g', h')] + D[\overline{B} \cdot (f'', g'', h'')]\}$$

$$= \frac{1}{\|T'\|}\Big\{ C[f'(g'h'' - g''h') - g'(f'h'' - f''h') + h'(f'g'' - f''g')]$$

$$+ D[f''(g'h'' - g''h') - g''(f'h'' - f''h') + h''(f'g'' - f''g')] \Big\}$$

$$= \frac{1}{\|T'\|}[C(0) + D(0)] = 0$$

CHAPTER 13
DIFFERENTIATING FUNCTIONS OF SEVERAL VARIABLES
13.1 Partial Derivatives and Differentiability (page 557)

1. $\frac{\partial z}{\partial x} = 2x$, $\frac{\partial z}{\partial y} = 2y$ 3. $\frac{\partial z}{\partial x} = 2xy^2 - 2y$, $\frac{\partial z}{\partial y} = 2x^2y - 2x$ 5. $\frac{\partial z}{\partial x} = 8xy^3 e^{x^2}$, $\frac{\partial z}{\partial y} = 12y^2 e^{x^2}$

7. $\frac{\partial z}{\partial x} = -\frac{y^2+1}{(xy-1)^2}$, $\frac{\partial z}{\partial y} = -\frac{x^2+1}{(xy-1)^2}$ 9. $\frac{\partial z}{\partial x} = \sin y \, e^{x\sin y}$, $\frac{\partial z}{\partial y} = x\cos y \, e^{x\sin y}$

11. $\frac{\partial z}{\partial x} = \frac{e^{x\cos y}}{x} + \ln x \cos y \, e^{x\cos y}$, $\frac{\partial z}{\partial y} = -(x\ln x \sin y)e^{x\cos y}$

13. $f_{xx} = 12$, $f_{yy} = 18$, $f_{xy} = f_{yx} = -8$

15. $f_{xx} = 4e^{2x+3y}$, $f_{yy} = 9e^{2x+3y}$, $f_{xy} = f_{yx} = 6e^{2x+3y}$

17. $f_{xx} = \frac{y^2}{(x^2+y^2)^{3/2}}$, $f_{yy} = \frac{x^2}{(x^2+y^2)^{3/2}}$, $f_{xy} = f_{yx} = \frac{-xy}{(x^2+y^2)^{3/2}}$

19. $f_{xx} = \frac{-\sin x}{\cos y}$, $f_{yy} = \frac{\sin x(\sin^2 y + 1)}{\cos^3 y}$, $f_{xy} = f_{yx} = \frac{\sin y \cos x}{\cos^2 y}$

21. $f_{xx} = -\frac{1}{(x-y)^2}$, $f_{yy} = \frac{-2y^2 + 2yx - x^2}{y^2(x-y)^2}$, $f_{xy} = f_{yx} = \frac{1}{(x-y)^2}$

23. $f_{xz} = f_{zx} = 3x^2y^2$, $f_{yyy} = f_{zzy} = f_{zyz} = 0$, $f_{xyz} = f_{zyx} = 6yx^2$

25. $f_{xz} = f_{zx} = (xzy^2 + y)e^{xyz}$, $f_{yyy} = x^3z^3e^{xyz}$, $f_{zzy} = f_{zyz} = (x^3y^2z + 2x^2y)e^{xyz}$,
   $f_{xyz} = f_{zyx} = (x^2y^2z^2 + 3xyz + 1)e^{xyz}$

27. $f_{xz} = f_{zx} = f_{yyy} = f_{zzy} = f_{zyz} = f_{zyx} = f_{xyz} = 0$

29. $f_x = 2xy$ and $f_y = x^2$ are continuous throughout the domain of $f(x,y) = x^2y - 9$.

31. $f_x = 2e^{2x+3y}$ and $f_y = 3e^{2x+3y}$ are continuous throughout the domain of $f(x,y) = e^{2x+3y}$.

33. $f_x = \frac{y}{xy-y^2} = \frac{1}{x-y}$ (as $y \neq 0$) and $f_y = \frac{x-2y}{xy-y^2}$ are continuous throughout the domain of
   $f(x,y) = \ln(xy-y^2)$.

35. $f_x = 3x^2y^2z^4$, $f_y = 2x^3yz^4$, and $f_z = 4x^3y^2z^3$ are continuous throughout the domain of
   $f(x,y,z) = x^3y^2z^4$.

37. $\frac{\partial z}{\partial x} = \frac{-2xz}{x^2+4y^2z}$, $\frac{\partial z}{\partial y} = -\frac{1+4yz^2}{x^2+4y^2z}$ 39. $\frac{\partial z}{\partial x} = -\frac{x^2+2yz}{z^2+2xy}$, $\frac{\partial z}{\partial y} = -\frac{y^2+2xz}{z^2+2xy}$ 41. xy^2

43. $f_x = g_y = 2x$ and $f_y = -g_x = -2y$ 45. $z_{xx} = -z_{yy} = \frac{-2xy}{(x^2+y^2)^2}$

47. $f_{xy} = [g(x)+h(y)]_{xy} = ([g(x)]_x + [h(y)]_x)_y = ([g(x)]_x + 0)_y = 0$

49. $x\frac{\partial z}{\partial x} + y\frac{\partial z}{\partial y} = x(4x^3 + 4xy^2) + y(4x^2y + 4y^3) = 4(x^4 + x^2y^2 + x^2y^2 + y^4) = 4z$

51. $\frac{\partial w}{\partial x} + \frac{\partial w}{\partial y} + \frac{\partial w}{\partial z} = 2xy + z^2 + x^2 + 2yz + y^2 + 2zx$ and $(x+y+z)^2 = (x+y+z)(x+y+z)$

same

$$= x(x+y+z) + y(x+y+z) + z(x+y+z)$$
$$= x^2 + xy + xz + yx + y^2 + yz + zx + zy + z^2$$

53. $z_{xx} - z_{yy} = [-\sin(x+y) - \sin(x-y)]_x - [-\sin(x+y) + \sin(x-y)]_y$

$$= -\cos(x+y) - \cos(x-y) - [-\cos(x+y) - \cos(x-y)] = 0$$

55. $V\frac{\partial P}{\partial V} = V\frac{\partial}{\partial V}\left(\frac{kT}{V}\right) = V\frac{\partial}{\partial V}(kTV^{-1}) = kTV(-V^{-2}) = -\frac{kT}{V} = -P$

$$V\frac{\partial P}{\partial V} + T\frac{\partial P}{\partial T} = -P + T\frac{\partial}{\partial T}\left(\frac{kT}{V}\right) = -P + T\boxed{\frac{k}{V}}\underset{PV\,=\,kT}{\longrightarrow} = -P + T\boxed{\frac{P}{T}} = 0$$

$$\left(\frac{\partial V}{\partial T}\right)\left(\frac{\partial T}{\partial P}\right)\left(\frac{\partial P}{\partial V}\right) = \left[\frac{\partial}{\partial T}(kTP^{-1})\right]\left[\frac{\partial}{\partial P}\left(\frac{1}{k}PV\right)\right]\left[\frac{\partial}{\partial V}(kTV^{-1})\right]$$

$$= \frac{k}{P} \cdot \frac{V}{k} \cdot \left[\frac{-kT}{V^2}\right] = \frac{V}{T} \cdot \frac{V}{k} \cdot \left[\frac{-kT}{V^2}\right] = -1$$

57. $\frac{\partial^2 R}{\partial R_1^2} \cdot \frac{\partial^2 R}{\partial R_2^2} = \frac{\partial}{\partial R_1}\left[R_2\frac{(R_1+R_2)-R_1}{(R_1+R_2)^2}\right] \cdot \frac{\partial}{\partial R_2}\left[R_1\frac{(R_1+R_2)-R_2}{(R_1+R_2)^2}\right]$

$$= \frac{\partial}{\partial R_1}[R_2^2(R_1+R_2)^{-2}] \cdot \frac{\partial}{\partial R_2}[R_1^2(R_1+R_2)^{-2}]$$

$$= \frac{-2R_2^2}{(R_1+R_2)^3} \cdot \frac{-2R_1^2}{(R_1+R_2)^3} = \frac{4(R_1R_2)^2}{(R_1+R_2)^6} \underset{\uparrow}{=} \frac{4R^2(R_1+R_2)^2}{(R_1+R_2)^6} = \frac{4R^2}{(R_1+R_2)^4}$$

$$R = \frac{R_1R_2}{R_1+R_2}$$

59. $-\frac{1}{2}$ **61.** $\cos 2t$ **63.** $1 + \cos 2t$

65. $z_t = 3\sin t^2 + (2st + 6t^2)\cos t^2$, $z_s = \sin t^2$ **67.** $w_t = s + 2se^{st} - 3\cos t$, $w_s = t + 2te^{st}$

13.2 Directional Derivatives, Gradient Vectors, and Tangent Planes (page 570)

1. $D_u f(x, y) = \lim\limits_{h \to 0} \dfrac{f(x + h\cos 45°, y + h\sin 45°) - f(x, y)}{h}$

$= \lim\limits_{h \to 0} \dfrac{f\left(x + \dfrac{1}{\sqrt{2}}h, y + \dfrac{1}{\sqrt{2}}h\right) - (x^2 + xy)}{h}$

$= \lim\limits_{h \to 0} \dfrac{\left(x + \dfrac{1}{\sqrt{2}}h\right)^2 + \left(x + \dfrac{1}{\sqrt{2}}h\right)\left(y + \dfrac{1}{\sqrt{2}}h\right) - (x^2 + xy)}{h}$

$= \lim\limits_{h \to 0} \dfrac{x^2 + \sqrt{2}hx + \dfrac{h^2}{2} + xy + \dfrac{hx}{\sqrt{2}} + \dfrac{hy}{\sqrt{2}} + \dfrac{h^2}{2} - x^2 - xy}{h}$

$= \lim\limits_{h \to 0} \dfrac{\sqrt{2}hx + \dfrac{h^2}{2} + \dfrac{hx}{\sqrt{2}} + \dfrac{hy}{\sqrt{2}} + \dfrac{h^2}{2}}{h} = \lim\limits_{h \to 0} \left(\sqrt{2}x + \dfrac{h}{2} + \dfrac{x}{\sqrt{2}} + \dfrac{y}{\sqrt{2}} + \dfrac{h}{2}\right)$

$= \sqrt{2}x + \dfrac{x}{\sqrt{2}} + \dfrac{y}{\sqrt{2}} = \dfrac{3x + y}{\sqrt{2}}$

3. $1 - 2\sqrt{3}$ **5.** $\dfrac{1}{2}(1 + \sqrt{3})$ **7.** $-\dfrac{1}{5}$ **9.** $-8\sqrt{2}$ **11.** 60   **13.** $-\dfrac{1}{16}$

15. $\dfrac{3}{5}$ **17.** $\dfrac{-3\sqrt{3}}{8}$ **19.** $\dfrac{1}{2\sqrt{2}}$ **21.** 3   **23.** e **25.** $30\sqrt{2}$ **27.** $\dfrac{13}{2}, -\dfrac{13}{2}$

29. $\sqrt{2}, -\sqrt{2}$ **31.** $\sqrt{5}, -\sqrt{5}$ **33.** $2, -2$ **35.** $3\sqrt{3}, -3\sqrt{3}$ **37.** $1, -1$

39. $2\sqrt{3}, -2\sqrt{3}$ **41.** $6x - 47y - z = -123$ **43.** $8x + 2y + z = 0$ **45.** $y - z = 0$

47. $3y - z = -1$ **49.** $4x + 3y + z = 10$ **51.** $2x - y + 2z = 0$ **53.** $x + y + z = e^3 - 1$

55. 300.00   **57.** -0.32 **59.** 0.23 cm^2 **61.** -0.32 **63.** 2.35 cm^3

65. $\nabla f(x_0, y_0) = f_x(x_0, y_0)\boldsymbol{i} + f_y(x_0, y_0)\boldsymbol{j} = \boldsymbol{0} \Rightarrow f_x(x_0, y_0) = f_y(x_0, y_0) = 0$. It follows that for

any $\boldsymbol{u} = a\boldsymbol{i} + b\boldsymbol{j}$: $D_u f(x_0, y_0) = af_x(x_0, y_0) + bf_y(x_0, y_0) = a \cdot 0 + b \cdot 0 = 0$.

67. For $f(x, y, z) = \dfrac{x^2}{a^2} + \dfrac{y^2}{b^2} - \dfrac{z^2}{c^2}$, $\nabla f(x, y, z) = \dfrac{2x}{a^2}\boldsymbol{i} + \dfrac{2y}{b^2}\boldsymbol{j} - \dfrac{2z}{c^2}\boldsymbol{k}$. It follows that

$\nabla f(x_0, y_0, z_0) = \dfrac{2x_0}{a^2}\boldsymbol{i} + \dfrac{2y_0}{b^2}\boldsymbol{j} - \dfrac{2z_0}{c^2}\boldsymbol{k}$ is a normal to the surface $\dfrac{x^2}{a^2} + \dfrac{y^2}{b^2} - \dfrac{z^2}{c^2} = 1$ at

(x_0, y_0, z_0), and that, therefore $\dfrac{2x_0}{a^2}(x - x_0) + \dfrac{2y_0}{b^2}(y - y_0) - \dfrac{2z_0}{c^2}(z - z_0) = 0$ is the tangent

plane at that point — a plane which can be rewritten in the form:

$$\dfrac{xx_0}{a^2} + \dfrac{yy_0}{b^2} - \dfrac{zz_0}{c^2} = \underbrace{\dfrac{x_0^2}{a^2} + \dfrac{y_0^2}{b^2} - \dfrac{z_0^2}{c^2}}_{\frac{x^2}{a^2} + \frac{y^2}{b^2} - \frac{z^2}{c^2} = 1} = 1$$

69. (a) $\nabla(rf) = (rf)_x\mathbf{i} + (rf)_y\mathbf{j} + (rf)_z\mathbf{k} = rf_x\mathbf{i} + rf_y\mathbf{j} + rf_z\mathbf{k} = r(f_x\mathbf{i} + f_y\mathbf{j} + f_z\mathbf{k}) = r\nabla f$

(b) $\nabla(f \pm g) = (f \pm g)_x\mathbf{i} + (f \pm g)_y\mathbf{j} + (f \pm g)_z\mathbf{k} = (f_x \pm g_x)\mathbf{i} + (f_y \pm g_y)\mathbf{j} + (f_z \pm g_z)\mathbf{k}$

$$= (f_x\mathbf{i} + f_y\mathbf{j} + f_z\mathbf{k}) \pm (g_x\mathbf{i} + g_y\mathbf{j} + g_z\mathbf{k}) = \nabla f \pm \nabla g$$

(c) $\nabla(fg) = (fg)_x\mathbf{i} + (fg)_y\mathbf{j} + (fg)_z\mathbf{k} = (fg_x + gf_x)\mathbf{i} + (fg_y + gf_y)\mathbf{j} + (fg_z + gf_z)\mathbf{k}$

$$= f(g_x\mathbf{i} + g_y\mathbf{j} + g_z\mathbf{k}) + g(f_x\mathbf{i} + f_y\mathbf{j} + g_z\mathbf{k}) = f\nabla g + g\nabla f$$

(d) $\nabla\left(\dfrac{f}{g}\right) = \left(\dfrac{f}{g}\right)_x\mathbf{i} + \left(\dfrac{f}{g}\right)_y\mathbf{j} + \left(\dfrac{f}{g}\right)_z\mathbf{k} = \dfrac{gf_x - fg_x}{g^2}\mathbf{i} + \dfrac{gf_y - fg_y}{g^2}\mathbf{j} + \dfrac{gf_z - fg_z}{g^2}\mathbf{k}$

$$= \dfrac{g(f_x\mathbf{i} + f_y\mathbf{j} + f_z\mathbf{k}) - f(g_x\mathbf{i} + g_y\mathbf{j} + g_y\mathbf{k})}{g^2} = \dfrac{g\nabla f - f\nabla g}{g^2}$$

13.3 Extreme Values (page 585)

1. Saddle point at $(5, 2)$. **3.** Saddle point at $(0, 0)$. **5.** Local maximum at $(-2, -2)$.

7. Local minimum at $(1, 1)$ and at $(-1, -1)$, saddle point at $(0, 0)$.

9. Local minimum at $(2, 1)$, saddle point at $(0, 0)$. **11.** Local minimum at $(2, -1)$.

13. None. **15.** Local maximum at $(-1, 0)$. **17.** Local maximum at $\left(\dfrac{\pi}{2}, \dfrac{\pi}{2}\right)$.

19. Absolute minimum: $f(3, 0) = -11$; absolute maximum: $f(3, 5) = 34$.

21. Absolute minimum: $f\left(\dfrac{1}{2}, \dfrac{1}{4}\right) = -\dfrac{1}{16}$; absolute maximum: $f\left(\dfrac{1}{4}, 2\right) = \dfrac{673}{64}$.

23. Absolute minimum: $f\left(\dfrac{2}{3}, \dfrac{1}{3}\right) = -\dfrac{13}{27}$; absolute maximum: $f(2, 0) = 6$.

25. Absolute minimum: $f(2, 0) = -3$; absolute maximum: $f\left(\dfrac{1}{2}, 0\right) = -\dfrac{3}{4}$.

27. Absolute minimum: $f(-2, 2) = -9$; absolute maximum: $f\left(1, \dfrac{3}{2}\right) = \dfrac{37}{4}$.

29. Absolute minimum: $f(0, -4) = -40$; absolute maximum: $f(\pm\sqrt{15}, 1) = 35$.

31. Minimum: $f\left(-\dfrac{1}{\sqrt{2}}, \dfrac{1}{\sqrt{2}}\right) = -\sqrt{2}$; maximum: $f\left(\dfrac{1}{\sqrt{2}}, \dfrac{1}{\sqrt{2}}\right) = \sqrt{2}$.

33. Minimum: $f(-3, 3) = f(3, -3) = 5$; maximum: $f(3, 3) = f(-3, -3) = 23$.

35. Minimum: $f(\pm 1, 0) = 1$; maximum: $f(0, \pm 1) = 2$.

37. Minimum: $f(4, 0) = 11$; maximum: $f(-2, \pm 2\sqrt{3}) = 47$.

39. Minimum: $f\left(-\dfrac{1}{\sqrt{2}}, -\dfrac{1}{\sqrt{2}}, -\sqrt{2}\right) = -3\sqrt{2}$; maximum: $f\left(\dfrac{1}{\sqrt{2}}, \dfrac{1}{\sqrt{2}}, \sqrt{2}\right) = 3\sqrt{2}$.

41. Minimum: $f(0, 0, 1) = f(0, 1, 0) = f(1, 0, 0) = 0$; maximum: $f\left(\dfrac{1}{3}, \dfrac{1}{3}, \dfrac{1}{3}\right) = \dfrac{1}{27}$.

43. Minimum: $f(0, 0) = 0$; maximum: $f(0, \pm 1) = 2$.

45. Minimum: $f(1, 0) = -7$; maximum: $f(-2, \pm 2\sqrt{3}) = 47$.

47. Minimum: $f\left(-\dfrac{2}{\sqrt{6}}, \dfrac{3}{\sqrt{6}}, \dfrac{1}{\sqrt{6}}\right) = -2\sqrt{6}$; maximum: $f\left(\dfrac{2}{\sqrt{6}}, -\dfrac{3}{\sqrt{6}}, -\dfrac{1}{\sqrt{6}}\right) = 2\sqrt{6}$.

49. Min: $f\left(-\sqrt{2}, -\dfrac{1}{\sqrt{2}}, \dfrac{1}{\sqrt{2}}\right) = f\left(\sqrt{2}, \dfrac{1}{\sqrt{2}}, -\dfrac{1}{\sqrt{2}}\right) = \dfrac{1}{2}$; max: $f\left(-\sqrt{2}, -\dfrac{1}{\sqrt{2}}, -\dfrac{1}{\sqrt{2}}\right) = f\left(\sqrt{2}, \dfrac{1}{\sqrt{2}}, \dfrac{1}{\sqrt{2}}\right) = \dfrac{3}{2}$.

51. Min: $f\left(\dfrac{2}{\sqrt{13}}, -\dfrac{3}{\sqrt{13}} - 2 + \dfrac{7}{\sqrt{13}}\right) = 4 - 2\sqrt{13}$; max: $f\left(-\dfrac{2}{\sqrt{13}}, \dfrac{3}{\sqrt{13}} - 2 - \dfrac{7}{\sqrt{13}}\right) = 4 + 2\sqrt{13}$.

53. $(0, 0, \pm 1)$ **55.** Square base of side $2 \cdot 4^{1/3}$ ft. and height of $6 \cdot 4^{1/3}$ ft.   **57.** $\dfrac{\sqrt{6}}{3}$

59. $2ab$ **61.** $x = 50, y = 150$ **63.** 10 in. by 20 in. by 15 in.

65. Base: $\sqrt{\dfrac{S}{3}}$ by $\sqrt{\dfrac{S}{3}}$; height $\dfrac{1}{2}\sqrt{\dfrac{S}{3}}$ **67.** $x = \dfrac{l}{3}, \alpha = 60°$

CHAPTER 14
VECTOR CALCULUS

14.1 Line Interals (page 601)

1. $\dfrac{5\sqrt{2}}{6}$ **3.** $2\pi + \dfrac{2}{3}$ **5.** $\dfrac{128}{5}$ **7.** $\dfrac{1}{4}(37\sqrt{37} - 1)$ **9.** $\dfrac{2\sqrt{2} + 9}{6}$ **11.** $\dfrac{20\sqrt{10} - 218}{27}$

13. 0 **15.** $-3\sqrt{10}\pi$ **17.** $4\pi + 8$ **19.** $45\pi + \dfrac{80}{3}\pi^3$ **21.** 0 **23.** e **25.** $\dfrac{19}{6} - \dfrac{e^4}{2}$

27. 1 **29.** 7 **31.** $\dfrac{6}{5} - \cos 1 + \sin 1$ **33.** $\dfrac{1069}{35}$ **35.** $2\pi^2$ **37.** $\dfrac{29}{60}$ **39.** $\dfrac{8}{3}$

14.2 Conservative Fields and Path Independence (page 614)

1. Not conservative   **3.** $f(x, y) = x^2 y^3 + y$ **5.** Not conservative   **7.** Not conservative

9. $f(x, y, z) = e^x \cos y + xyz + \dfrac{1}{2}z^2$ **11.** $f(x, y, z) = xy^2 + ye^{3z}$ **13.** $-\pi$ **15.** 20

17. $3e + \dfrac{1}{3}e^2 - \dfrac{737}{216}$ **19.** 2 **21.** $e^{3\pi} + 1$ **23.** $-\dfrac{2}{3}$ **25.** $e^2 + 4e + 1$ **27.** $2x^2 y$

29. $2xe^y$ **31.** $\dfrac{1}{2}y^4 i + 2xy^3 j + zk$ **33.** 0 **35.** 1

37. Since $f(x, y)$ and $h(x, y)$ are both potential functions of the same conservative vector field, $\nabla f = \nabla h$, which means that (1): $f_x(x, y) = h_x(x, y)$ and (2): $f_y(x, y) = h_y(x, y)$.
  Let $k(x, y) = f(x, y) - h(x, y)$. From (1): $k_x(x, y) = 0 \Rightarrow k(x, y) = k(y)$. From (2):
  $k'(y) = 0 \Rightarrow k(y) = C$. It follows that $f(x, y) - h(x, y) = C$.

14.3 Green's Theorem (page 623)

1. 2π **3.** $\dfrac{1}{2}$ **5.** $-\dfrac{1}{4}$ **7.** $\dfrac{\pi}{2}$ **9.** $\dfrac{2}{3}$ **11.** -24π **13.** 0 **15.** $-\dfrac{4}{3}$ **17.** $1 - \sin 1$

19. $-\dfrac{1}{12}$ **21.** 12π **23.** πr^2 **25.** $\dfrac{1}{12}$ **27.** $3\pi a^2$ **29.** π **31.** 0

14.4 Curl and Div (page 632)

1. $(y+1)k$ **3.** $(-y\sin xy - x\cos xy)k$ **5.** $xi - (y - y^2)j + (-3x^2 - 2yz)k$

7. $(-2y - xy)i + xj + yzk$ **9.** $(x + ye^{yz})i - \left(y - \dfrac{1}{x+z}\right)j$

11. $-x(\cos xy + \cos zx)i + y(\cos xy + \cos yz)j + z(\cos zx - \cos yz)k$

13. $-x\cos xyi + y(\cos xy + \sec^2 yz)j - (y\sin xy + z\sec^2 yz)k$ **15.** $2xy$ **17.** $y + z - 2x$

19. $e^{x-z} - e^{z-y}$ **21.** $-y\sin xy - z\cos yz$ **23.** Show curl$(F) = 0$ **25.** Show curl$(F) = 0$
35. Not meaningful   **37.** Meaningful   **39.** Meaningful   **41.** Meaningful

14.5 Surface Integral (page 641)

1. $4\sqrt{2}\pi$ **3.** 38 **5.** $\dfrac{\pi}{6}(17\sqrt{17} - 5\sqrt{5})$ **7.** 10π **9.** $\dfrac{4}{15}(275 - 36\sqrt{6} - 49\sqrt{7})$

11. $\dfrac{1}{3}(26\sqrt{26} - 10\sqrt{10})$ **13.** $\dfrac{13\sqrt{2}}{3}$ **15.** $12\pi + 36$ **17.** $\dfrac{\pi\sqrt{2}}{32}$ **19.** $\dfrac{1}{4}(37\sqrt{37} - 1)$

21. $\dfrac{\sqrt{3}}{12}ka^4$ **23.** $-\dfrac{10\sqrt{6}}{3}$ **25.** $\dfrac{4\pi}{3}$ **27.** $\dfrac{3}{2}$ **29.** 128π

14.6 Stoke's Theorem (page 650

1. -324π **3.** -112π **5.** $\dfrac{3\pi}{2}$ **7.** 12π **9.** -1 **11.** 0 **13.** 5 **15.** -18π

17. π **19.** $-\pi$ **21.** -2π **23.** 0 **25.** $-\dfrac{16}{3}$ **27.** 2π

29. Since a normal to the plane $x + y + z = 1$ is $i + j + k$, $n = \dfrac{1}{\sqrt{3}}(i + j + k)$, and:

$$\text{curl}(F) = \det\begin{vmatrix} i & j & k \\ \dfrac{\partial}{\partial x} & \dfrac{\partial}{\partial y} & \dfrac{\partial}{\partial z} \\ z & -2x & 3y \end{vmatrix} = 3i + j - 2k$$

$$\iint_S \text{curl}\,F \cdot n\,dS = \iint_S (3i + j - 2k) \cdot \dfrac{1}{\sqrt{3}}(i + j + k)dS = \dfrac{2}{\sqrt{3}}\iint_S dS = \dfrac{2}{\sqrt{3}}(\text{surface area of } S)$$

$$= \dfrac{2}{\sqrt{3}}(\text{area of the region enclosed by } C)$$

14.7 The Divergence Theorem (page 657)

1. $\dfrac{12\pi a^5}{5}$ **3.** $\dfrac{2\pi}{3}$ **5.** $\dfrac{81\pi}{2}$ **7.** 256π **9.** 32π **11.** $\dfrac{184}{35}$ **13.** 3 **15.** 32π

17. $F = c \Rightarrow \text{div}(F) = 0 \Rightarrow \iint_S F \cdot n\,dS = \iiint_E \text{div}(F)dV = \iiint_E 0\,dV = 0$